OXFORD STATISTICAL SCIENCE SERIES

SERIES EDITORS

OXFORD STATISTICAL SCIENCE SERIES

1. A. C. Atkinson: *Plots, transformations, and regression*
2. M. Stone: *Coordinate-free multivariable statistics*
3. W. J. Krzanowski: *Principles of multivariate analysis: a user's perspective*
4. M. Aitkin, D. Anderson, B. Francis, and J. Hinde: *Statistical modelling in GLIM*
5. Peter J. Diggle: *Time series: a biostatistical introduction*
6. Howell Tong: *Non-linear time series: a dynamical system approach*
7. V. P. Godambe: *Estimating functions*
8. A. C. Atkinson and A. N. Donev: *Optimum experimental designs*
9. U. N. Bhat and I. V. Basawa: *Queuing and related models*
10. J. K. Lindsey: *Models for repeated measurements (see new edition)*
11. N. T. Longford: *Random coefficient models*
12. P. J. Brown: *Measurement, regression, and calibration*
13. Peter J. Diggle, Kung-Yee Liang, and Scott L. Zeger: *Analysis of longitudinal data (see new edition)*
14. J. I. Ansell and M. J. Phillips: *Practical methods for reliability data analysis*
15. J. K. Lindsey: *Modelling frequency and count data*
16. J. L. Jensen: *Saddlepoint approximations*
17. Steffen L. Lauritzen: *Graphical models*
18. A. W. Bowman and A. Azzalini: *Applied smoothing methods for data analysis*
19. J. K. Lindsey: *Models for repeated measurements* second edition
20. Michael Evans and Tim Swartz: *Approximating integrals via Monte Carlo and deterministic methods*
21. D. F. Andrews and J. E. Stafford: *Symbolic computation for statistical inference*
22. T. A. Severini: *Likelihood methods in statistics*
23. W. J. Krzanowski: *Principles of multivariate analysis: a user's perspective* updated edition
24. J. Durbin and S. J. Koopman: *Time series analysis by State Space Models*
25. Peter J. Diggle, Patrick Heagerty, Kung-Yee Liang, Scott L. Zeger: *Analysis of Longitudinal Data 2/e*
26. J. K. Lindsey: *Nonlinear models in medical statistics*
27. Peter J. Green, Nils L. Hjort & Sylvia Richardson: *Highly structured stochastic systems*
28. Margaret S. Pepe: *Statistical evolution of medical tests*
29. Christopher G. Small and Jinfang Wang: *Numerical methods for nonlinear estimating equations*
30. John C. Gower and Garmt B. Dijksterhuis: *Procrustes problems*

Procrustes Problems

J. C. GOWER
Statistics Department
Open University,
Milton Keynes

G. B. DIJKSTERHUIS
Consumer and Market Insight
Agrotechnology and Food Innovations B.V.
Wageningen University and Research Centre
Wageningen, the Netherlands
and
Department of Marketing and Marketing Research
Faculty of Economics
University of Groningen

OXFORD
UNIVERSITY PRESS

OXFORD
UNIVERSITY PRESS

Great Clarendon Street, Oxford OX2 6DP

It furthers the University's objective of excellence in research, scholarship, and education by publishing worldwide in

Oxford New York

Auckland Bangkok Buenos Aires Cape Town Chennai
Dar es Salaam Delhi Hong Kong Istanbul Karachi Kolkata
Kuala Lumpur Madrid Melbourne Mexico City Mumbai Nairobi
São Paulo Shanghai Taipei Tokyo Toronto

Oxford is a registered trade mark of Oxford University Press in the UK and in certain other countries

Published in the United States by Oxford University Press Inc., New York

First published 2004

A catalogue record for this title is available from the British Library

Library of Congress Cataloging in Publication Data
(Data available)

ISBN 0 19 851058 6

10 9 8 7 6 5 4 3 2 1

Typeset by Newgen Imaging Systems (P) Ltd., Chennai, India
Printed in Great Britain
on acid-free paper by
Biddles Ltd, www.biddles.co.uk

Procrustes *(the subduer), son of Poseidon, kept an inn benefiting from what he claimed to be a wonderful all-fitting bed. He lopped off excessive limbage from tall guests and either flattened short guests by hammering or stretched them by racking. The victim fitted the bed perfectly but, regrettably, died. To exclude the embarrassment of an initially exact-fitting guest, variants of the legend allow Procrustes two, different-sized, beds. Ultimately, in a crackdown on robbers and monsters, the young Theseus fitted Procrustes to his own bed.*

Contents

Preface

In 1990 we started writing an article about Procrustes analysis, prompted by some apparent misconceptions about Procrustes methods, especially relating to confusions between projections and generalised rotations but also concerning comparisons across different goodness-of-fit criteria. This article grew and grew, and we realised that there was much more to say about Procrustes analyses than would fit into a single article; a book was the only option.

The literature on Procrustes, and related problems was confusing, and even where to draw the boundaries of the subject was a problem. We decided that we were mainly, but not exclusively, concerned with matching geometrical configurations of points. After writing an introduction, we intended to start with a chapter that discussed the main types of Procrustes methods for two data sets, but soon found that there was actually material for six chapters. These are Chapters 3–8 that now form the heart of the book. Some of the material here is rather theoretical, but technical aspects are mainly confined to an appendix. Issues that initially we thought were rather straightforward, such as scaling and weighting, turned out to be quite intricate.

The process of writing proved susceptible to the machinations of Procrustes, the evil figure from Greek mythology. The scope of the matter seemed to be continuously stretching before our eyes, and writing time stretched correspondingly. Eventually, we decided to take the other Procrustean action, to come to a completion by chopping things. As a result there are fewer examples and applications of the methods than we would have liked. There are several new ideas in the book that are not fully explored and which students of multivariate analysis are invited to take up, test and further develop. Dispersed throughout the book are algorithms, some old, some new, given in an obvious pseudo code, and often in a very abbreviated form. Some are illustrated with small data sets, but we realise that their worth is in their serious application. We hope that readers will put the algorithms to the test and develop them further.

Sometimes we feel that this book should be subtitled *All you need to know and a lot that you do not need to know about Procrustes analysis*. However, this would be wrong on both counts. On the one hand we are aware that many useful Procrustes methods are covered only superficially—the important specialised area of shape analysis is one such example. On the other hand, we are confident that many of the seemingly more esoteric methods will find applications.

Acknowledgements

Many people and institutions helped us in many different ways. The Data Theory group at the University of Leiden, the Netherlands, is thanked for hosting JCG in 1990 and 1991 when the cooperation between JCG and GBD, which culminated in this book, began. We thank especially Willem Heiser and Jacqueline Meulman.

Colleagues of JCG, especially Frank Critchley, at the Statistics Department of the Open University, Milton Keynes are thanked for their advice and support. Some parts of this work were carried out when GBD was working on the Føtek programme supported by the Danish Dairy Research Foundation and the Danish Government. Colleagues at the Sensory Science Group of the Department of Dairy and Food science, at the Veterinary and Agricultural University, Frederiksberg, Denmark, in particular Harald and Magni Martens, are thanked for stimulating discussions on data analysis and far beyond.

Part of this work was carried out at the ZA-EUROLAB at the Zentralarchiv für Empirische Sozialforschung (ZA), Cologne. The ZA is a Large Scale Facility supported by the Improving Human Potential (IHP) Programme—Access to Research Infrastructures of the European Union. At this institute we thank Ekkehard Mochmann, Jörg Blasius, Reiner Mauer, and Ingvill Mochman.

The LMC, Centre for advanced food studies, Copenhagen, Denmark is thanked for providing a grant enabling JCG to visit GBD in Copenhagen in March 2001. Further we thank individuals who supported us, professionally or privately: Gillian Arnold, Jos ten Berge, Derek Byrne, Alex Collins, Janet Gower, Michael Greenacre, Renger Jellema, Henk Kiers, Richard Nicholls, Karen van der Vlies, Gerjo de Vries.

1
Introduction

There are three elements in the Procrustes story, the unfortunate traveller, who we might label $\mathbf{X}_1$, the Procrustean bed $\mathbf{X}_2$, and the treatment, $\mathbf{T}$, meted out of racking, hammering, or amputation. The simplest algebraic statement of a Procrustes problem seeks a matrix $\mathbf{T}$ that minimises

$$\|\mathbf{X}_1\mathbf{T} - \mathbf{X}_2\| \tag{1.1}$$

over $\mathbf{T}$ $(P_1 \times P_2)$, for given $\mathbf{X}_1$ $(N \times P_1)$ and $\mathbf{X}_2$ $(N \times P_2)$.* Different variants of this problem depend on the shapes of $\mathbf{X}_1$ and $\mathbf{X}_2$ and any restrictions put on admissible forms of $\mathbf{T}$. In this book we shall be especially concerned with the cases where $\mathbf{T}$ is orthogonal, which we denote by $\mathbf{Q}$, orthonormal denoted by $\mathbf{P}$ or a matrix of direction cosines denoted by $\mathbf{C}$.

There is considerable confusion in the literature between the terms orthogonal and orthonormal, so we state at the outset *our* meanings for these terms. The prefix *ortho* refers to the inner product of two vectors being zero (geometrically they represent perpendicular directions) and *normal* refers to vectors of length one. Thus an orthonormal matrix ${}_{P_1}\mathbf{P}_{P_2}$ satisfies ${}_{P_2}\mathbf{P}'{}_{P_1}\mathbf{P}_{P_2} = {}_{P_2}\mathbf{I}_{P_2}$. More precisely $\mathbf{P}$, so defined, is column-orthonormal; we may also have row-orthonormality, defined by ${}_{P_1}\mathbf{P}_{P_2}\mathbf{P}'{}_{P_1} = {}_{P_1}\mathbf{I}_{P_1}$. By an orthogonal matrix $\mathbf{Q}$, we mean a square matrix, in which case $P_1 = P_2 = P$, say, which is both row and column orthonormal, so that $\mathbf{Q}\mathbf{Q}' = \mathbf{Q}'\mathbf{Q} = {}_P\mathbf{I}_P$. Actually, for square matrices, row orthonormality implies column orthonormality and vice versa; thus a $P \times P$ orthonormal matrix is an orthogonal matrix. The term orthogonal for square orthonormal matrices is very common in the literature but it is a little misleading because from the etymological point of view we would understand only the diagonality of $\mathbf{Q}\mathbf{Q}'$ and $\mathbf{Q}'\mathbf{Q}$ and not the unit-length. However, orthogonal matrices are of such importance that some special terminology is needed to distinguish them from non-square orthonormal matrices. In the literature, what we call orthonormal matrices are sometimes described as orthogonal, which is etymologically correct but a source of confusion. Another source of confusion in the literature is that quite general matrices $\mathbf{T}$ may be referred to as representing *rotations* whereas, strictly speaking, rotation only refers to orthogonal matrices and then not to all orthogonal matrices

* At this stage notation should be self-explanatory; for convenience of reference we have gathered together in Section 1.5 all the notational devices used in this book.

(see Appendix B). We reserve the term rotation to refer only to operations by orthogonal matrices. Yet a further source of confusion is that we shall often use orthonormal matrices $\mathbf{P}$ to denote the geometrical concept of orthogonal projection. In the mathematical literature, a projection matrix is square and idempotent ($\mathbf{T}^2 = \mathbf{T}$ such as when $\mathbf{T} = \mathbf{PP}'$ and $\mathbf{P}$ is column orthonormal); we explain our looser use of the projection terminology in Appendix C.

If the columns of a matrix $\mathbf{C}$ give direction cosines of a set of oblique axes then $\text{diag}(\mathbf{C}'\mathbf{C}) = \mathbf{I}$ and the off-diagonal elements give the cosines of the angles between the axes. Thus, the columns of $\mathbf{C}$ are normalised to unit length but are not orthogonal. The use of oblique axes is explained in Appendix D.

In this book we shall be concerned with two ways of looking at the transformations $\mathbf{T}$. In the first we shall be primarily concerned with the jth column $\mathbf{X}_1\mathbf{t}_j$ and how this relates to the jth column of $\mathbf{X}_2$. These may be regarded as giving values on a jth coordinate axis. The axes form a complete orthogonal set of P_1 axes when $\mathbf{T} = \mathbf{Q}$, a subset of P_2 orthogonal axes when $\mathbf{T} = \mathbf{P}$ and a set of oblique axes when $\mathbf{T} = \mathbf{C}$. The focus is on the coordinate values themselves. In the second way of looking at the transformations the focus is on the configuration of points generated by the coordinates and there is, at most, secondary interest in the coordinate values themselves. Most of the original work in Procrustes analysis was concerned with the coordinate interpretation of matching factor loadings. Recent work tends to be more concerned with matching configurations.

1.1 Historical review

The minimisation problem (1.1) is called a Procrustes problem because $\mathbf{X}_1$ is transformed by $\mathbf{T}$ to fit the 'bed' $\mathbf{X}_2$ (see the frontispage for the legend of Procrustes). The terminology is due to Hurley and Cattell (1962) for whom the matrices have sizes given by $\mathbf{X}_1$, $\mathbf{X}_2$, $\mathbf{T}$, as defined above, and where $\mathbf{T}$ is any real matrix. This, of course, is the multivariate multiple regression problem, estimating $\mathbf{T}$ by:

$$\mathbf{T} = \left(\mathbf{X}_1'\mathbf{X}_1\right)^{-1} \mathbf{X}_1'\mathbf{X}_2. \tag{1.2}$$

The objective of Hurley and Cattell (1962) was to relate a factor structure $\mathbf{X}_1$, obtained by factor analysis, to a specified hypothetical target matrix $\mathbf{X}_2$ by finding an estimated transformation matrix $\mathbf{T}$ that would transform the obtained factor structure to fit the hypothesised structure. Ideally, Hurley and Cattell would have liked to estimate a matrix $\mathbf{C}$ giving the directions of a set of oblique axes. They were unable to do this, so estimated $\mathbf{T}$ in unconstrained form as in equation (1.2) and normalised its columns, thus giving a matrix $\mathbf{C}$ interpreted as a matrix of direction cosines. The reason that they coined the term Procrustes was that their Procrustes programme 'lends itself to the brutal feat of making almost any data fit almost any hypothesis!' (p. 260). One might draw attention to the additional brutality of normalising the columns of $\mathbf{T}$ rather than directly fitting a matrix $\mathbf{C}$ of direction cosines.

Procrustes methods in this context, though not the name itself, go back at least to Mosier (1939). Indeed Mosier formulated the correct least squares normal equations for oblique Procrustes analysis (Section 6.1). Writing many years before the computing revolution, Mosier remarks 'Since [the equations] are not homogeneous their solution is excessively laborious, though possible, and the following considerations justify an approximation.' Mosier's approximation seems to be none other than the normalised **T** solution proposed by Hurley and Cattell 23 years later. The oblique Procrustes problem was not fully solved until Browne (1967) gave a thorough analysis that led to a practical algorithm; Cramer (1974) and ten Berge and Knevels (1977) supplied some improvements.

The orthogonal Procrustes problem arises when **T** is an orthogonal matrix **Q**. The minimisation of (1.1) when $\mathbf{T} = \mathbf{Q}$ seems first to have been discussed and solved by Green (1952). Green confined himself to the case where $\mathbf{X}_1$ and $\mathbf{X}_2$ were both of full column rank; Schönemann (1966) relaxed this condition. Schönemann and Carroll (1970) associated a scaling factor with one of the matrices. Independently, Gower had derived the same results without publication but had applied them in Gower (1971) and had taken the first steps towards a generalised orthogonal Procrustes analysis (GPA or GOPA), where $\mathbf{X}_1$ and $\mathbf{X}_2$ are replaced by K sets $\mathbf{X}_1, \ldots, \mathbf{X}_K$, by forming a $K \times K$ table whose i, jth element is the residual sum-of-squares $\|\mathbf{X}_i\mathbf{Q}_{ij} - \mathbf{X}_j\|$, this table being further analysed by Multidimensional Scaling (see Section 9.3 on pairwise methods). This may have been the first instance where the matrices $\mathbf{X}_1, \mathbf{X}_2, \ldots, \mathbf{X}_K$ are regarded as configuration matrices rather than as coordinate matrices of factor loadings.

Still in the factor loading tradition, Kristof and Wingersky (1971) considered minimising the multi-set criterion $\sum_{i<j}^{K} \|\mathbf{X}_i\mathbf{Q}_i - \mathbf{X}_j\mathbf{Q}_j\|$ to determine the orthogonal matrices and to give an average 'factor matrix' $\mathbf{G} = K^{-1}\sum_{k=1}^{K}\mathbf{X}_k\mathbf{Q}_k$. Gower (1975) introduced GPA, minimising the same criterion as Kristof and Wingersky (1971) but introducing scaling factors for each set and viewing the whole problem as one of matching configurations. The equivalent to the average factor matrix **G** was initially called a *consensus* matrix but now is generally termed a *group average* matrix, using the same terminology as for a similar construct found in individual differences scaling (INDSCAL, Carroll and Chang 1971, see also Chapter 13). Gower also introduced an analysis of variance framework (Chapter 10). ten Berge (1977*a*) found an analytical result for the scaling factors (Section 9.1.2) which suggested an improved algorithm for GPA.

Although he described his method as *orthogonal Procrustes,* Cliff (1966) seems to have been the first to address a projection Procrustes problem where $\mathbf{T} = \mathbf{P}$. He used the inner product criterion: maximise $\mathrm{tr}(\mathbf{X}_2'\mathbf{X}_1\mathbf{P})$, and showed that this had a simple algebraic solution similar to that of the two-set orthogonal Procrustes problem, with which it coincides when $P_1 = P_2$ and with which it has a link even when $P_2 < P_1$ (Section 5.4). Green and Gower (1979) addressed the same problem but used the least-squares criterion. This does not have an algebraic solution but they suggested an algorithm for finding **P** (Section 5.4), which when $P_2 = 1$ offers an alternative, but less efficient, solution to the Browne (1967) oblique Procrustes algorithm. ten Berge and Knol (1984) discussed the Generalised Projection Procrustes Problem for K sets and with scaling factors, using both the least-squares and inner-product criteria (Sections 9.2.2 and 9.2.3). Peay (1988) also addressed the Generalised Projection Procrustes problem but rather than minimising the residual sum-of-squares, maximised the sum-of-squares of the group average (Section 9.2.3). We shall see that although in GPA the least-squares, inner products and maximal group average criteria are all equivalent, this equivalence does not extend to projection problems.

A variant of the Generalised Procrustes problem seems to have originated with PINDIS (Lingoes and Borg 1978) where transformations are attached to a given group-average, rather than to the individual configurations. This type of approach is reminiscent of the treatment of the group average in INDSCAL. These, and related methods, are discussed in Chapter 13.

In Table 1.1 an overview of the different Procrustes methods encountered in the literature is given, based on the optimisation criterion and the number of sets of data matrices involved in the analyses (two or more than two), and whether or not isotropic scaling is discussed.

Table 1.1 indicates only the tip of an iceberg, as is evident from a glance at our list of references. ten Berge (1977*b*) identified a taxonomy of 36 Procrustes problems and many new variants have been discussed since then; happily not all of them seem to be of value to data analysis. From Table 1.1 there are four types of transformation matrix if we distinguish row and column orthonormality, three criteria, two forms with and without scaling, and forms which handle two, or K sets giving a total of $4 \times 3 \times 2 \times 2 = 48$ possibilities. This number explodes if we take into account one and two sided problems, various anisotropic scaling possibilities (see Chapter 8), row and column weighting, double transformations, considerations concerning the order of the matrices and the handling of missing values.

1.2 Current trends

As we have seen, the origins of Procrustes analysis are concerned with matching factor loading matrices. This is not a main topic of this book, but we discuss some of the main issues in Chapter 6. From what we have said so far, it would seem that Procrustes analysis is solely an off-shoot of psychometrics. However, in the last 20 to 30 years similar problems have arisen in other fields of application and these have coloured more recent developments. In this section we mention some of the different strands that now stimulate research into Procrustean methods.

Rather than mathematical constructs like factors, in many of the more recent applications the matrices $\mathbf{X}_k$ derive from direct experimental observations on sets of variables. Perhaps after standardisation, this form of data naturally suggests viewing the cases as a set of points in a multidimensional Euclidean space, one point for each case and one dimension for each variable (Appendix A). The resulting cloud of points is now commonly referred to as a configuration, so we are concerned with matching configurations. The configurations may derive from multidimensional

Table 1.1. Chronology and coverage of some papers relating to Procrustes Analysis.

Year	Author	**Q**	**P**	**C**	2	*K*	*s*	LS	IP	GA	Comment
1939	Mosier			✓	✓			✓			Approximate solution
1952	Green	✓			✓			✓			Full rank matrices
1962	Hurley, Cattell			✓	✓			✓			'Procrustes' overlap with Mosier
1966	Schönemann	✓			✓			✓			Deficient rank case
1966	Cliff		✓		✓				✓		Two-sided
1967	Browne			✓	✓			✓			
1968	Schönemann	✓			✓			✓			Two-sided
1970	Schönemann, Carroll	✓			✓		*i*	✓			
1971	Gower	✓				✓		✓			Pairwise
1971	Kristof, Wingersky	✓				✓		✓			Simultaneous
1975	Gower	✓				✓	*i*	✓			Simultaneous Anova framework
1977*a*	ten Berge	✓				✓	*i*		✓		
1978	Borg, Lingoes		✓			✓	*a*	✓			PINDIS
1979	Green, Gower		✓			✓		✓			
1984	ten Berge, Knol		✓		✓	✓	*i*	✓	✓		
1988	Peay		✓			✓	*i*			✓	
1991	Commandeur		✓				*a*	✓			Missing rows
1991	Dijksterhuis, Gower	✓	✓		✓	✓	*i*	✓		✓	Review
1993	ten Berge, Kiers Commandeur	✓				✓	*i*	✓			Missing cells
1995	Gower	✓	✓		✓	✓	*i*	✓		✓	Review

Note: The meaning of the matrices **Q**, **P**, and **C** is as in the text; 2 stands for two sets and K for K-sets methods; the column s contains an i for isotropic and an a for anisotropic scaling; LS: Least Squares; IP: Inner Product; GA: Group Average, are abbreviations used for the optimisation criterion.

scaling of distance matrices, whose elements are themselves derived from the variables. Then, the dimensions of the configurations refer only indirectly to the variables; indeed, the different configurations occurring in the same analysis may be based on different variables. What is important is that the same row of each configuration matrix refers to the same case, otherwise matching makes no sense. We shall be concerned with the relative orientations of configurations but not so much with the associated coordinate values that are the focus of interest in factor analysis.

Since the mid-1970s, Procrustes methods have been widely used in aspects of food research concerned with sensory evaluations (Banfield and Harries 1975, Harries and Macfie 1976) and later extending into applications in the general field of the sensory and consumer sciences (Dijksterhuis 1997).

One of us came to Procrustes analysis via an unusual route at a time when he was much involved with hierarchical cluster analysis. Realising that dendrograms generate ultrametric distances and that ultrametric distances were embeddable in Euclidean space, he considered Procrustean rotation of the Euclidean ultrametric configuration to best fit the Euclidean configuration of the data (Gower and Banfield 1975). Later he realised that Procrustes methodology had direct application to data analysis, giving an application concerned with the physical anthropology of fossil skulls (Gower 1971). In that example, preliminary canonical variate analyses produced configurations for different parts of the skulls. The configurations were rotated to fit in pairs, to give a symmetric matrix whose elements $\|\mathbf{X}_i\mathbf{Q} - \mathbf{X}_j\| = \|\mathbf{X}_j\mathbf{Q}' - \mathbf{X}_i\|$ gave the 'distances' between different regions of the fossil skulls. A multidimensional scaling of this distance matrix gave visualisations in which pairs of adjacent points had close Procrustean fits. This gives a pairwise method for comparing more than two matrices (Section 9.3).

Rather than comparing configurations in pairs, Generalised Procrustes methods, developed in the early 1970s, sought to fit all configurations simultaneously; this turned out to be the same as fitting to a group average configuration. GPA (Gower 1975), has become popular in the sensory sciences. It is applied as an individual difference method to adjust for the differences in scoring between the assessors in a sensory panel (see, for example, Williams and Langron 1984, Arnold and Williams 1986). The paper by Banfield and Harries (1975) is probably the first in which Procrustes Analysis is applied in a sensory context. These authors used pairwise Procrustes Analysis to match the scores of pairs of judges. Harries and MacFie (1976) used a variant of this method to study the judging of meat by a several assessors. The use of pairwise methods differs from the simultaneous fitting of GPA as used in most later studies of sensory data. In Gower (1975), GPA is applied to the raw data sets from three assessors visually scoring properties of beef carcasses.

Up to this point most research was done in a least-squares context and, apart from the factor analysis literature with its emphasis on oblique rotations, was concerned with orthogonal matrices. More recent work based on projection methods using rectangular orthonormal matrices has several origins. Projection is justified when matching factors, in which case one is looking for orthogonal or oblique directions in a space of lower dimension than the full set of factors. However, projection is sometimes seen as an alternative to orthogonal rotation methods for comparing configurations, but projection can distort configurations and lead to subsequent misconceptions in interpretation. Geometrically, rotation and projection are essentially different things and when matching configurations it is wrong to regard them as variant methods with the same, or even similar, objectives. A further potential source of confusion is that although the least-squares and inner-product criteria are equivalent for orthogonal matrices, they differ for projection methods and so lead to different estimates. We hope that the following pages will help clarify the distinction between orthogonal matching and projection matching that, in our opinion, has so often obfuscated the current literature.

Ever since the days of d'Arcy Thompson*, biologists have been interested in transformation grids. The idea is to impose a rectangular grid on, say, a two-dimensional view of a human skull, and then study the transformed positions of the 'same' points in other anthropoid skulls, thus inducing a non-linear transformation of the original rectangular grid. Transformation grids embody a concept similar to that of matching configurations but more general than the linear transformations induced by **Q**, **P**, and **C**. Interest has moved overtly into the Procrustes domain, matching configurations described by landmark points. Shape analysis is discussed briefly in Section 4.6.2 and again in Section 14.1.4. More recently, non-linear transformation methods have been developed—see Section 14.1.5. Nowadays Procrustes methods are used in molecular biology in matching molecular structures, in ecology matching environments described by different species groups, and in consumer research, matching individual consumers' perceptions of products.

1.3 Overview of the topics in the book

In Chapter 2 we discuss initial transformations that may be useful before embarking on a Procrustes analysis proper.

Chapter 3 is concerned with the two sets Procrustes problem in its one-sided form (1.1) and in its two-sided form:

$$\|\mathbf{X}_1\mathbf{T}_1 - \mathbf{X}_2\mathbf{T}_2\|. \tag{1.3}$$

We start by discussing two-set criteria with general forms of $\mathbf{T}$. We also consider problems of translating $\mathbf{X}_1$ and $\mathbf{X}_2$, pointing out that in this book we nearly always assume that $\mathbf{X}_1$ and $\mathbf{X}_2$ are translated to a common origin, which is most conveniently achieved by arranging that both matrices have zero column sums. Next we discuss the principal forms of $\mathbf{T}$, starting in Chapter 4 with the case where $\mathbf{T}$ is an orthogonal matrix $\mathbf{Q}$; then $\mathbf{X}_1$ and $\mathbf{X}_2$ must have the same number P of columns. If the rows of $\mathbf{X}_1$ and $\mathbf{X}_2$ are regarded as giving the coordinates of points, then the orthogonal transformation leaves the distances between the points of each configuration unchanged; then the orthogonal Procrustes problem can be considered as rotating the configuration $\mathbf{X}_1$ to match the configuration $\mathbf{X}_2$. We also consider the special cases of orthogonal matrices that represent reflections in a plane (Householder transforms) and rotations in a plane (Jacobi rotations). In Chapter 5 we consider projection Procrustes where $\mathbf{T}$ is an orthonormal matrix $\mathbf{P}$ which has an interpretation of rotating an orthogonal projection of the P_1-dimensional configuration $\mathbf{X}_1$ to match a P_2-dimensional configuration $\mathbf{X}_2$. In the two-sided version $\mathbf{T}_1$ and $\mathbf{T}_2$ may have Q columns where $Q \leq P_1 P_2$; often, $Q = \min(P_1, P_2)$.

* d'Arcy Thompson's book *On Growth and Form* (1917) gave a seminal discussion of how to describe and measure shape and growth in biological organisms.

Chapter 6 considers oblique Procrustes problems where **T** is, or is a function of, a direction cosine matrix **C**. There are four possibilities and we develop a new unified approach, which determines the optimal direction of one axis conditional on the directions of the other axes. Then, for all four methods, one iterates over all axes until convergence. All the cases have clear geometrical interpretations.

Chapter 7 briefly reviews other choices of **T**, such as when **T** is of given rank or is a permutation, symmetric, or positive matrix. Finally, we discuss double Procrustes problems of the form:

$$\|\mathbf{S}\mathbf{X}_1\mathbf{T} - \mathbf{X}_2\| \tag{1.4}$$

where **S** and **T** are orthogonal; **S** and **T** diagonal are considered under the discussion of anisotropic scaling in Chapter 8. Clearly, **S** and **T** may be restricted in other ways, but these appear to be of little interest and are not discussed here.

In Chapter 8, we introduce weighting. One form of weighting is when the *rows* of $\mathbf{X}_1$ are weighted. It is also possible to weight the columns. In these cases the weighting is expressed in terms of diagonal matrices. The most general form of weighting is when every cell of $\mathbf{X}_1$ gets separate weighting. Missing values may be specified by giving rows, columns, or cells zero weights. Next, we consider the important case of isotropic scaling where a scaling factor can be applied to the whole of the matrix $\mathbf{X}_1$. This allows for the common situation where the relative sizes of $\mathbf{X}_1$ and $\mathbf{X}_2$ are unknown. Then we introduce anisotropic *scaling*, represented by diagonal matrices **S** and **T** in equation (1.4). Unlike weighting-matrices, which are given, scaling matrices are to be estimated. However, in iterative algorithms, the current estimates of scaling matrices may be treated as given weights while estimating updates of transformation matrices or further scaling matrices.

Chapter 9 is concerned with generalisations where the two sets of configurations $\mathbf{X}_1$ and $\mathbf{X}_2$ are replaced by K sets, $\mathbf{X}_1, \mathbf{X}_2, \ldots, \mathbf{X}_K$, each with its own transformation matrix $\mathbf{T}_1, \ldots, \mathbf{T}_K$. All the variants of two-sets Procrustes problems generalise; we have to discuss different choices of $\mathbf{T}_k$, scaling, weighting, and the optimisation criteria. The group average configuration $\mathbf{G} = K^{-1}\sum_{k=1}^{K}\mathbf{X}_k\mathbf{T}_k$ of the transformed configurations plays a central role. We defer until Chapter 13 a discussion of related models, some of which are close to the material of Chapter 9.

Chapter 10 continues the discussion of the multi-set problem, presenting several forms of analysis of variance which have only a rudimentary form for two sets Procrustes problem but which give more detailed information with K sets. Terms in this analysis of variance help with interpretation and throw more light on the possible choices of criteria suitable for fitting Procrustes models by least squares.

So far we have been concerned entirely with visualisations of the cases (row information), but when available, it is useful to incorporate information on variables. This is the concern of biplot methodology discussed in Chapter 11.

What little is known about probability models for Procrustes problems is discussed in Chapter 12. For most practical purposes we shall have to rely on data sampling procedures like permutation tests, jack-knifing, or boot-strapping.

The multi-set Procrustes problem may be viewed as a part of the family of three-mode multidimensional scaling methods. In Chapter 13 we discuss links with other three-mode methods, especially PINDIS, INDSCAL, STATIS, SMACOF, ALSCAL, and generalisations of canonical correlation.

Some application areas of Procrustes analysis are discussed in Chapter 14.

1.4 Closed form and algorithmic solutions

Some problems discussed in this book yield closed form solutions while many are open only to algorithmic solutions. Outline algorithms are displayed in boxed form throughout the text. Classically, closed-form means expressible in terms of *known* functions, but what is a known function? Certainly, it is a function that is well-defined and whose properties have been explored and published in the literature, but ultimately, it means a function whose values can be calculated. In the days before computers, this usually meant a function that had been, or could have been, tabulated but nowadays an algorithm for computing the values of the function suffices and is in many ways more convenient than tabulation. Even before computers, some acceptable functions were not amenable to tabulation and an algorithmic solution was essential—the calculation of the inverse or eigenvalues and vectors of a matrix provide two such examples. Ultimately, a known function is one whose values can be computed as functions of previously defined functions. Just as new functions may be expressed in terms of previously studied functions so new algorithms for their computation may incorporate existing algorithms. Thus, the old distinction between closed-form and algorithmic solutions has lost much of its force. If there remains a difference between the classical and the computer-age study of this aspect of functions, it is that nowadays less effort seems to be put into establishing the properties of new computable functions, but this is not a difference in principle. The impetus of the computer is such that emphasis is on the, almost daily, discovery of new c … e effort should be directed into establishing the …

GowerDijksterhuis.XLS
Notation Sheet.

1.5 Notation

In this section the notation that is used throughout the book is introduced. When deviations from this notation occur, it is made clear in the text. The order of a matrix is indicated in subscript characters; when no order is specified, the appropriate order, needed for the indicated operation, is assumed. In general, scalars are printed in italics and matrices and vectors in bold (upper and lower case, repectively).

Thus, by ${}_p\mathbf{A}_q$ we indicate that $\mathbf{A}$ has p rows and q columns. The notation ${}_p\mathbf{A}_q\mathbf{B}_r$ confirms the commensurability of the number q of columns of $\mathbf{A}$ with the number of rows q of $\mathbf{B}$.

By $\|\mathbf{A}\|$ we mean the Euclidean/Frobenius norm trace$(\mathbf{A}'\mathbf{A})$, the sum-of-squares of the elements of $\mathbf{A}$. For other norms we use a suffix, for example, $\|\mathbf{A}\|_1$ represents the L_1 norm.

Norms

Frobenius $\|A\|_F^2 = \sum_{i=1}^{m}\sum_{j=1}^{n}|a_{ij}|^2 = \text{trace}(A^*A) = \sum_{i=1}^{\min\{m,n\}}\sigma_i^2$ aka Euclidean norm (not to be confused with the L^2 norm [aka Eucl

Trace $\|A\|_{tr} = \text{trace}(\sqrt{A^*A}) = \sum_{i=1}^{\min\{m,n\}}\sigma_i$

A	point
B	point
$\mathbf{A}$	matrix used only to illustrate notation
$\mathbf{B}$	inner product matrix, used in INDSCAL context
$\mathbf{C}$	matrix of direction cosines
d	distance observed (usually with suffices, d_{ij})
$\mathbf{e}_j$	unit vector with 1 in the jth position, zero elsewhere
δ	distance approximated (usually with suffices, δ_{ij})
$\mathbf{D}$	distance matrix, observed
Δ	distance matrix, approximated
$\mathbf{G}$	group average configuration matrix
i	running index for cases, usually associated with the rows of a configuration or data matrix, normally $i = 1, \ldots, N$
$\mathbf{I}$	identity matrix of appropriate order
j	running index for the columns of a configuration or data matrix; columns may refer to variables
k	running index used for data sets, usually $k = 1, \ldots, K$
l	running index used for data sets
K	number of data sets
$\mathbf{m}$	translation vector
$\mathbf{M}$	matrix of pairwise residual sums-of-squares
N	number of cases, rows
$\mathbf{N}$	centring matrix, $\mathbf{I} - \mathbf{11}'/N$
O	origin
P	number of columns when $P_1 = P_2 = \cdots = P_K$
P_k	number of columns for the kth set
$\mathbf{P}$	orthonormal projection matrix
$\mathbf{Q}$	orthogonal matrix
R	number of dimensions for display, often $R = 2$ is chosen
$\mathbf{R}$	pre-scaling matrix
s, s_k	isotropic scaling parameters
$\mathbf{S}$	matrix with elements trace $(\mathbf{X}_k'\mathbf{X}_l)$
$\mathbf{S}$	post-scaling matrix (different meaning in STATIS)
S	Procrustes least-squares loss, under rotation
S^*	Procrustes least-squares loss, under rotation and isotropic scaling
$\mathbf{T}$	general transformation matrix
$\mathbf{v}$	in STATIS
$\mathbf{W}$	weight matrix
$\mathbf{X}_k$	configuration matrix for the kth set
$\mathbf{Y}_k$	observed data matrix for the kth set
$\mathbf{Z}$	inner-product matrix $\mathbf{X}_2'\mathbf{X}_1$
$\mathbf{Z}_{kl}$	inner-product matrix for ordered sets k and l : $\mathbf{X}_k'\mathbf{X}_l$
$\mathbf{1}$	column vector of ones

2
Initial transformations

In this chapter, several matters will be discussed concerning the scaling of data, prior to a Procrustes Analysis. Data-scaling and configuration-scaling are the terms adopted for the many kinds of transformations that may be deemed desirable before embarking on the actual Procrustes matching. Their aim is to eliminate possible incommensurabilities of variables within the individual data sets (data-scaling) and size differences between data sets (configuration-scaling). Although some choices of data-scaling have clear justification one must admit that other choices are largely subjective. The form and source of the data should help decide what, if any, data-scaling is needed. We consider three main types of data: (i) sets of coordinates derived from some form of multidimensional scaling (MDS) or, in the case of shape studies, directly measured landmark coordinates (see, for example, Lele and Richtsmeier 2001), (ii) data matrices whose columns refer to different variables and (iii) sets of loadings derived from factor analysis.

Figure 2.1 shows possible links between *configurations* (i) of N points with coordinates in R dimensions given as the rows of a matrix denoted by ${}_N\mathbf{X}_R$ and data matrices (ii) for N cases and P variables, denoted by the symbol ${}_N\mathbf{Y}_P$. The data-scaling considerations (ii) also apply to a data matrix that is to be used for factor analysis but now the dimensions of the factor loadings are given by ${}_P\mathbf{X}_F$ where $F < P$, is the number of factors (see Chapter 6).

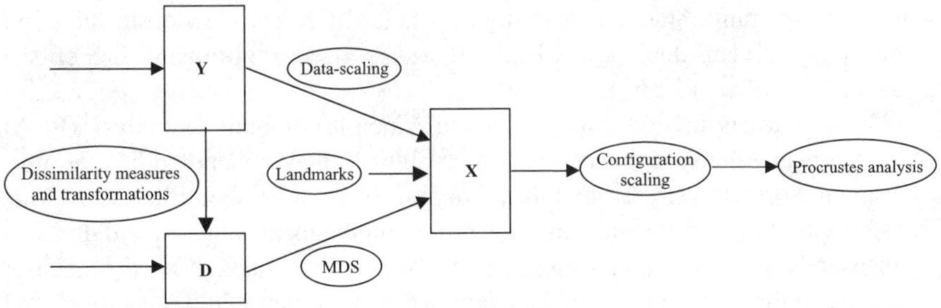

Fig. 2.1. Schematic representation of the relation between matrices **X**, **D**, and **Y**. Rectangles contain data types; **Y** is a data matrix, **D** is a distance matrix, and **X** is a configuration matrix; ellipses contain operations and measures described in the text.

Figure 2.1 shows that the most direct route to a Procrustes analysis is when the configuration **X** is directly measured, as with landmark data. The next most direct route is when **X** is clearly related to the raw data **Y** (with $R = P$); in extreme cases we may have $\mathbf{X} = \mathbf{Y}$. A less direct route is indicated in the Figure where a distance matrix ${}_N\mathbf{D}_N$, derived from a data matrix **Y** or given by direct experimental observation, is approximated by an MDS, yielding a configuration **X**. To be more precise, the distances between pairs of rows of **X** approximate the values in the distance matrix **D**. Here R is the chosen dimensionality of the approximation; often $R = 2$. For further information about MDS see Cox and Cox (2001), Kruskal and Wish (1978). When configurations arise from MDS, data-scaling is usually straightforward, but when they are derived directly from raw data, problems of data-scaling may be acute. Data-scaling for variables and distance matrices is discussed in Section 2.1.

Procrustes analysis compares two, or more, configurations $\mathbf{X}_1$ and $\mathbf{X}_2$ which may be derived by different methodologies. For example $\mathbf{X}_1$ may represent an MDS solution and $\mathbf{X}_2$ a pre-scaled data matrix. It is quite possible for the two configurations to be derived from the same **Y** by different methods of analysis. Even when the same data-scaling is used for **Y** (or for $\mathbf{Y}_1$ and $\mathbf{Y}_2$) different methodologies are likely to induce size differences in $\mathbf{X}_1$ and $\mathbf{X}_2$ that should be eliminated before proceeding to a Procrustes analysis. This is the concern of configuration-scaling discussed in Section 2.2.

2.1 Data-scaling of variables

The greatest need for data-scaling is when the columns of a data matrix **Y** represent different variables whose incommensurabilities of measurement scales need to be eliminated before giving acceptable coordinates of **X**. Of course, there are occasions when users are confident that variables are measured in the same units and call for no further scaling, so then $\mathbf{X} = \mathbf{Y}$. Table 2.1 presents an example of seven variables from different chemical/physical measurements on a set of apples (Dijksterhuis 1994). Clearly the variables in Table 2.1 are incommensurate; note the differences in mean, standard deviation and range. MOIST has a high mean value and large range/standard deviation, while ITHICK has a low mean value and small range/standard deviation. Clearly, there is a need to eliminate such effects by some form of standardisation.

The situation is directly analogous to the Principle components analysis (PCA) of **Y**, where any incommensurability of variables is usually handled by their normalisation, so that every variable has unit sum-of-squares about the mean. This transformation is accomplished through removing the mean value $\bar{y}_j$ and dividing by the standard deviation, s_j, of a vector of observations, making x_{ij} the distance of a score to the mean, measured in standard deviation units. In the notation used throughout this book:

$$x_{ij} = \frac{y_{ij} - \bar{y}_j}{s_j}. \tag{2.1}$$

Table 2.1. Mean, standard deviation, range, minimum and maximum of a number of different variables measured on apples.

Variable	Unit	Mean	Std. dev	Range	Minimum	Maximum
PRED	mm	4.43	0.981	4	3	7
PGREEN	mm	4.41	0.919	3.8	2.8	6.6
MOIST	g	24.5	13.1	39.65	10.18	49.83
DRYMAT	g	15.4	0.989	4.96	13.26	18.22
ACID	pH	7.60	1.16	5.42	4.77	10.19
ITHICK	mm	1.66	0.279	1.25	1.23	2.480
KATAC		12.1	2.17	9.9	7.2	17.1

Abbreviations:
PRED, PGREEN: penetrometer at red, respectively, green side of the apple; MOIST: expelled moisture; DRYMAT: amount of dry matter; ACID: acidity; ITHICK: INSTRON-thickness; KATAC: katalase activity.

Alternative standardisations may be obtained by dividing by the maximum value, by the range, or any other statistic reflecting spread of values of a variable. The considerations that apply to PCA apply equally to data-scaling of **Y** prior to a Procrustes Analysis.

For variables measured on ratio scales, with no negative values, an alternative strategy for eliminating incommensurabilities is to take logarithms:

$$x_{ij} = \log(y_{ij}). \tag{2.2}$$

Then the effect of measurement scales merely becomes an additive constant which is automatically eliminated when centring to each variable (see Section 3.2). When the distributions of some of the variables are very skewed, taking a logarithm can be advantageous; the base of the logarithm is immaterial.

2.1.1 CANONICAL ANALYSIS AS A FORM OF DATA-SCALING

With 'open-ended' quantitative variables (i.e. variables with no assigned upper limit), one can always normalise to unit Sum-of-squares (about the mean) as described above. When data on the *same* variables are available for K different sets, another possibility (see Dijksterhuis and Gower 1991/2) is to regard each set as a group in a canonical variate analysis and replace observed variables by the canonical variables, thus eliminating the effects of within group correlations and providing values of **X** that are invariant to the measurement scales used for the variables. This type of canonical variate analysis is unusual, because the same cases occur in all the sets, as when the same food products (the cases) are assessed by different judges (the sets), but no allowance should be made for this or interesting aspects of the data could be eliminated.

2.1.2 PRIOR MULTIVARIATE ANALYSES

The previous section was concerned with using a multivariate method, canonical analysis, to eliminate the effect of scales in the *observed* data sets $\mathbf{Y}_k$. All kinds of multivariate methods (MDS, factor analysis, PCA, etc.) may also be used to produce *clean* configurations prior to a Procrustes analysis. Of particular importance are quantification methods, such as nonlinear principal components analysis, that replace categorical variables by normalised quantitative scores. Quantifications of variables may be treated like a numerical data matrix **Y** and possibly used to generate **X** by a further MDS step. Thus, in Figure 2.1, the data-scaling path from **Y** to **X** now represents some form of multivariate analysis; for example, factor analysis fits in here.

Factor analysis and MDS both produce a configuration **X** in fewer dimensions than those of **Y**. Sometimes a more overt attitude is taken to dimension reduction, as when the individual data sets $\mathbf{Y}_k$ are separately analysed, by PCA, and *cleaned up* by retaining only a few dimensions $R < P_k$ for the subsequent Procrustes analysis. The information not included in the P_k–R dimensions of a set $\mathbf{X}_k$ will, of course, not contribute to the matching of the $\mathbf{X}_k$. A configuration of points common to all $\mathbf{X}_k$, but only visible in the higher P_k–R dimensions, can never be found, and a potential source of agreement between the K sets is lost. Thus, one has to be aware of a risk of sacrificing potential fit. Unless one has good reason to believe that the excluded dimensions represent 'noise', it is advisable that in such situations one retains all dimensions of the obtained configurations for use in the Procrustes analysis. Analysis of variance, as discussed in Chapter 10, helps one disentangle signal from noise, thus allowing decisions on dimension reduction to be made after the Procrustes analysis.

2.1.3 DISTANCE MATRICES

We have seen that a distance matrix **D** may be computed from more basic data given in a units × variables data matrix **Y**, possibly itself derived from quantifying categorical information as described above. Concerns about incommensurabilities between variables may persist but are usually taken care of by using a variety of distance/dissimilarity/similarity coefficients (e.g. Gordon 1999). The way these coefficients are defined depends on the kinds of variables included in **Y**, but details need not detain us here because we can assume, rightly or wrongly, that the coefficients themselves adequately handle any incommensurability. Then, we may proceed with a multidimensional scaling of **D**, so computed, to give an acceptable configuration **X**.

Another form of MDS seeks optimal transforms of **Y** leading to **D** and thence a configuration **X**. Permissible transformations include polynomial, monotonic and, especially, transformations of categorical and ordered categorical variables to give quantitative scales (Meulman 1992). Ordinal variables may be easily constructed from quantitative variables by defining grouping ranges and then requantified by using a method such as nonlinear principal components analysis. Thus, an appeal to ordinality is one way of eliminating incommensurabilities between the measurement scales.

2.2 Configuration-scaling

The configuration represented by $\mathbf{X}$ is the raw material of a Procrustes analysis. In Section 2.1 we have discussed several issues that help ensure that $\mathbf{X}$ is not influenced adversely by the measurement scales of $\mathbf{Y}$. Of course, we need more than one configuration, $\mathbf{X}_1$ and $\mathbf{X}_2$, say, and then the question arises as to whether $\mathbf{X}_1$ and $\mathbf{X}_2$ are themselves commensurable. For example, two sets of factor loadings may be differently standardised, in which case it would be sensible to take steps to put them on the same footing. Similarly, MDS configurations are not necessarily compatibly scaled. For example, $\mathbf{X}_1$ could be standardised to have total sum-of-squares of unity, and $\mathbf{X}_2$ could be scaled to lie in a unit square. Obviously $\mathbf{X}_1$ and $\mathbf{X}_2$ should be put on an equal footing before Procrustean comparisons can be made. In Figure 2.1 we have used the term *configuration-scaling* to distinguish the scaling of configurations as discussed in this section from the scaling of more basic data, termed data-scaling and discussed above in Section 2.1. The most simple strategy is to make $\mathbf{X}_1$ and $\mathbf{X}_2$ the same size. When $\mathbf{X}_1$ and $\mathbf{X}_2$ are both centred, *same size* is taken to mean that both are scaled to have the same sum-of-squares, which may be taken to be unity, about their centroids, that is, $\mathrm{tr}(\mathbf{X}_1'\mathbf{X}_1) = \mathrm{tr}(\mathbf{X}_2'\mathbf{X}_2) = 1$. Another way of expressing size is in terms of the square distances between the rows of the configuration matrices. Writing $\boldsymbol{\Delta}_1$ and $\boldsymbol{\Delta}_2$ for the matrices of squared distances generated by the rows of $\mathbf{X}_1$ and $\mathbf{X}_2$, then equal size may be defined by requiring that $\mathbf{1}'\boldsymbol{\Delta}_1\mathbf{1} = \mathbf{1}'\boldsymbol{\Delta}_2\mathbf{1}$ which implies that $\mathrm{tr}(\mathbf{X}_1'\mathbf{X}_1) = \mathrm{tr}(\mathbf{X}_2'\mathbf{X}_2)$ and again we find that it would be sensible to scale both to have the same total sum-of-squares $\mathrm{tr}(\mathbf{X}_1'\mathbf{X}_1) = \mathrm{tr}(\mathbf{X}_2'\mathbf{X}_2) = 1$. We have chosen a standardised sum-of-squares of unity, but any other value would be equally acceptable. We shall see in Chapter 3 that for many forms of Procrustes analysis the configurations $\mathbf{X}$ should be centred to have zero means. The argument based on equalising the sizes of $\boldsymbol{\Delta}_1$ and $\boldsymbol{\Delta}_2$ suggests that a scaling of unit sum-of-squares about the centroid is sensible even without recourse to considerations of centering.

2.2.1 DIFFERING NUMBERS OF VARIABLES

Free choice profiling (FCP, Arnold and Williams 1985), permits judges considerable freedom to choose what, and how many, rating scales they will use. This freedom almost guarantees that different and different numbers of variables are supplied by each judge. Further, different judges may define scales of different lengths and use the scales differentially—some people will use the whole range, and others use only an idiosyncratic part of the scale. Even variables with fixed ranges (such as 0–9 scales) may be used differentially. When different uses of scale are of primary interest, data-scaling would be wrong, but in other circumstances attempts should be made to correct for differential uses. As above, this can be done by scaling each variable for each judge to have unit sum-of-squares about its mean as in the normalisation equation (2.1).

A special difficulty arises when $\mathbf{Y}_1$ and $\mathbf{Y}_2$ have different numbers of variables. The normalisation (2.1) retains the same structure for the configurations $\mathbf{X}_1$ and $\mathbf{X}_2$ and now we shall have $\text{tr}(\mathbf{X}_1'\mathbf{X}_1) = P_1$ while $\text{tr}(\mathbf{X}_2'\mathbf{X}_2) = P_2$ so that sizes differ. To equalise the sizes, divide each $\mathbf{X}_k$ by $\sqrt{P_k}$. This is the same as dividing each variable in the kth data set by $\sqrt{P_k}$, which puts the size of the matrix $\mathbf{X}_k$ on a *per variable* basis (Dijksterhuis and Gower 1991/2). By this, we mean that not only does $\text{tr}(\mathbf{X}_k'\mathbf{X}_k) = 1$ but also each column of $\mathbf{X}_k$ has the same sum-of-squares, $1/P_k$. We term this process P_k scaling.

2.2.2 ISOTROPIC AND ANISOTROPIC SCALING

To complicate things further it will be seen in Section 3.3 that Procrustes Analysis itself can handle certain forms of scaling. For example, when comparing two configurations of different sizes, Procrustes analysis can scale one isotropically to fit the other. It turns out that even if the two configurations each have been scaled to have unit sums-of-squares, the estimated isotropic scaling factor can produce apparent further improvement. In our view, if the configurations have been scaled to the same size, there is little justification for additional estimation of isotropic scaling, but one could omit the configuration-scaling and let the isotropic scaling take care of size differences. When comparing two configurations whose coordinates are given by data-scaled variables, instead of configuration-scaling we could include a diagonal matrix of anisotropic scaling factors (see Section 8.3) to allow for incommensurabilities, but this would not cope well with variables that were very incommensurable with one another.

Our overall recommendation is to use configuration-scaling to adjust for all foreseeable size differences and only use the isotropic or anisotropic scaling in the Procrustes analysis if there are special reasons for doing so. One such reason would be when differences in the use of scales for different variables are of substantive interest and deserve separate estimation. See Section 8.3 for further comment on the interaction between configuration-scaling and isotropic scale estimation.

2.3 Types of data sets

We have mentioned that Procrustes Analysis is concerned with matching two *or more* configurations. Much of what we have said above carries over directly into the data-scaling and configuration-scaling of several configurations, but it may be useful to explore in a little more detail what happens with K sets. First we shall consider K configurations $\mathbf{X}_k, k = 1, \ldots, K$, then K data matrices $\mathbf{Y}_k$, $k = 1, \ldots, K$ and finally K distance matrices $\mathbf{D}_k$, $k = 1, \ldots, K$.

Figure 2.2 shows K configurations. What we have said about configuration-scaling of one or two matrices of configuration coordinates extends immediately to K sets. In Procrustes analysis the ith row of each configuration must refer to the same object, so each must have the same number N of rows. However, the

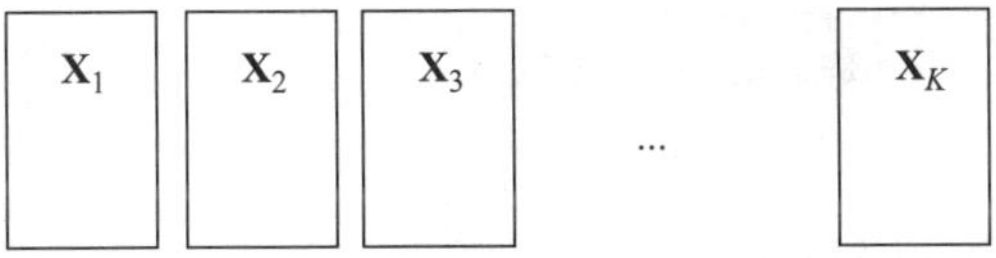

Fig. 2.2. Configuration matrices for K data sets, $\mathbf{X}_k, k = 1, \ldots, K$.

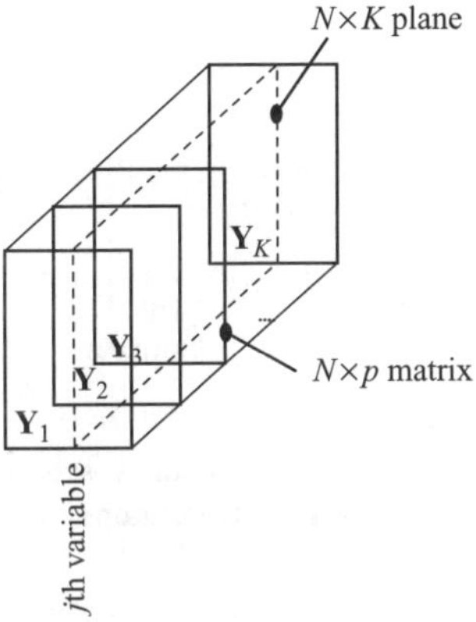

Fig. 2.3. K data matrices, $\mathbf{Y}_k$, $k = 1, \ldots, K$, with common variables.

columns refer to dimensions and the number of dimensions may differ from one configuration to the other. In general it is not legitimate to regard the jth dimension of one configuration as being in any way related to the jth dimension of another, even when both have the same number of columns. A possible exception is when the dimensions are ranked from most to least variable, as is common in PCA and factor analysis. Then, when the ranked dimensions are deemed to have substantive meanings (reification), one might possibly consider matching dimensions.

With K data matrices we have to consider two distinct situations. The first is shown in Figure 2.3 in which the jth variable is common to each data matrix. Note that it is not merely the number of variables but the variables themselves that match. Useful information is very likely to be contained in these matching variables and its analysis is the concern of classical multivariate methods (e.g. Hand and Taylor 1987). After the data matrices $\mathbf{Y}_k$ have been data-scaled and/or analysed by MDS, information on the variables is totally lost unless special steps are taken—see below. Even when data-scaling is used without MDS, Procrustean methods ignore any remaining information on the scaled variables, treating them as unmatched dimensions. It may seem that Procrustes analysis is ignoring much information but this would not be an appropriate assessment. Rather, Procrustes analysis leaves you free to do the classical analysis and provides information *additional* to that contained in the mean values of the variables. We have mentioned that special

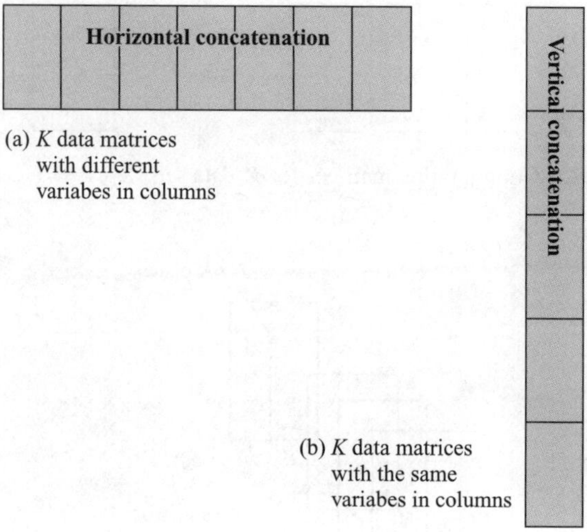

Fig. 2.4. Permissible concatenation directions of K sets of data matrices: (a) different variables in columns, and (b) same variables in columns; recall that in both cases the rows match.

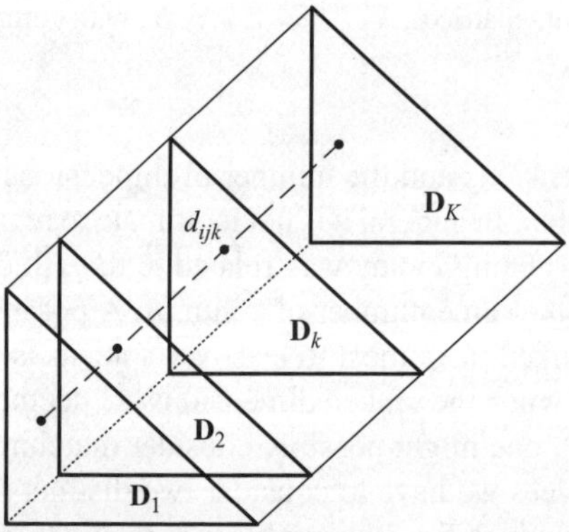

Fig. 2.5. K distance matrices $\mathbf{D}_k$, $k = 1, \ldots, K$. The dashed line indicates points giving the distances between i and j for the kth set.

steps may be taken to include information on the variables. This is the concern of biplot methodology (see Gower and Hand 1995) discussed in Chapter 11.

The second kind of collection of K data matrices, typically arising from FCP, is when the jth variable of each $\mathbf{Y}_k$ does not match, so that Figure 2.3 is inapplicable. Figure 2.4 gives a visual impression contrasting the two situations. In Figure 2.4, (a) is possible for both types of configuration matrix, since all that is required is that rows match. However, (b) is applicable only when the columns (i.e. the variables) match. In Figure 2.4(a) it is appropriate to normalise each variable together with scaling each variable of the kth set by dividing by the square root of P_k (see Section 2.2.1).

In Chapter 13 we shall also meet K sets of distance matrices $\mathbf{D}_k$, $k = 1, \ldots, K$ as in Figure 2.5. These also occur in non-Procrustean methods such as INDSCAL but we shall mainly be concerned with Procrustean operations on K configuration matrices derived by MDS from the $\mathbf{D}_k$.

We should also mention genuine three-way tables, where a single variable is classified by three categorical variables. Such tables may be represented as in Figure 2.3 but, unlike data-matrices, can be sliced in three directions to give either I, J or K sets of two-way arrays which, at least formally, could be analysed by Procrustean methods. We rarely regard such arrays as configuration matrices and would rather use classical methods such as generalised linear models (e.g. McCullagh and Nelder 1989), multivariate Anova (e.g. Hand and Taylor 1987) and some three mode models (e.g. Coppi and Bolasco 1989) as offering more appropriate methodology.

3 Two-set Procrustes problems—generalities

The next few chapters are concerned with various forms of two-set Procrustes problems. In this chapter we discuss some generalities which are applicable to all two-set problems, while the following chapters are concerned with the details of the main kinds of constraint on **T**—orthogonal transformations **Q** (Chapter 4), orthonormal transformations **P** (Chapter 5) and transformations to oblique axes **C** (Chapter 6). Chapter 7 covers some little used transformations including a discussion of double Procrustes problems. Chapter 8 discusses various forms of scaling and weighting and their relationships. Simple illustrative examples are used throughout, while Chapter 9 contains some examples of *K*-sets Procrustes analysis of real data.

3.1 Introduction

In this chapter we discuss problems where ${}_N(\mathbf{X}_1)_{P_1}$ and ${}_N(\mathbf{X}_2)_{P_2}$ are given configurations and we wish to find a transformation matrix ${}_{P_1}\mathbf{T}_{P_2}$ such that $\mathbf{X}_1\mathbf{T}$ in some sense best matches $\mathbf{X}_2$. Several ways in which best match may be measured are discussed in this section.

3.1.1 LEAST-SQUARES CRITERIA

The most commonly used criterion is least squares, which defines the best match as being given by the matrix **T** which minimises:

$$S = \|\mathbf{X}_1\mathbf{T} - \mathbf{X}_2\|. \tag{3.1}$$

Normally, **T** will be constrained in some way and in some applications must be square with $P_1 = P_2 = P$ but this is not assumed unless specifically stated. In this introductory section, the constraint, if any, on **T** is not specified; later chapters consider specific constraints.

In (3.1), $\mathbf{X}_1$ is matched to $\mathbf{X}_2$, the target matrix. Sometimes it is more appropriate to treat $\mathbf{X}_1$ and $\mathbf{X}_2$ symmetrically. There are two ways of doing this.

The first is the *two-sided* variant of (3.1) which seeks to find transformation matrices $\mathbf{T}_1$ and $\mathbf{T}_2$, each with *R* columns, such that $\mathbf{X}_1\mathbf{T}_1$ best matches $\mathbf{X}_2\mathbf{T}_2$, so rather than (3.1) we require the minimum of:

$$S = \|\mathbf{X}_1\mathbf{T}_1 - \mathbf{X}_2\mathbf{T}_2\|. \tag{3.2}$$

Note, that without constraints on $\mathbf{T}_1$ and/or $\mathbf{T}_2$ (3.2) has the trivial null solution $\mathbf{T}_1 = \mathbf{0}, \mathbf{T}_2 = \mathbf{0}$. A further variant consists of the *double* Procrustes problems expressed as $\|\mathbf{T}_2\mathbf{X}_1\mathbf{T}_1 - \mathbf{X}_2\|$ which gets some mention in Chapters 7 and 8.

The second way of expressing a symmetric relationship between $\mathbf{X}_1$ and $\mathbf{X}_2$ is to note that (3.2) may be written as:

$$\tfrac{1}{4}S = \|\mathbf{X}_1\mathbf{T} - \mathbf{G}\| = \|\mathbf{X}_2\mathbf{T} - \mathbf{G}\|,$$

where $\mathbf{G} = \frac{1}{2}(\mathbf{X}_1\mathbf{T} + \mathbf{X}_2\mathbf{T})$. It follows that:

$$\frac{1}{2}S = \sum_{k=1}^{2} \|\mathbf{X}_k\mathbf{T} - \mathbf{G}\|.$$

This is merely a rewriting of (3.2) but, as we shall see in Chapter 9, it suggests a simple way of generalising Procrustes problems to more than two sets, where there is no concept of a target matrix.

3.1.2 CORRELATIONAL AND INNER-PRODUCT CRITERIA

Criteria other than least squares may be chosen, several of which occur in the following. Here we mention the more important of these, which have correlational forms. The first of these is the RV coefficient, due to Escoufier (1970) and Robert and Escoufier (1976) the two forms of which are defined as:

$$r_V^2 = \frac{\{\mathrm{trace}(\mathbf{X}_2'\mathbf{X}_1\mathbf{T})\}^2}{\mathrm{trace}\{(\mathbf{X}_1\mathbf{T})'(\mathbf{X}_1\mathbf{T})\}\mathrm{trace}\{\mathbf{X}_2'\mathbf{X}_2\}} \tag{3.3}$$

and

$$r_V^2 = \frac{\{\mathrm{trace}\{(\mathbf{X}_2\mathbf{T}_2)'(\mathbf{X}_1\mathbf{T}_1)\}\}^2}{\mathrm{trace}\{(\mathbf{X}_1\mathbf{T}_1)'(\mathbf{X}_1\mathbf{T}_1)\}\mathrm{trace}\{(\mathbf{X}_2\mathbf{T}_2)'(\mathbf{X}_2\mathbf{T}_2)\}}. \tag{3.4}$$

Effectively, these are uncentered correlations between the elements of $\mathbf{X}_1\mathbf{T}_1$ and $\mathbf{X}_2$ (or $\mathbf{X}_2\mathbf{T}_2$ in the two-sided form), where each of the two matrices is assumed to be strung out column by column to form a single vector. When $\|\mathbf{X}_1\| = \|\mathbf{X}_1\mathbf{T}\|$ for (3.3) and when $\|\mathbf{X}_1\| = \|\mathbf{X}_1\mathbf{T}_1\|$ and $\|\mathbf{X}_2\mathbf{T}_2\| = \|\mathbf{X}_2\|$ for (3.4), the denominators are independent of $\mathbf{T}$ and then only the numerator is effective, leading to the inner-product criteria:

$$\mathrm{trace}(\mathbf{X}_2'\mathbf{X}_1\mathbf{T}) \tag{3.5}$$

and

$$\mathrm{trace}(\mathbf{T}_2'\mathbf{X}_2'\mathbf{X}_1\mathbf{T}_1). \tag{3.6}$$

It is hard to justify the use of the inner product as a matching criterion when the correlational interpretation is not available; normally this confines the use of the inner product forms when $\mathbf{T}_1$ and $\mathbf{T}_2$ are orthogonal matrices (see Chapter 4), although we shall see in Chapter 5 an example where the transformed matrices are constrained so that $\|\mathbf{X}_1\mathbf{T}_1\| = 1$ and $\|\mathbf{X}_2\mathbf{T}_2\| = 1$. Criteria (3.3) and (3.4) differ from the usual definitions of product–moment correlation in that no correction for the mean is specified but we shall see below, when discussing translation, that such corrections have a place in modified forms of (3.3) and (3.4).

An alternative correlational criterion is Tucker's (1951) coefficient of congruence. Effectively, this evaluates the uncentered correlations between corresponding columns of $\mathbf{X}_1\mathbf{T}$ and $\mathbf{X}_2$ and sums them. Algebraically this becomes:

$$r_T = \text{trace}\left[\text{diag}(\mathbf{X}_2'\mathbf{X}_2)^{-1/2}(\mathbf{X}_2'\mathbf{X}_1\mathbf{T}_1)\text{diag}(\mathbf{T}_1'\mathbf{X}_1'\mathbf{X}_1\mathbf{T}_1)^{-1/2}\right], \tag{3.7}$$

with a similar two-sided variant. With normalised matrices this too coincides with the inner-product criterion.

3.1.3 ROBUST CRITERIA

The statistical literature increasingly is interested in using the L_1 norm, the sum of the absolute residuals, as a criterion for fitting models. This is because it is more robust, in the sense that it gives less weight to large residuals than do least-squares methods. Another possibility is to minimise the median residual (Siegel and Benson 1982, see Section 4.8.2). These are just two examples of robust estimation, a subject which now has an extensive statistical literature (see, for example, Andrews *et al.* 1972 and Huber 1981). Many robust estimation methods are formulated in a least-squares context by applying weights to the residuals of each observation. These weights are derived as functions of the residuals chosen so that small residuals get high weights and big residuals get low weights. The weights will themselves generate new estimates of the residuals, so the whole process is iterative, known as Iterative (Re)weighted Least Squares (IRLS). From an applied standpoint, the particular choice of weighting function may not be critical. The use of IRLS adds a further level of iteration to Procrustes algorithms. In principle these ideas may be applied to all Procrustes problems but we know of few applications; Verboon and Heiser (1992) give one example that we discuss in Section 4.8.2.

3.1.4 QUANTIFICATIONS

Procrustes methods may be applied when $\mathbf{X}_1$ and $\mathbf{X}_2$ are replaced by categorical data represented by indicator matrices $\mathbf{G}_1$ and $\mathbf{G}_2$. In the least-squares context one may proceed by searching for numerical category scores $\mathbf{Z}_1$ and $\mathbf{Z}_2$ which minimise $\|\mathbf{G}_1\mathbf{Z}_1\mathbf{T} - \mathbf{G}_2\mathbf{Z}_2\|$. For ordered categorical variables, order constraints can be put on the scores. There is a considerable literature on the quantification of categorical variables (see Gifi 1990). In principle, similar methods would apply to any choice of criterion. Little work has been done applying these concepts to Procrustes analysis but see van Buuren and Dijksterhuis (1988) for an example of this approach.

3.1.5 MATCHING OF ROWS AND COLUMNS

Initially we are concerned with the least-squares problems (3.1) and (3.2), regarding the rows of $\mathbf{X}_1$ as giving the coordinates of N points in P dimensions, and similarly for $\mathbf{X}_2$. Then (3.1) represents a matching of two configurations, each of N points. For matching to be meaningful, the ith point of each configuration should refer to the same entity—typically—the same sample or the same case or the same product. Although the *rows* of $\mathbf{X}_1$ and $\mathbf{X}_2$ must correspond in this way, the *columns* may refer to different observed variables or may be *derived* from different variables. The latter is the more usual, as different variables are normally measured on different scales (incommensurable), so that (3.1), although algebraically defined, would be statistically meaningless. Indeed, problems of commensurability may persist even when the columns of $\mathbf{X}_1$ and $\mathbf{X}_2$ *do* refer to the same variables (see Section 2.2.1). Commonly, $\mathbf{X}_1$ and $\mathbf{X}_2$ will have been derived from some form of multidimensional scaling (see Chapter 2) and then there is no problem of commensurability* apart from a possible arbitrary scaling factor that allows for MDS software not to scale its output coordinates to a standard size. Nevertheless, even with MDS configurations, there is rarely any good reason for believing that the columns of $\mathbf{X}_1$ and $\mathbf{X}_2$ match in any substantive sense. The fundamental reason for this is that the dimensions of configurations are generally unordered so that, for example, the term *first dimension* is meaningless. It follows that one should rarely try to match the corresponding dimensions of two configurations. The exception is when an ordering is imposed, for example, when dimensions are given in order of decreasing eigenvalues, as in many methods of multivariate analysis. Then, dimensions could be matched, provided that there is good reason (reification) to believe that the ordering has a substantive interpretation.

3.2 Translation

When it is the shapes of configurations $\mathbf{X}_1$ and $\mathbf{X}_2$ that are of most interest, the origins to which they are referred are irrelevant. Hence, it is usual, though not essential, to consider the better fits of the configurations that can be obtained by allowing for shape-preserving translations of origin $\mathbf{X}_1 - \mathbf{1a}_1'$ and $\mathbf{X}_2 - \mathbf{1a}_2'$. How the details of handling translations depend on the goodness of fit criterion used and on the precise form of the Procrustean model are discussed in this section.

* It would be truer to say that any incommensurability between variables has been subsumed when defining (dis)similarity coefficients used by the MDS.

3.2.1 SIMPLE TRANSLATION

If we translate $\mathbf{X}_1$ and $\mathbf{X}_2$ (3.1) becomes $\|(\mathbf{X}_1 - \mathbf{1a}_1')\mathbf{T} - (\mathbf{X}_2 - \mathbf{1a}_2')\|$, which may be written:

$$\|\mathbf{X}_1\mathbf{T} - \mathbf{1a}' - \mathbf{X}_2\|, \tag{3.8}$$

where $\mathbf{a}' = \mathbf{a}_1'\mathbf{T} - \mathbf{a}_2'$. The reason why translation can be expressed in terms of a single parameter $\mathbf{a}$ is because it is only the relative positions of the origins that have any operational significance. Thus we have only to minimise (3.8) over $\mathbf{T}$ and a single translational parameter $\mathbf{a}$. In (3.8) $\mathbf{a}$ is a P_2-vector, so the term $\mathbf{1a}'$ represents a translation or constant displacement for every row of $\mathbf{X}_1\mathbf{T} - \mathbf{X}_2$. We note that $\mathbf{X}_1\mathbf{T}$ represents a linear transformation of $\mathbf{X}_1$ while $\mathbf{X}_1\mathbf{T} - \mathbf{1a}'$ represents an affine transformation. Of course, $\|\mathbf{A} - \mathbf{1a}'\|$ is minimised when $\mathbf{a}' = \mathbf{1}'\mathbf{A}/N$, the means of the columns of $\mathbf{A}$. A simple way of effecting this translation on the elements of $\mathbf{A} = \mathbf{X}_1\mathbf{T} - \mathbf{X}_2$ is to remove the column-means from $\mathbf{X}_1$ and $\mathbf{X}_2$ separately, that is, express the data in terms of deviations from their column-means, a process known as centring. Algebraically, centring involves evaluating:

$$\mathbf{A} - \mathbf{1a}' = \mathbf{A} - \mathbf{11}'\mathbf{A}/N = (\mathbf{I} - \mathbf{11}'/N)\mathbf{A} = \mathbf{NA}, \quad \text{where } \mathbf{N} = (\mathbf{I} - \mathbf{11}'/N).$$

We term $\mathbf{N}$ a centring matrix. With this notation we see that $(\mathbf{NX})\mathbf{T} = \mathbf{N}(\mathbf{XT})$, showing that the result is the same whether we centre $\mathbf{X}$ before transformation or centre the transformed $\mathbf{XT}$. The simplest thing to do is to centre before transformation.

With the inner-product criterion (3.5) both $\mathbf{X}_1$ and $\mathbf{X}_2$ may need separate translations $\mathbf{a}_1$ and $\mathbf{a}_2$ so that (3.5) becomes:

$$\text{trace}(\mathbf{X}_2 - \mathbf{1a}_2)'(\mathbf{X}_1\mathbf{T} - \mathbf{1a}_1'), \tag{3.9}$$

or the equivalent alternative parameterisation:

$$\text{trace}(\mathbf{X}_2 - \mathbf{1a}_2')'(\mathbf{X}_1 - \mathbf{1a}_1')\mathbf{T}.$$

Obviously (3.9) can be made indefinitely large by choosing large (negative) values of $\mathbf{a}_1$ and $\mathbf{a}_2$ showing that, in this context, translation is not a useful concept. However, progress can be made with the correlational form (3.3), which now becomes:

$$r^2 = \frac{\{\text{trace}(\mathbf{X}_2 - \mathbf{1a}_2')'(\mathbf{X}_1\mathbf{T} - \mathbf{1a}_1')\}^2}{\{\text{trace}(\mathbf{X}_1\mathbf{T} - \mathbf{1a}_1')'(\mathbf{X}_1\mathbf{T} - \mathbf{1a}_1')\}\{\text{trace}(\mathbf{X}_2 - \mathbf{1a}_2')'(\mathbf{X}_2 - \mathbf{1a}_2')\}}$$
$$= \frac{C^2}{\text{AB}} \text{ (say).}$$

Differentiating with respect to $\mathbf{a}_1$ gives:

$$2\text{ABC}\frac{\partial C}{\partial \mathbf{a}_1} = C^2 B \frac{\partial A}{\partial \mathbf{a}_1} \quad \text{giving} \quad A(\mathbf{1}'\mathbf{X}_2 - N\mathbf{a}_2') = -C(\mathbf{1}'\mathbf{X}_1\mathbf{T} - N\mathbf{a}_1').$$

Similarly, differentiating with respect to $\mathbf{a}_2$ gives:

$$B(\mathbf{1}'\mathbf{X}_1\mathbf{T} - N\mathbf{a}_1') = -C(\mathbf{1}'\mathbf{X}_2 - N\mathbf{a}_2').$$

Combining these equations yields that either $C^2 = AB$, in which case $r^2 = 1$, or $\mathbf{1}'\mathbf{X}_1\mathbf{T} - N\mathbf{a}_1' = 0$ and $\mathbf{1}'\mathbf{X}_2 - N\mathbf{a}_2' = 0$. Unit correlation implies that $\mathbf{X}_1\mathbf{T} - \mathbf{1a}_1'$ must be proportional to $\mathbf{X}_2 - \mathbf{1a}_2'$ an unlikely event. Therefore we are left with the other solution which replaces $\mathbf{X}_1\mathbf{T} - \mathbf{1a}_1'$ by $\mathbf{NX}_1\mathbf{T}$ and $\mathbf{X}_2 - \mathbf{1a}_2'$ by $\mathbf{NX}_2$ as for the least-squares criterion. Again, $\mathbf{a}_1$ and $\mathbf{a}_2$ are eliminated by centring the columns of $\mathbf{X}_1$ and $\mathbf{X}_2$. The same thing is found when allowing for translation terms in the two-sided criteria (3.2) and (3.4). Thus, both least squares and correlational criteria lead to the same way of handling translations by centring. Therefore, throughout this book, unless otherwise specified, we shall assume that all matrices $\mathbf{X}_k$ are centred. Of course, if one does not wish to allow for translational effects, then the matrices should not be centred. With centred matrices we have that $\mathbf{1}'\mathbf{X}_1 = \mathbf{1}'\mathbf{X}_2 = \mathbf{0}$ and then all translational effects are estimated as zero. Then (3.1) is the same as (3.8) and (3.3) is the same as (3.9). Geometrically, the N points whose coordinates are given by the rows of $\mathbf{X}_1$ have the same centroid as the N points whose coordinates are given by the rows of $\mathbf{X}_2$. An important situation, where simple centring does not suffice, is when $\mathbf{X}_1$ is preceded by a weighting matrix; this is discussed in Chapter 8.

Before closing this section on eliminating translational effects, we emphasise once again (see Section 2.3) that there are many instances, especially when the columns of $\mathbf{X}_1$ and $\mathbf{X}_2$ refer to variables, where the main, perhaps only, interest is in the differences between the means.

3.3 Isotropic scaling

The solution to the minimisation problem (3.1) depends on what, if any, constraints are put on the form permitted for $\mathbf{T}$. A scaling factor s which requires estimation may be included and then $\|s\mathbf{X}_1\mathbf{T} - \mathbf{X}_2\|$ is to be minimised. In unconstrained problems, s may be absorbed into $\mathbf{T}$; in constrained problems, the least-squares estimate of s is given by:

$$\hat{s} = \frac{\text{trace}(\mathbf{X}_2'\mathbf{X}_1\mathbf{T})}{\text{trace}(\mathbf{T}'\mathbf{X}_1'\mathbf{X}_1\mathbf{T})}. \tag{3.10}$$

Thus, the scaling parameter may be estimated trivially, whatever constraint may be imposed on $\mathbf{T}$. When, as here, the same scaling factor is attached to all elements of $\mathbf{X}_1$, the scaling is said to be *isotropic*; anisotropic scaling will be discussed in Section 8.3. Substituting for $\hat{s}$ gives:

$$\|\hat{s}\mathbf{X}_1\mathbf{T} - \mathbf{X}_2\| = \|\mathbf{X}_2\|[1 - r_V^2(\mathbf{X}_1\mathbf{T}_1, \mathbf{X}_2)],$$

showing the link between estimating an isotropic scaling coefficient by least squares and maximising the correlation between $\mathbf{X}_2$ and the transformed $\mathbf{X}_1$. Thus,

The following is a general strategy for handling the case $P_1 > P_2$. The strategy is presented in the context of least squares but similar strategies are available for other criteria. We proceed by adding P_1-P_2 extra columns to $\mathbf{T}$ and $\mathbf{X}_2$ and consider the minimisation of:

$$S_i = \|\mathbf{X}_1(\mathbf{T}_i, \mathbf{R}_i) - (\mathbf{X}_2, \mathbf{X}_1\mathbf{R}_{i-1})\|, \tag{3.12}$$

where we assume that $\mathbf{R}_{i-1}$ is known; initially, we could set $\mathbf{X}_1\mathbf{R}_0$ to be the last P_1-P_2 columns of $\mathbf{X}_1$. Now $P_1 = P_2$ and this constrained minimisation can be handled to give estimates of $\mathbf{T}_i$ and $\mathbf{R}_i$. We have that:

$$S_{i+1} = \|\mathbf{X}_1(\mathbf{T}_{i+1}, \mathbf{R}_{i+1}) - (\mathbf{X}_2, \mathbf{X}_1\mathbf{R}_i)\| < \|\mathbf{X}_1(\mathbf{T}_i, \mathbf{R}_i) - (\mathbf{X}_2, \mathbf{X}_1\mathbf{R}_i)\|, \tag{3.13}$$

because the left-hand-side minimises S_{i+1}. Also,

$$\|\mathbf{X}_1(\mathbf{T}_i, \mathbf{R}_i) - (\mathbf{X}_2, \mathbf{X}_1\mathbf{R}_i)\| < \|\mathbf{X}_1(\mathbf{T}_i, \mathbf{R}_i) - (\mathbf{X}_2, \mathbf{X}_1\mathbf{R}_{i-1})\| = S_i, \tag{3.14}$$

because the final P_1-P_2 columns are zero on the left-hand-side and non-zero on the right. Putting together (3.13) and (3.14) we see that $S_{i+1} < S_i$ so the residual sum-of-squares has decreased. It follows that the sequence S_i converges. At convergence, $S_i = \|\mathbf{X}_1(\mathbf{T}, \mathbf{R}) - (\mathbf{X}_2, \mathbf{X}_1\mathbf{R})\| = \|\mathbf{X}_1\mathbf{T} - \mathbf{X}_2\|$, so we have found at least a local minimum of the original criterion with $P_1 > P_2$.

Algorithm 5.1 is a special case of the above procedure, where $\mathbf{T}$ is orthonormal and $(\mathbf{T}, \mathbf{R})$ is an orthogonal matrix.

The above case, where $\mathbf{T}$ has fewer columns than rows, is not uncommon. More rarely, we may require that $P_1 < P_2$. Then, we may write:

$$\|\mathbf{X}_1\mathbf{T} - \mathbf{X}_2\| = \left\|(\mathbf{X}_1, \mathbf{0})\begin{pmatrix}\mathbf{T}\\ \mathbf{T}^*\end{pmatrix} - \mathbf{X}_2\right\|,$$

where $\mathbf{0}$ represents P_2-P_1 null columns appended to $\mathbf{X}_1$ and $\binom{\mathbf{T}}{\mathbf{T}^*}$ is square. Thus, the minimisation of $\|\mathbf{X}_1\mathbf{T} - \mathbf{X}_2\|$ may be converted to operate on a square matrix. This is a simpler problem than when $P_1 > P_2$ but one has to ensure that the estimate of $\binom{\mathbf{T}}{\mathbf{T}^*}$ is consistent with the constraints put on $\mathbf{T}$. An algorithmic application of this approach is discussed in Section 4.5 (see also Section 5.7 and Appendix C.1).

3.6 A K-sets problem with two-sets solutions

We explore K-sets Procrustes problems in Chapters 9 and 13. However, there is one apparent K-sets problem that we can handle in the two-sets framework. This is when we consider the Procrustes problem of minimising:

$$S = \sum_{k=1}^{K} \|\mathbf{C}_k\mathbf{T} - \mathbf{B}_k\|, \tag{3.15}$$

where $\mathbf{B}_k$ and $\mathbf{C}_k$, $k = 1, 2, \ldots, K$ are all given. At first sight, this is a K-sets problem but we may expand the criterion as:

$$
\begin{aligned}
S &= \operatorname{trace}\left[\mathbf{T}'\left(\sum_{k=1}^{K}\mathbf{C}_k'\mathbf{C}_k\right)\mathbf{T} - 2\mathbf{T}'\left(\sum_{k=1}^{K}\mathbf{C}_k'\mathbf{B}_k\right) + \left(\sum_{k=1}^{K}\mathbf{B}_k'\mathbf{B}_k\right)\right] \\
&= \left\| \left(\sum_{k=1}^{K}\mathbf{C}_k'\mathbf{C}_k\right)^{1/2}\mathbf{T} - \left(\sum_{k=1}^{K}\mathbf{C}_k'\mathbf{C}_k\right)^{-1/2}\left(\sum_{k=1}^{K}\mathbf{C}_k'\mathbf{B}_k\right)\right\| \\
&\quad + \operatorname{trace}\left[\left(\sum_{k=1}^{K}\mathbf{B}_k'\mathbf{B}_k\right) - \left(\sum_{k=1}^{K}\mathbf{B}_k'\mathbf{C}_k\right)\left(\sum_{k=1}^{K}\mathbf{C}_k'\mathbf{C}_k\right)^{-1}\left(\sum_{k=1}^{K}\mathbf{C}_k'\mathbf{B}_k\right)\right],
\end{aligned}
\tag{3.16}
$$

where the matrix square roots are the unique positive definite symmetric square root of the indicated positive definite symmetric matrices. Only the first term of (3.16) involves $\mathbf{T}$, so the minimisation of (3.15) turns out to be a simple one-sided two-sets Procrustes problem (3.1) with $\mathbf{X}_1 = \left(\sum_{k=1}^{K}\mathbf{C}_k'\mathbf{C}_k\right)^{1/2}$ and $\mathbf{X}_2 = \left(\sum_{k=1}^{K}\mathbf{C}_k'\mathbf{C}_k\right)^{-1/2}\left(\sum_{k=1}^{K}\mathbf{C}_k'\mathbf{B}_k\right)$.

4
Orthogonal Procrustes problems

In Orthogonal Procrustes Analysis, $\mathbf{T}$ is constrained to be an orthogonal matrix (see Section 1.3) so is square with $P_1 = P_2 = P$. An orthogonal matrix represents a generalised rotation* (see Appendix B) so we require the rotation of $\mathbf{X}_1$, in P-dimensional space, that best fits $\mathbf{X}_2$. Figure 4.1 illustrates the procedure in the simple case where $\mathbf{X}_1$ and $\mathbf{X}_2$ are both triangles.

As in Section 1.3 we use $\mathbf{Q}$ to denote an orthogonal matrix and then the constrained version of (3.1) requires the minimum of $\|\mathbf{X}_1\mathbf{Q} - \mathbf{X}_2\|$. We have that:

$$\|\mathbf{X}_1\mathbf{Q} - \mathbf{X}_2\| = \text{trace}\,(\mathbf{X}_1'\mathbf{X}_1 + \mathbf{X}_2'\mathbf{X}_2) - 2\,\text{trace}\,(\mathbf{X}_2'\mathbf{X}_1\mathbf{Q}).$$

The first part of the right-hand-side does not contain $\mathbf{Q}$, so we require the maximum of $\text{trace}(\mathbf{X}_2'\mathbf{X}_1\mathbf{Q})$, showing that in this case the least-squares (3.1) and inner-products (3.3) criteria must give the same result. Further, because orthogonal transformations leave size unchanged, $\|\mathbf{X}_1\mathbf{Q}\| = \|\mathbf{X}_1\|$ and the correlational interpretation of the inner-product criterion is valid. It also follows that the same orthogonal matrix $\mathbf{Q}$ maximises $\|\mathbf{X}_1\mathbf{Q} + \mathbf{X}_2\|$ because this leads to the same inner-product criterion. We may write $\|\mathbf{X}_1\mathbf{Q} + \mathbf{X}_2\|$ as $4\|\mathbf{G}\|$ where $\mathbf{G}$ is the average configuration $\frac{1}{2}(\mathbf{X}_1\mathbf{Q} + \mathbf{X}_2)$; that is, orthogonal Procrustes analysis finds the rotation that gives the biggest average configuration among all possible generalised rotations, a result which will be of interest later (Sections 5.6 and 9.1.9). When we consider the minimisation of $\|\mathbf{X}_1\mathbf{Q}_1 - \mathbf{X}_2\mathbf{Q}_2\|$, the orthogonal version of (3.3), we see that this is the same as minimising $\|\mathbf{X}_1\mathbf{Q} - \mathbf{X}_2\|$ where $\mathbf{Q} = \mathbf{Q}_1\mathbf{Q}_2'$ is itself orthogonal; similarly for the maximisation of $\|\mathbf{X}_1\mathbf{Q}_1 + \mathbf{X}_2\mathbf{Q}_2\|$. This means that rotating $\mathbf{X}_1$ through $\mathbf{Q}_1$ and $\mathbf{X}_2$ through $\mathbf{Q}_2$ is the same as rotating $\mathbf{X}_1$ through $\mathbf{Q}_1\mathbf{Q}_2'$ and leaving $\mathbf{X}_2$ fixed. Thus, one-sided and two-sided orthogonal Procrustes problems are the same. That so many different criteria give the same solution is a feature of orthogonal Procrustes analysis, which singles it out from all the other Procrustes problems considered in this book.

* Recall that generalised rotations include the possibility of reflection (see Section 4.3 and Appendix). Note also that in the literature the term *rotation* is sometimes used as a synonym for transformation and need not refer to an orthogonal matrix. In this book a rotation is always expressed by an orthogonal matrix.

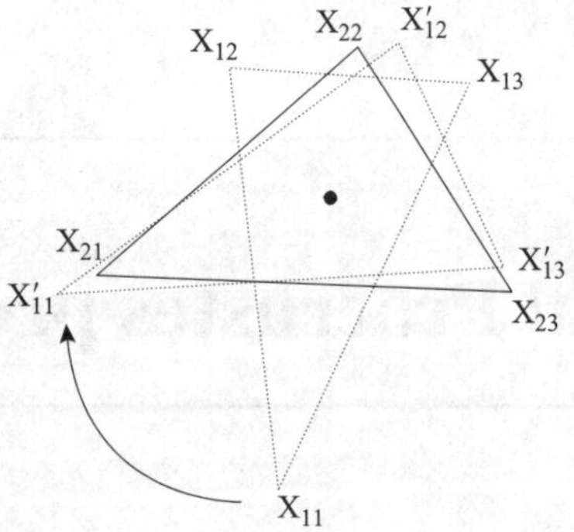

Fig. 4.1. $X_{11}X_{12}X_{13}$ and $X_{21}X_{22}X_{23}$ are two dissimilar triangles with the same centroid. $X'_{11}X'_{12}X'_{13}$ shows the position of the first triangle after it has been rotated to best fit according to the orthogonal Procrustes criterion. The arrow gives the sense (here, clockwise) of rotation.

4.1 Solution of the orthogonal Procrustes problem

In this paragraph, we derive the well-known analytical solution to the orthogonal Procrustes problem. Expressing $\mathbf{X}_2'\mathbf{X}_1$ in terms of its singular value decomposition $\mathbf{X}_2'\mathbf{X}_1 = \mathbf{U\Sigma}V'$ gives:

$$\text{trace}(\mathbf{X}_2'\mathbf{X}_1\mathbf{Q}) = \text{trace}(\mathbf{U\Sigma V'Q}) = \text{trace}(\mathbf{\Sigma V'QU}) = \text{trace}(\mathbf{\Sigma H}),$$

where $\mathbf{H} = \mathbf{V'QU}$, being the product of orthogonal matrices, is itself orthogonal. We have that:

$$\text{trace}(\mathbf{\Sigma H}) = \sum_{i=1}^{P} h_{ii}\sigma_i, \tag{4.1}$$

which, because the singular values σ_i are non-negative, is maximum when $h_{ii} = 1$ for $i = 1, 2, \ldots, P$, the maximal value attained by the elements of an orthogonal matrix. Thus at the maximum $\mathbf{H} = \mathbf{I}$ giving, $\mathbf{I} = \mathbf{V'QU}$ and

$$\mathbf{Q} = \mathbf{VU}'. \tag{4.2}$$

The residual sum-of-squares is given by:

$$\|\mathbf{X}_1\mathbf{Q} - \mathbf{X}_2\| = \|\mathbf{X}_1\| + \|\mathbf{X}_2\| - 2\,\text{trace}(\mathbf{\Sigma}), \tag{4.3}$$

which may be used as the basis for a simple analysis of variance (ANOVA).

A special case of (4.2) is when $\mathbf{X}_1 = \mathbf{I}$, then $\mathbf{Q}$ gives the best orthogonal matrix approximation to $\mathbf{X}_2$. As before $\mathbf{Q} = \mathbf{VU}'$ is given by (4.2) but now $\mathbf{U}$ and $\mathbf{V}$ are the singular vectors of $\mathbf{X}_2'$; this result was required by Gibson (1962).

An example of a two sets orthogonal Procrustes analysis is given in Section 4.4.

4.2 Necessary and sufficient conditions for optimality

We have shown that trace($\mathbf{X}_2'\mathbf{X}_1\mathbf{Q}$) is maximised when $\mathbf{Q} = \mathbf{VU}'$ where $\mathbf{U\Sigma V}'$ is the singular value decomposition of $\mathbf{X}_2'\mathbf{X}_1$. It follows that:

$$\mathbf{X}_2'\mathbf{X}_1 = \mathbf{U\Sigma V'VU'} = \mathbf{U\Sigma U}',$$

showing that at the maximum, $\mathbf{X}_2'\mathbf{X}_1\mathbf{Q}$ is a symmetric positive-semi-definite (p.s.d.) matrix. We now show that the converse is true: if $\mathbf{X}_2'\mathbf{X}_1\mathbf{Q}$ is a symmetric p.s.d matrix for some orthogonal matrix $\mathbf{Q}$, then the trace is maximised. This follows from noting that if $\mathbf{X}_2'\mathbf{X}_1\mathbf{Q}$ is a symmetric p.s.d. then we have the spectral decomposition:

$$\mathbf{X}_2'\mathbf{X}_1\mathbf{Q} = \mathbf{L\Lambda L}',$$

where $\mathbf{L}$ is orthogonal and $\mathbf{\Lambda}$ is diagonal with non-negative eigenvalues. It follows that:

$$\mathbf{X}_2'\mathbf{X}_1 = \mathbf{L\Lambda(QL)}',$$

giving the SVD of $\mathbf{X}_2'\mathbf{X}_1$. The singular value decomposition is unique, unless some of the singular values repeat. It follows from Section 4.1 that trace($\mathbf{X}_2'\mathbf{X}_1$)$\mathbf{Q}^*$ is maximised when:

$$\mathbf{Q}^* = (\mathbf{QL})\mathbf{L}' = \mathbf{Q}.$$

This establishes that if $\mathbf{X}_2'\mathbf{X}_1\mathbf{Q}$ is symmetric and p.s.d. then trace($\mathbf{X}_2'\mathbf{X}_1\mathbf{Q}$) is maximised.

It is instructive to consider the case where $\mathbf{X}_2'\mathbf{X}_1\mathbf{Q}$ is symmetric but not necessarily p.s.d. Then, $\mathbf{X}_2'\mathbf{X}_1\mathbf{Q} = \mathbf{VJ\Sigma V}'$ where $\mathbf{J}$ is a diagonal matrix giving the signs ($+1$ or -1) to be associated with the positive diagonal matrix $\mathbf{\Sigma}$. Then:

$$\mathbf{X}_2'\mathbf{X}_1 = \mathbf{VJ\Sigma V'Q}',$$

is in the form of a singular value decomposition of $\mathbf{X}_2'\mathbf{X}_1$ with row singular vectors $\mathbf{VJ}$ and column singular vectors $\mathbf{QV}$. The singular value decomposition is unique (unless some of the singular values repeat). It follows that trace($\mathbf{X}_2'\mathbf{X}_1\mathbf{Q}^*$) is maximised not by setting $\mathbf{Q}^* = \mathbf{Q}$ but by the orthogonal matrix $\mathbf{Q}^* = (\mathbf{QV})(\mathbf{JV}')$. This establishes that symmetricity is not enough; p.s.d. of $\mathbf{X}_2'\mathbf{X}_1\mathbf{Q}$ is also essential.

In summary, we have shown that trace($\mathbf{X}_2'\mathbf{X}_1\mathbf{Q}$) is maximised if and only if $\mathbf{X}_2'\mathbf{X}_1\mathbf{Q}$ is symmetric and p.s.d. (ten Berge 1977a attributed to Fischer and Poppert 1965). The 'if' (i.e. sufficiency) part of this result is useful, for if we find an orthogonal matrix $\mathbf{Q}$ such that $\mathbf{X}_2'\mathbf{X}_1\mathbf{Q}$ is symmetric p.s.d. then we know that $\mathbf{Q}$ minimises $\|\mathbf{X}_1\mathbf{Q} - \mathbf{X}_2\|$ (see 4.1).

Even without the p.s.d. property, we have that $\mathbf{Q}^* = \mathbf{QVJV}'$. Thus, if $\mathbf{X}_2'\mathbf{X}_1\mathbf{Q}$ is symmetric but not p.s.d., the trace may be maximised merely by replacing $\mathbf{Q}$ by $\mathbf{Q}^*$.

4.3 Scaling

If, now, it is desired to fit an isotropic scaling factor s to minimise $\|s\mathbf{X}_1\mathbf{Q} - \mathbf{X}_2\|$ then the estimate of $\mathbf{Q}$ is unchanged and directly from (3.10):

$$s = \text{trace}(\mathbf{X}_2'\mathbf{X}_1\mathbf{Q})/\|\mathbf{X}_1\| = \text{trace}\,\mathbf{\Sigma}/\|\mathbf{X}_1\|, \tag{4.4}$$

with residual sum-of-squares $\|\mathbf{X}_2\| - (\text{trace}\,\mathbf{\Sigma})^2/\|\mathbf{X}_1\|$. Thus, the inverse transformation does not estimate scaling by the inverse of s but by $s\|\mathbf{X}_1\|/\|\mathbf{X}_2\|$. When $\|\mathbf{X}_1\| = \|\mathbf{X}_2\|$, a commonly used form of pre-scaling (Section 2.2), then the estimated residual sum-of-squares and the scaling for fitting $\mathbf{X}_1$ to $\mathbf{X}_2$ are the *same* as that for fitting $\mathbf{X}_2$ to $\mathbf{X}_1$, showing that even in this special case the scaling does not invert, as might be expected, but remains unchanged. This asymmetry can be circumvented by using the special case of estimation of scaling with K sets when $K = 2$ (see Section 9.2.1); the variant of Procrustes analysis described by Kendall (1984) is the even more special case for $p = 2$ and restricted to exclude reflections (see Section 4.6.2).

Because singular values are non-negative (4.4) shows that the estimate s_i must be positive.

4.4 Example of an orthogonal rotation of two sets, including a scaling factor

Lele and Richtsmeier (2001) measured six landmarks on Macqaque skulls. One set of landmarks came from juvenile skulls, the other set from adults. The data we use are coordinates derived by Lele and Richtsmeier from their 'mean form' matrix, an average of matrices of distances between the landmarks. Our $\mathbf{X}_1$ contains the coordinates of six such landmarks, from the juveniles, in three dimensions, $\mathbf{X}_2$ contains analogous coordinates from adult skulls.

$$\mathbf{X}_1 = \begin{pmatrix} 0.918 & 0.726 & -0.266 \\ 1.070 & -0.529 & -0.420 \\ 0.556 & -0.648 & 0.345 \\ 0.199 & 0.551 & 0.628 \\ -1.400 & -0.292 & 0.199 \\ -1.343 & 0.192 & -0.488 \end{pmatrix}, \quad \mathbf{X}_2 = \begin{pmatrix} 0.860 & -1.194 & 0.455 \\ 2.192 & 0.750 & 0.392 \\ 1.461 & 0.700 & -0.577 \\ -0.424 & -0.288 & -0.648 \\ -2.141 & 0.917 & -0.297 \\ -1.947 & 0.114 & 0.657 \end{pmatrix}.$$

It seems natural to perform an orthogonal rotation to fit $\mathbf{X}_1$ to $\mathbf{X}_2$ using the solution from Section 4.1, that is, a rotation matrix $\mathbf{Q} = \mathbf{V}\mathbf{U}'$ where $\mathbf{X}_2'\mathbf{X}_1 = \mathbf{U}\mathbf{\Sigma}\mathbf{V}'$, its SVD. In addition we estimate a scaling factor, as in (4.4) because we expect a size difference between the juvenile and adult skulls.

We obtain

$$\mathbf{Q} = \begin{pmatrix} 0.9693 & -0.2423 & -0.0411 \\ -0.2396 & -0.9690 & 0.0605 \\ -0.0545 & -0.0488 & -0.9973 \end{pmatrix}$$

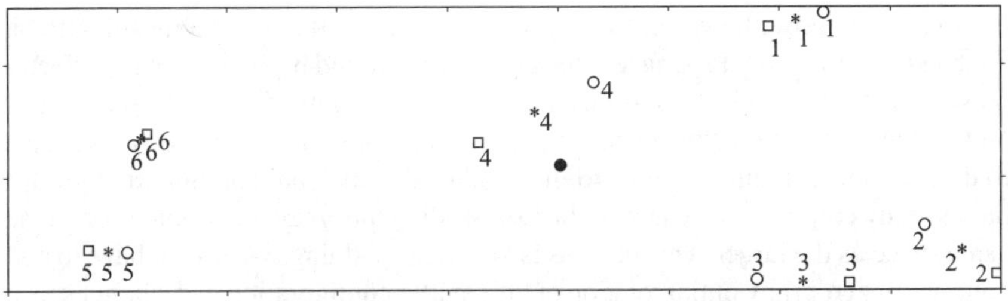

Fig. 4.2. Positions of the six landmarks in $\mathbf{X}_1$ (circles), $\mathbf{X}_2$ (squares) and their average (stars). N.B. vertical axis has scale reversed

and

$$
s\mathbf{X}_1\mathbf{Q} = \begin{pmatrix} 1.0996 & -1.3743 & 0.4087 \\ 1.7868 & 0.4123 & 0.5161 \\ 1.0168 & 0.7172 & -0.6115 \\ 0.0401 & -0.9225 & -0.9050 \\ -1.9541 & 0.9220 & -0.2387 \\ -1.9891 & 0.2456 & 0.8334 \end{pmatrix}
$$

with a scaling factor of 1.5055.

The residual sum-of-squares is 1.3572. As expected, the scaling factor is quite large; without it the two configurations did not match well. The results may be summarised in the following analysis of variance:

Source of variation	Algebraic formulae	Sums of squares
Fitted	$2s\ \text{trace}(\boldsymbol{\Sigma})$	39.8492
Residual	$\lVert s\mathbf{X}_1\mathbf{Q} - \mathbf{X}_2 \rVert$	1.3572
Total	$s^2 \lVert \mathbf{X}_1 \rVert + \lVert \mathbf{X}_2 \rVert$	41.2064

which is obtained from (4.3) but with $\mathbf{X}_1$ replaced by $s\mathbf{X}_1$.

To obtain a two-dimensional display of the solution, we express the results relative to the first two principal components of the average of $s\mathbf{X}_1\mathbf{Q}$ and $\mathbf{X}_2$. The two-dimensional approximation accounts for 91% of the total variation. From trace(Σ)

From Figure 4.2 where the scaling is included, we see that the match is good but that landmark number 4 gives the poorest fit. Presumably, this landmark represents a feature of the skull that differs most between adults and juveniles.

4.5 Different dimensionalities in $\mathbf{X}_1$ and $\mathbf{X}_2$

In the above, it has been assumed that $\mathbf{X}_1$ and $\mathbf{X}_2$ have the same number P of columns. This has been essential because $\mathbf{Q}$, being an orthogonal matrix, must be square. Suppose, however, that P_1 and P_2 differ, Is there still a meaning to be attached to orthogonal Procrustes analysis? There should be, because it is perfectly reasonable to rotate three points on a line to fit three points in, say, two dimensions and clearly this notion extends to more points and more dimensions. All we have to do is to add columns of zeros to the smaller-dimensional configuration until it has as many columns as the larger, thus embedding the smaller configuration in the same space as the larger. This process is known as padding. We do not have to add columns of zeros; any initial rotation of the smaller configuration in the larger space will do. For example, the three points with one-dimensional coordinates (-15), (5), (10) determine exactly the same linear configuration if represented by the two-dimensional coordinates $(-9, -12)$, $(3, 4)$, $(6, 8)$; only the orientation differs. It makes no difference what initial orientation the one-dimensional configuration has relative to the two-dimensional configuration, except of course that the actual value of $\mathbf{Q}$ will depend on it. What is important is that the *relative* orientation of the two configurations *after* rotation to best fit is invariant to the initial orientation. See Appendix C.1 and Section 5.7 for further discussion on the interpretation of padding.

4.6 Constrained orthogonal Procrustes problems

In this section we discuss some special cases of orthogonal transformations. In Section 4.6.1 we discuss how to ensure that $\mathbf{Q}$ represents a proper rotation or includes a reflection. Recall from Appendix B that when $\det \mathbf{Q} = 1$ we have a rotation and when $\det \mathbf{Q} = -1$ we have a reflection. Section 4.6.2 gives some results for the special case of two dimensions ($P = 2$). Section 4.6.3 handles simple Householder reflections and Section 4.6.4 handles rotation in a best two-dimensional plane.

4.6.1 ROTATIONS AND REFLECTIONS

When determining the orthogonal matrix $\mathbf{Q}$ which minimises $\|\mathbf{X}_1\mathbf{Q} - \mathbf{X}_2\|$ we arrive at $\mathbf{Q} = \mathbf{V}\mathbf{U}'$ where $\mathbf{U}\boldsymbol{\Sigma}\mathbf{V}'$ is the SVD of $\mathbf{Z} = \mathbf{X}_2'\mathbf{X}_1$. In deriving this result, there was no constraint on the orthogonal matrices found, which may be of type $\det \mathbf{Q} = 1$ or $\det \mathbf{Q} = -1$ (see Appendix B). Suppose we wish to specify the determinant of $\mathbf{Q}$. We may wish to restrict matching to exclude reflections (i.e. require $\det \mathbf{Q} = 1$) or insist that only reflections are considered (i.e. require $\det \mathbf{Q} = -1$). When $\mathbf{X}_1$ and $\mathbf{X}_2$ are two-dimensional, simple direct solutions may be obtained as described in Section 4.6.2. To tackle these constrained versions of orthogonal Procrustes analysis in general, we require a new representation of the SVD of $\mathbf{Z}$ which we shall write as:

$$\mathbf{Z} = \mathbf{U}\boldsymbol{\Sigma}^*\mathbf{V}^{*\prime}, \quad \text{where} \quad (\det \mathbf{U})(\det \mathbf{V}^*) = 1 \text{ and } |\sigma_i^*| \geq |\sigma_{i-1}^*|. \tag{4.5}$$

This is merely a rewriting of the usual SVD result but to accommodate the positivity of the product of the determinants of $\mathbf{U}$ and $\mathbf{V}^*$, we have to allow for the possibility that the singular values have different signs, although their absolute values continue to be presented in non-decreasing order. Unlike the classical form of the SVD, this representation is not unique but it may be made so by ensuring that all singular values are non-negative except, possibly, the final singular value σ_P^* which may be negative. This form is easily derived from the classical form by inspecting $(\det \mathbf{U})(\det \mathbf{V})$ and, if it is negative, changing the sign of the final column of $\mathbf{V}$ (or $\mathbf{U}$), to give $\mathbf{V}^*$, and making $\sigma_P^* = -\sigma_P$. This representation of the SVD will be written SVD*.

Solution of the generalised rotation/reflection problem With this new notation, the constrained maximum of $\text{trace}(\mathbf{X}_2'\mathbf{X}_1\mathbf{Q}) = \text{trace}(\mathbf{ZQ})$ requires the maximum of $\text{trace}(\mathbf{\Sigma}^*\mathbf{V}^{*\prime}\mathbf{QU}) = \text{trace}(\mathbf{\Sigma}^*\mathbf{H})$ where $\mathbf{H} = \mathbf{V}^{*\prime}\mathbf{QU}$, and we merely have to ensure that $\det \mathbf{Q} = \det \mathbf{H}$ has the required sign. By setting $h_{ii} = 1$ for $i = 1, 2, \ldots, P-1$ we have that $\det \mathbf{H} = h_{PP}$. Thus if we require $\det \mathbf{Q} = 1$, we should set $h_{PP} = 1$ (giving $\mathbf{H} = \mathbf{I}$) and if we require $\det \mathbf{Q} = -1$ we should set $h_{PP} = -1$ (giving $\mathbf{H} = \mathbf{I} - 2\mathbf{e}_P\mathbf{e}_P'$). We shall write $\mathbf{J} = \mathbf{I} - 2\mathbf{e}_P\mathbf{e}_P'$, a diagonal orthogonal matrix with a unit diagonal except that the final position has -1. We have that:

$$\text{trace}(\mathbf{ZQ}) = \text{trace}(\mathbf{ZV}^*\mathbf{HU}') = \text{trace}(\mathbf{\Sigma}^*\mathbf{H}) = \sum_{i=1}^{p-1} \sigma_i + h_{PP}\sigma_P^*. \tag{4.6}$$

When $h_{PP} = \text{sign}(\sigma_P^*)$, (4.6) gives the same maximum to $\text{trace}(\mathbf{X}_2'\mathbf{X}_1\mathbf{Q})$ as for the unconstrained problem, but when $h_{PP} = -\text{sign}(\sigma_P^*)$ the maximum is decreased to $\text{trace}(\mathbf{\Sigma}^*\mathbf{H}) = \sum_{i=1}^{P-1} \sigma_i - \sigma_P$. Thus, to solve the constrained problem, set:

$$\mathbf{Q} = \mathbf{V}^*\mathbf{HU}', \tag{4.7}$$

where $\mathbf{H} = \mathbf{I}$ for a rotation and $\mathbf{H} = \mathbf{J}$ for a reflection.

The effect of constraints on the necessary and sufficient conditions Note that $\mathbf{ZQ} = \mathbf{U\Sigma}^*\mathbf{HU}'$, is symmetric, so symmetricity is a necessary condition for the validity of the constrained solution. If $\mathbf{ZQ}$ is symmetric and positive definite, we know from Section 4.2, that $\mathbf{Q}$ must be the matrix that minimises the unconstrained Procrustes criterion. However, the sign of $\det \mathbf{Q}$ need not be that required in a constrained minimisation. If $\mathbf{ZQ}$ is symmetric with only its *smallest*, in absolute value, eigenvalue negative, then $\mathbf{Q}$ is a candidate for an optimal constrained solution. Then, for some orthogonal $\mathbf{U}$ and positive diagonal $\mathbf{\Sigma}^*$ we have:

$$\mathbf{ZQ} = \mathbf{UJ\Sigma U}' \tag{4.8}$$

and

$$\mathbf{Z} = \mathbf{UJ\Sigma U}'\mathbf{Q}'. \tag{4.9}$$

Consider first the case where det $\mathbf{Q} = 1$. Then the SVD* of (4.9) requires:

$$\mathbf{\Sigma}^* = \mathbf{\Sigma J} \quad \text{and} \quad \mathbf{V}^* = \mathbf{QU}$$

for which $(\det \mathbf{U})(\det \mathbf{V}^*) = (\det \mathbf{U})^2(\det \mathbf{Q}) = 1$ as is required for a SVD*. The optimal constrained orthogonal matrix $\mathbf{Q}_0$, say, is given by (4.7). Thus

when we require (i) det $\mathbf{Q}_0 = 1$, then $\mathbf{H} = \mathbf{I}$ and $\mathbf{Q}_0 = \mathbf{QUIU}' = \mathbf{Q}$,

and when we require (ii) det $\mathbf{Q}_0 = -1$, then $\mathbf{H} = \mathbf{J}$ and $\mathbf{Q}_0 = \mathbf{QUJU}'$.

In case (ii) $\mathbf{ZQ}_0 = (\mathbf{U\Sigma JU'Q'})(\mathbf{QUJU'}) = \mathbf{U\Sigma U'}$ which is symmetric and positive definite. It follows from the necessary and sufficient conditions of Section 4.2 that $\mathbf{Q}_0 = \mathbf{QUJU}'$ is the unconstrained solution. In case (i) $\mathbf{Q}$ is the constrained solution.

Next consider the case where det $\mathbf{Q} = -1$. Then the SVD* of (4.9) requires:

$$\mathbf{\Sigma}^* = \mathbf{\Sigma} \quad \text{and} \quad \mathbf{V}^* = \mathbf{QUJ}$$

for which $(\det \mathbf{U})(\det \mathbf{V}^*) = (\det \mathbf{U})^2(\det \mathbf{Q})(\det \mathbf{J}) = 1$ as is required for a SVD*. As above, the optimal constrained orthogonal matrix $\mathbf{Q}$, is given by (4.7). Thus:

when we require (i) det $\mathbf{Q}_0 = 1$, then $\mathbf{H} = \mathbf{I}$ and $\mathbf{Q}_0 = \mathbf{QUJIU}' = \mathbf{QUJU}'$,

and when we require (ii) det $\mathbf{Q}_0 = -1$, then $\mathbf{H} = \mathbf{J}$ and $\mathbf{Q}_0 = \mathbf{QUJ}^2\mathbf{U}' = \mathbf{QUU}' = \mathbf{Q}$.

In case (i) $\mathbf{ZQ}_0 = (\mathbf{U\Sigma JU'Q'})(\mathbf{QUJU'}) = \mathbf{U\Sigma U'}$ which is symmetric and positive definite. It follows from the necessary and sufficient conditions of Section 4.2 that $\mathbf{Q}_0 = \mathbf{QUJU}'$ is the unconstrained solution. In case (ii) $\mathbf{Q}$ is the constrained solution.

In summary, if $\mathbf{ZQ} = \mathbf{UJ\Sigma U}'$ and one requires an optimal $\mathbf{Q}_0$ whose determinant is constrained to have positive or negative sign, then $\mathbf{Q}_0 = \mathbf{Q}$ when sign(det $\mathbf{Q}$) is the same as the required sign, else $\mathbf{Q}_0 = \mathbf{QUJU}'$, which coincides with the unconstrained estimate. Thus, not only is it necessary that $\mathbf{ZQ}$ be symmetric for $\mathbf{Q}$ to be an optimal solution to the constrained orthogonal Procrustes problem but also it is sufficient to identify the correct orthogonal matrix either as $\mathbf{Q}$ itself or as the simple unique transformation $\mathbf{Q} = \mathbf{UJU}'$. In the latter case, the constrained and unconstrained solutions coincide. These results extend the necessary and sufficient conditions of Section 4.2.

4.6.2 THE TWO-DIMENSIONAL CASE

The case $P = 2$ is of some importance, especially in shape analysis (see, for example, Kendall 1984 and Dryden and Mardia 1991). Because $\mathbf{Q}$ is now a 2×2 orthogonal matrix, explicit expressions could be found for the singular values of $\mathbf{Z} = \mathbf{X}_2'\mathbf{X}_1$ but more direct approaches are simpler. We consider separately the cases of rotation and reflection in two dimensions. Using the notation $\mathbf{e}_1' = (1, 0)$, $\mathbf{e}_2' = (0, 1)$, $\mathbf{J} = (-\mathbf{e}_2, \ \mathbf{e}_1)$, $c = \cos(\theta)$ and $s = \sin(\theta)$, we may write:

(i) Rotation: $\mathbf{Q} = c\mathbf{I} + s\mathbf{J}$.

(ii) Reflection: $\mathbf{Q} = \mathbf{I} - 2\mathbf{uu}'$, where $\mathbf{u} = c\mathbf{e}_1 + s\mathbf{e}_2$.

In both cases, $\mathbf{Q}$ depends on the single parameter θ and we may directly maximise trace($\mathbf{ZQ}$) with respect to θ for $\mathbf{Q}$ given by (i) and (ii). Here we adopt an alternative approach, appealing to the condition that $\mathbf{ZQ}$ must be symmetric (Sections 4.2 and 4.6.1). For (i) this gives:

$$c\mathbf{Z} + s\mathbf{ZJ} = c\mathbf{Z}' - s\mathbf{JZ}' \quad \text{or} \quad c(\mathbf{Z}' - \mathbf{Z}) = s(\mathbf{ZJ} + \mathbf{JZ}').$$

Equating the elements on both sides in row one and column two immediately gives: $c(z_{21} - z_{12}) = s(z_{11} + z_{22})$ so the angle of rotation is given by:

$$\tan(\theta) = \frac{z_{21} - z_{12}}{z_{11} + z_{22}}.$$

Similarly, the symmetry of $\mathbf{ZQ}$ associated with (ii) gives:

$$\mathbf{Z} - \mathbf{Z}' = 2(\mathbf{Zuu}' - \mathbf{uu}'\mathbf{Z}')$$

and equating the elements on both sides in the first row and second column gives, after some elementary rearrangement:

$$\tan(2\theta) = \frac{z_{21} + z_{12}}{z_{11} - z_{22}}.$$

These results give the best rotation (reflection) even when the reflection (rotation) gives a smaller residual sum-of-squares. In each case, assuming only the symmetricity of $\mathbf{ZQ}$ has led to a unique result. This is consistent with the stronger version of Section 4.2 mentioned at the end of 4.6.1, that it is sufficient for $\mathbf{ZQ}$ to be symmetric to guarantee an optimal restricted solution.

The previously found formulae for the residual sum-of-squares (4.3) and isotropic scaling factor (4.4) remain valid. Both require trace($\mathbf{\Sigma}$) = trace($\mathbf{ZQ}$) which has an explicit form when $P = 2$. We may calculate these by inserting the values of $\mathbf{Q}$ found above but it is interesting to derive the result directly from observing that:

$$(\sigma_1 + \sigma_2)^2 = (\sigma_1^2 + \sigma_2^2) + 2\sigma_1\sigma_2,$$

that is, $(\operatorname{trace}\mathbf{\Sigma})^2 = \operatorname{trace}(\mathbf{Z}'\mathbf{Z}) + 2\sigma_1\sigma_2$.
Now $\det\mathbf{Z} = \det(\mathbf{U\Sigma V}') = (\det\mathbf{Q})\sigma_1\sigma_2$, so that:

$$(\operatorname{trace}\mathbf{\Sigma})^2 = \operatorname{trace}(\mathbf{Z}'\mathbf{Z}) \pm 2\det\mathbf{Z},$$

where the positive sign signifies a rotation and the negative sign a reflection. In terms of the elements of $\mathbf{Z}$ this gives for a rotation (see Section 9.2.1):

$$\text{trace}(\boldsymbol{\Sigma}) = \left\{(z_{11} + z_{22})^2 + (z_{21} - z_{12})^2\right\}^{1/2},$$

and for a reflection:

$$\text{trace}(\boldsymbol{\Sigma}) = \left\{(z_{11} - z_{22})^2 + (z_{21} + z_{12})^2\right\}^{1/2}.$$

4.6.3 BEST HOUSEHOLDER REFLECTION

An even stronger constraint on the orthogonal matrix Procrustes problem is to minimise $\|\mathbf{X}_1\mathbf{Q} - \mathbf{X}_2\|$ where $\mathbf{Q} = \mathbf{I} - 2\mathbf{u}\mathbf{u}'$, a Householder transform representing a reflection in the plane normal to the unit vector $\mathbf{u}$ (see Appendix equation (B.6)). As previously, we require the maximum of trace$(\mathbf{X}_2'\mathbf{X}_1\mathbf{Q})$. Now, writing $\mathbf{Z} = \mathbf{X}_2'\mathbf{X}_1$ we have:

$$\text{trace}(\mathbf{X}_2'\mathbf{X}_1\mathbf{Q}) = \text{trace}(\mathbf{Z}(\mathbf{I} - 2\mathbf{u}\mathbf{u}')),$$

which is maximised when $\mathbf{u}'\mathbf{Z}\mathbf{u}$ is minimised. This has the well-known solution of choosing $\mathbf{u}$ to be the eigenvector corresponding to the smallest eigenvalue of the symmetric matrix $\frac{1}{2}(\mathbf{Z} + \mathbf{Z}')$. We refer to such eigenvectors as being *subordinate*.

4.6.4 BEST PLANE ROTATION

Next we may require $\mathbf{Q}$ to be a plane rotation. Plane rotations are discussed in Appendix B.2, where it is shown that a rotation in the plane spanned by vectors $\mathbf{u}$ and $\mathbf{v}$ may be parameterised in several ways. It seems that the most simple parameterisation for handling the Procrustes problem is given by case (b) of Table B.1, which amounts to expressing $\mathbf{Q}$ as the product of two Householder reflections $\mathbf{Q} = (\mathbf{I} - 2\mathbf{u}\mathbf{u}')(\mathbf{I} - 2\mathbf{v}\mathbf{v}')$; the angle of rotation is $2\cos^{-1}(\mathbf{u}'\mathbf{v})$. The least-squares criterion becomes:

$$\mathrm{C} = \|\mathbf{X}_1(\mathbf{I} - 2\mathbf{u}\mathbf{u}')(\mathbf{I} - 2\mathbf{v}\mathbf{v}') - \mathbf{X}_2\|,$$

which, because $(\mathbf{I} - 2\mathbf{v}\mathbf{v}')$ is a symmetric orthogonal matrix, may be written:

$$\mathrm{C} = \|\mathbf{X}_1(\mathbf{I} - 2\mathbf{u}\mathbf{u}') - \mathbf{X}_2(\mathbf{I} - 2\mathbf{v}\mathbf{v}')\|,$$

showing that the plane rotation of $\mathbf{X}_1$ is equivalent to the two-sided Procrustes problem with separate reflections of $\mathbf{X}_1$ and $\mathbf{X}_2$. As previously, we require the maximum of trace$(\mathbf{Z}\mathbf{Q})$ = trace$(\mathbf{Z}(\mathbf{I} - 2\mathbf{u}\mathbf{u}')(\mathbf{I} - 2\mathbf{v}\mathbf{v}'))$. This does not have a closed form solution but for fixed $\mathbf{v}$, is precisely the Householder Procrustes problem just discussed but with $\mathbf{Z}$ replaced by $\mathbf{Z}^* = (\mathbf{I} - 2\mathbf{v}\mathbf{v}')\mathbf{Z}$. Hence for fixed $\mathbf{v}$, $\mathbf{u}$ is given as the subordinate eigenvector corresponding to the minimal eigenvalue λ of the symmetric matrix:

$$\mathbf{A} = \tfrac{1}{2}(\mathbf{Z} + \mathbf{Z}') - (\mathbf{v}\mathbf{v}'\mathbf{Z} + \mathbf{Z}'\mathbf{v}\mathbf{v}').$$

Now with $\mathbf{Q} = \mathbf{I} - 2\mathbf{uu}'$ we have:

$$\text{trace}\,\mathbf{Z}^*\mathbf{Q} = \text{trace}\,\mathbf{Z} - 2\mathbf{v}'\mathbf{Z}\mathbf{v} - 2\lambda,$$

which is the greatest possible value for the given $\mathbf{v}$. Similarly, for fixed $\mathbf{u}$, $\mathbf{v}$ is given as the subordinate eigenvector corresponding to the minimal eigenvalue μ of the symmetric matrix:

$$\mathbf{B} = \tfrac{1}{2}(\mathbf{Z} + \mathbf{Z}') - (\mathbf{uu}'\mathbf{Z}' + \mathbf{Zuu}'),$$

reducing C even further.

Thus, we have a value of $\mathbf{u}$ and a value of $\mathbf{v}$ which define a plane rotation and give a value to the criterion C. If now, a new value of $\mathbf{A}$ is found using this value of $\mathbf{v}$, giving in turn a new setting of $\mathbf{u}$, then these must combine to reduce C, or at least leave it unchanged. Similarly, using this new $\mathbf{u}$ we may redefine $\mathbf{B}$ to find a new $\mathbf{v}$ to get further improvement. Thus we may now solve new eigenvalue problems for $\mathbf{A}$ and $\mathbf{B}$ alternately, each step reducing the residual sum-of-squares C until convergence to the final values of $\mathbf{u}$ and $\mathbf{v}$. The angle of rotation is $2\cos^{-1}(\mathbf{u}'\mathbf{v})$. Note that the definitions of $\mathbf{Z}$ in $\mathbf{A}$ and $\mathbf{B}$ do not use up-dated versions of $\mathbf{X}_1$ and $\mathbf{X}_2$, as that would amount to transforming them by products of Householder transforms, giving the general orthogonal matrices of the orthogonal Procrustes problem of Section 4.1 rather than the plane rotations required here.

The first round of iteration deserves separate analysis. Consider the perfect fit, $\mathbf{X}_1 = \mathbf{X}_2 = \mathbf{X}$, say, then the best reflection is given by the subordinate eigenvector $\mathbf{u}$, with eigenvalue λ, of the positive definite matrix $\mathbf{X}'\mathbf{X}$ and then $\mathbf{X}_1(\mathbf{I} - 2\mathbf{uu}')$ no longer has perfect fit with $\mathbf{X}_2$. However, a second reflection requires the subordinate eigenvector of $\mathbf{X}'\mathbf{X}(\mathbf{I} - 2\mathbf{uu}')$ which is again $\mathbf{u}$ but now with eigenvalue $-\lambda$. This restores the initial position because $\mathbf{X}_1(\mathbf{I} - 2\mathbf{uu}')(\mathbf{I} - 2\mathbf{uu}') = \mathbf{X}_1$. Thus C worsens after determining $\mathbf{u}$ from $\mathbf{A}$ and improves again after determining $\mathbf{v}$ from $\mathbf{B}$; it is not clear to us whether in all cases C *necessarily* improves after determining $\mathbf{v}$ in the first round. Not that it matters, because the criterion improves uniformly in the second and subsequent iterations.
This gives the following algorithm.

Algorithm 4.1

Step 1 Initialise by setting $\mathbf{u} = \mathbf{v} = \mathbf{0}$.

Step 2 Find λ and $\mathbf{u}$ for matrix $\mathbf{A}$. Update $\mathbf{B}$.

Step 3 Find μ and $\mathbf{v}$ for matrix $\mathbf{B}$. Update $\mathbf{A}$.

Step 4 Test for convergence; if $\lambda + \mu$ is not within threshold of its previous value return to step 2.

Step 5 Evaluate $\mathbf{X}_1(\mathbf{I} - 2\mathbf{uu}')(\mathbf{I} - 2\mathbf{vv}')$ and the angle of rotation, $2\cos^{-1}(\mathbf{u}'\mathbf{v})$.

As an illustration of this algorithm we use data from Harman (1976) quoted by Gruvaeus (1970, table 1, p. 498) who provides a target matrix in his table 3, p. 500.

$\mathbf{X}_1$ (Harman)

0.90	−0.09	−0.03
0.83	0.09	−0.04
0.87	−0.01	0.07
0.55	0.79	−0.07
0.56	0.65	0.04
0.63	0.60	0.03
0.28	0.27	0.45
0.38	0.20	0.63
0.38	0.19	0.77

Target $\mathbf{X}_2$ (Gruvaeus)

0.98	0	0
0.76	0	0
0.86	0	0
0	1.00	0
0	0.84	0
0	0.77	0
0	0	0.55
0	0	0.76
0	0	0.93

With $\varepsilon = 0.0001$, convergence was achieved after only three iterations, the values of λ and μ are given in the following table.

Iteration	λ	μ
1	1.2266	−1.4166
2	−1.4836	−1.4166
3	−1.4835	−1.4166

The final settings of **u**,**v**: defining the best plane were:

u=	**v**=
0.6218	0.7536
−0.5992	−0.4369
−0.5042	−0.4912

and the plane-rotation matrix **Q** was

0.9242	0.3409	0.1725
−0.3629	0.9243	0.1180
−0.1192	−0.1716	0.9779

This is close to a unit matrix, as might have been expected with initial similarity between the data and the target, the rotated form of which was:

$\mathbf{X}_1\mathbf{Q}$:

0.8680	0.2287	0.1153
0.7392	0.3730	0.1147
0.7993	0.2753	0.2173
0.2299	0.9297	0.1196
0.2768	0.7848	0.2124
0.3609	0.7642	0.2088
0.1071	0.2678	0.5202
0.2035	0.2063	0.7052
0.1904	0.1730	0.8410

The angle of rotation was 0.4197 radians, equivalent to 24.05°.

4.7 Best rank-R fit, principal components analysis and the Eckart–Young theorem

The material given in this section is a little outside the main theme of this book but we include it because of the importance of the Eckart–Young theorem in much of the following and because of an interesting link with orthogonal Procrustes analysis. We are now concerned with minimising $\|\mathbf{X}_1 - \mathbf{X}_2\|$ where $\mathbf{X}_2$ is constrained to have *given* rank R. In this section, note that there is no transformation matrix $\mathbf{T}$ and, apart from its rank, $\mathbf{X}_2$ is not given but is to be determined. Both $\mathbf{X}_1$ and $\mathbf{X}_2$ have P columns. First we consider the problem where $\mathbf{X}_2$ is specified to lie in a *given* R-dimensional subspace spanned by the columns of some orthonormal matrix ${}_P\mathbf{P}_R$. Thus, for some matrix $\mathbf{A}$, $\mathbf{X}_2 = \mathbf{APP}'$. We have that:

$$\|\mathbf{APP}' - \mathbf{X}_1\| = \|\mathbf{X}_1\mathbf{PP}' - \mathbf{X}_1\| + \|\mathbf{APP}' - \mathbf{X}_1\mathbf{PP}'\|$$

so that the minimum over $\mathbf{A}$ of the left-hand-side occurs when $\|\mathbf{APP}' - \mathbf{X}_1\mathbf{PP}'\|$ is minimal. Trivially the minimum sum-of-squares is zero when $\mathbf{APP}' = \mathbf{X}_1\mathbf{PP}'$ so we have established that $\mathbf{X}_2 = \mathbf{X}_1\mathbf{PP}'$, the orthogonal projection of $\mathbf{X}_1$ onto the given subspace. The residual sum-of-squares is $\|\mathbf{X}_1 - \mathbf{X}_1\mathbf{PP}'\|$.

Next we ask for which R-dimensional subspace is there a global minimum. We have:

$$\|\mathbf{X}_1 - \mathbf{X}_1\mathbf{PP}'\| = \|\mathbf{X}_1(\mathbf{I} - \mathbf{PP}')\| = \text{trace}\{\mathbf{X}_1(\mathbf{I} - \mathbf{PP}')\mathbf{X}_1'\},$$

so we require the maximum of:

$$\text{trace}(\mathbf{X}_1\mathbf{PP}'\mathbf{X}_1') = \text{trace}(\mathbf{X}_1'\mathbf{X}_1\mathbf{PP}') = \text{trace}(\mathbf{V\Lambda V}'\mathbf{PP}') = \text{trace}(\mathbf{L}'\mathbf{\Lambda L})$$

where $\mathbf{X}_1'\mathbf{X}_1 = \mathbf{V\Lambda V}'$ is a spectral decomposition of a positive definite symmetric matrix and ${}_P\mathbf{L}_R = \mathbf{V}'\mathbf{P}$ is an orthonormal matrix. Writing $\mathbf{l}_i$ for the ith column of $\mathbf{L}$ we have:

$$\text{trace}(\mathbf{X}_1\mathbf{PP}'\mathbf{X}_1') = \text{trace}\left(\sum_{i=1}^{R} \lambda_i \mathbf{l}_i \mathbf{l}_i'\right) = \sum_{i=1}^{R} \lambda_i \mathbf{l}_i' \mathbf{l}_i.$$

Now the eigenvalues λ_i are non-negative and may be assumed to be given in decreasing order of magnitude. Also, because $\mathbf{L}$ is orthonormal, the maximum value in any column is unity, when all the other values in that column are zero. Thus, the trace has a maximum of $\lambda_1 + \lambda_2 + \cdots + \lambda_R$ when $l_{ii} = 1$ for $i = 1, 2, \ldots, R$ and $l_{ij} = 0$ whenever $i \neq j$. It follows that $\mathbf{P} = \mathbf{VL} = \mathbf{V}_R$, the first R columns of $\mathbf{V}$. We have shown that the residual sum-of-squares $\|\mathbf{X}_1 - \mathbf{X}_1\mathbf{PP}'\|$ is minimised for all projections onto R-dimensional subspaces when $\mathbf{P} = \mathbf{V}_R$ or, equivalently, the fitted sum-of-squares in the subspace is maximised. That is, we have established the basic property of principal components analysis (PCA).

Another way of expressing the above result is that $\mathbf{X}_1\mathbf{V}_R\mathbf{V}_R'$ is the best least-squares rank R approximation to $\mathbf{X}_1$. Now, from the singular value decomposition $\mathbf{X}_1 = \mathbf{U\Sigma V}'$ we have that:

$$\mathbf{X}_1\mathbf{V}_R\mathbf{V}_R' = \mathbf{U\Sigma V}'\mathbf{V}_R\mathbf{V}_R' = \mathbf{U\Sigma}_R\mathbf{V}_R',$$

where $\mathbf{\Sigma}_R$ represents the first R columns of the diagonal matrix $\mathbf{\Sigma}$ of singular values. This is the usual form of the Eckart–Young theorem (Eckart and Young 1936) which is fundamental to so many least-squares matrix approximation problems. The Eckart–Young theorem is often confused with the singular value decomposition of a matrix but the latter has been known in algebra from 1873 (Beltrami 1873, Jordan 1874). It is the important least-squares property that was established by Eckart and Young. A simple derivation of the singular value decomposition is available in Gower and Hand (1996, p. 236).

We close this section by establishing a link between Orthogonal Procrustes Analysis and PCA. Geometrically, the link is obvious, for PCA maximises the orthogonal projections $\mathbf{X}_1\mathbf{V}_R\mathbf{V}'_R$ of $\mathbf{X}_1$ onto any R-dimensional subspace and minimises the residual sum-of-squares $\|\mathbf{X}_1\left(\mathbf{I}-\mathbf{V}_R\mathbf{V}'_R\right)\|$. Any rotation of $\mathbf{X}_1\mathbf{V}_R\mathbf{V}'_R$ cannot give a smaller residual sum-of-squares. But the residual sum-of-squares between the two configurations is just the Orthogonal Procrustes least-squares criterion. Thus, PCA gives the best R-dimensional Orthogonal Procrustes fit to $\mathbf{X}_1$, all other R-dimensional fits must be worse.

To prove this result analytically we need to find an orthogonal matrix $\mathbf{Q}$ that minimises:

$$\|\mathbf{X}_1\mathbf{Q}-\mathbf{X}_1\mathbf{V}_R\mathbf{V}'_R\|.$$

From Section 4.1 we know that this requires the singular value decomposition of $\mathbf{X}'_1\mathbf{X}_1\mathbf{V}_R\mathbf{V}'_R = \mathbf{V}\mathbf{\Lambda}\mathbf{V}'\mathbf{V}_R\mathbf{V}'_R = \mathbf{V}\mathbf{\Lambda}_R\mathbf{V}'_R$. This may be expressed as the singular-value decomposition $\mathbf{V}\mathbf{\Lambda}_0\mathbf{V}'$ where $\mathbf{\Lambda}_0 = (\mathbf{\Lambda}_R, \mathbf{0})$ is diagonal with its final P–R columns set to zero. From (4.2) we see that $\mathbf{Q} = \mathbf{V}\mathbf{V}' = \mathbf{I}$ so the best rotation is given by the identity matrix, which has no effect.

4.8 Other criteria

All the other criteria mentioned in Chapter 3 are potentially available. Indeed, we have seen that with orthogonal Procrustes problems, the least-squares, inner-product, maximal group average criteria and the RV coefficient all coincide. When $\|\mathbf{X}_1\| = \|\mathbf{X}_2\|$ even the Tucker's congruence coefficient (see Section 3.1) coincides with the inner product, and so with the others. This is a welcome simplification that we shall see does not extend to non-orthogonal transformations.

4.8.1 ORTHOGONAL PROCRUSTES BY MAXIMISING CONGRUENCE

However, there remains the possibility that $\|\mathbf{X}_1\| \neq \|\mathbf{X}_2\|$ and we do not wish to normalise them nor scale them isotropically to allow for size differences but we do wish to maximise congruence. Brokken (1983) has examined the maximisation of the Tucker coefficient of congruence (equation (3.7)) in conjunction with orthogonal transformations. The problem has no closed form solution so he uses a Newton optimisation algorithm to estimate the optimal matrix $\mathbf{Q}$ and resulting maximal congruence. His findings are rather surprising. Having tried several examples, Brokken reports that the 'most striking finding is … the success of the orthogonal Procrustes rotation [which gives results that] are usually not at a maximum [but are] nonetheless very close to the maximum'. Indeed, Brokken reports some occasions where orthogonal Procrustes by least squares gives higher congruence coefficients than the algorithm that is meant to maximise them. The likely explanation is that there is some deficiency in the algorithm—see Appendix E.2 for some remarks on potential dangers, in a different context, of using the Newton–Raphson algorithm. We are surprised that orthogonal Procrustes has done so well in optimising congruence and wonder if this might be a consequence of using data that were nearly standardised, in which case the two criteria will be very similar. With grossly unstandardised data, we would expect that orthogonal Procrustes estimated by least squares and by optimising congruence to give quite different results.

4.8.2 ROBUST ORTHOGONAL PROCRUSTES

Robust methods of estimation get little discussion in this book, not because they are unimportant but because, so far, they have received little attention. We are aware only of the papers by Verboon and Heiser (1992) and Siegel and Benson (1982). The material in this section relies heavily on that of Verboon and Heiser ending with a brief reference to the approach of Siegel and Benson.

The idea of robust methods is to down-weight cases with large residuals relative to cases with small residuals. Following Huber (1981), rather than minimising a sum-of-squares of residuals r_i, we minimise $\sum_{i=1}^{n} f(r_i)$ where:

$$\begin{aligned} f(r_i) &= r_i^2 \quad \text{when } r_i \le c, \\ f(r_i) &= 2cr_i - c^2 \quad \text{when } r_i > c, \end{aligned} \tag{4.10}$$

where c is a given *tuning constant*, discussed below. Thus, when $r_i > c$, the residuals have less effect than they would have with ordinary least-squares (Figure 4.3(a) illustrates the weighting function). Of course, the residuals are not known and have to be given initial estimates. After fitting an appropriate model, new estimates of the residuals will be available. Thus, we can try to minimise (4.10) by using an iterative (re)weighted least-squares algorithm (IRLS)

In the orthogonal Procrustes context, the residuals (whose sum-of-squares is minimised) are the distances between a rotated point and its target, for example, between X_{11}' and X_{21} in Figure 4.1. The residuals are the sums-of-squares of the rows of $\mathbf{R} = \mathbf{X}_1\mathbf{Q} - \mathbf{X}_2$. Thus r_i^2 is the ith diagonal value of $\mathbf{RR}'$. It is easy to formulate the criterion (4.5) in a weighted least-squares form. All we have to do is to define weights w_i such that $f(r_i) = w_i r_i^2$. This gives:

$$\begin{aligned} w_i &= 1 \quad \text{when } r_i \le c, \\ w_i &= 2c/r_i - c^2/r_i^2 \quad \text{when } r_i < c, \end{aligned}$$

so we have to maximise trace $\mathbf{WRR}'$. Weighted forms of Procrustes criteria are discussed in Chapter 8 but this form is particularly simple because it may be written $\|\mathbf{W}^{1/2}\mathbf{R}\|$ or, equivalently:

$$\|\mathbf{W}^{1/2}\mathbf{X}_1\mathbf{Q} - \mathbf{W}^{1/2}\mathbf{X}_2\|, \tag{4.11}$$

whose minimisation is an ordinary orthogonal Procrustes problem with $\mathbf{X}_1$ replaced by $W^{1/2}\mathbf{X}_1$ and $\mathbf{X}_2$ replaced by $\mathbf{W}^{1/2}\mathbf{X}_2$. Once $\mathbf{Q}$ has been estimated, there will be a new estimate of $\mathbf{R}$ and so to the IRLS algorithm. The problem with this approach is that the convergence properties of the algorithm are uncertain unless the weights are carefully chosen. Verboon and Heiser (1992) have developed a function $g(r_i)$ that majorises (4.10), defined by:

$$\begin{aligned} g(r_i) &= w_i r_i^2 \quad \text{when } r_{0i} \le c, \\ g(r_i) &= w_i r_i^2 + c r_{0i} - c^2 \quad \text{when } r_{0i} > c. \end{aligned} \tag{4.12}$$

A function $g(x)$ that majorises $f(x)$ is one that for all values of $x = x_0$ satisfies:

$$\begin{aligned} g(x) &\ge f(x) \quad \text{for all } x, \\ g(x_0) &= f(x_0), \end{aligned} \tag{4.13}$$

from which it follows that $f'(x_0) = g'(x_0)$. The important thing is that when the majorising function $g(x)$ is sufficiently simple, the value x_0' at its minimum may be easily calculated. Thus, $f(x_0') \le g(x_0') \le g(x_0) = f(x_0)$, so we have reduced the value of the function $f(x)$ and we may repeat the minimisation with $x_0 = x_0'$.

Because Huber's function is convex and bounded below by zero, the IRLS based on any majorising function must converge to the global minimum (Verboon and Heiser 1992). Verboon and Heiser showed that $g(r_i)$ given by (4.12) majorises $f(r_i)$ given by (4.10) when:

$$\begin{aligned} w_i &= 1 \quad \text{when } r_{0i} \le c, \\ w_i &= c/r_{0i} \quad \text{when } r_{0i} > c, \end{aligned} \tag{4.14}$$

where r_{0i} is the ith residual given by a current value of $\mathbf{Q}$. With these settings, it is immediate that $g(r_0) = f(r_0)$ but to establish that $g(r_0) \ge f(r_0)$ needs some care, considering the mutually exclusive cases (i) $r_i < c, r_{0i} < c$, (ii) $r_i < c, r_{0i} > c$, (iii) $r_i > c, r_{0i} < c$, and (iv) $r_i > c, r_{0i} > c$. Clearly, when $g(r_i)$ majorises $f(r_i)$ for $i = 1, 2, \ldots, n$ then $\sum_{i=1}^n g(r_i)$ majorises $\sum_{i=1}^n (r_i)$.

The interesting thing about (4.12) is that both terms involving r_i are the same for all values of r_0. It follows that we have only to minimise trace$\mathbf{WRR}'$ which is the same as (4.11) but now with the weights defined by (4.14). This leads to a reliable algorithm that converges to a global minimum of the Huber function. The choice of c is clearly crucial. Verboon and Heiser, basing their advice on Huber (1981),

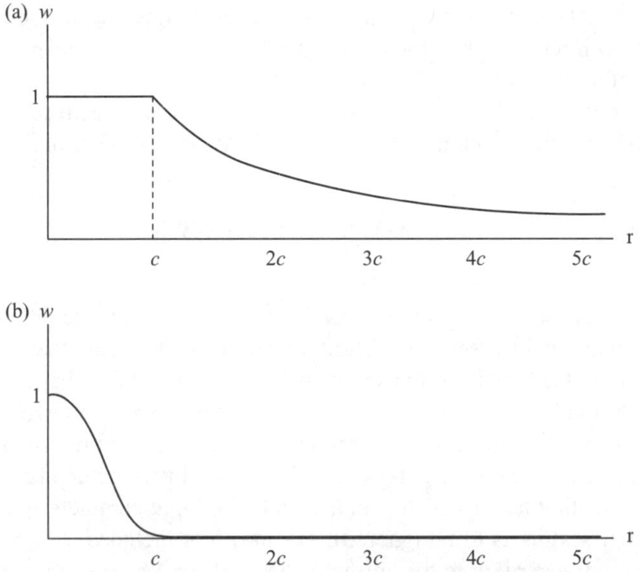

Fig. 4.3. The weighting of residuals for (a) Huber's (1981) robust estimator and (b) the form used by Mosteller and Tukey (1971).

suggest the following. For a vector $\mathbf{x}$ we define $\text{MAD}(\mathbf{x}) = \text{med}|x_i - \text{med}(x)|$. For a matrix $\mathbf{X}$ with p columns we define S to be the average of the p MAD estimates for the columns. Then we choose c to satisfy $S < c < 2S$, the exact value within this range being a matter of choice or experiment.

Verboon and Heiser (1992) have also considered a second weighting function, due to Mosteller and Tukey (1971). This leads to a majorisation function which is minimised by weights:

$$
\begin{aligned}
w_i &= \left(1 - \left(\frac{r_{0i}}{c}\right)^2\right)^2 \quad \text{when } r_{0i} \le c, \\
w_i &= 0 \quad \text{when } r_{0i} > c.
\end{aligned}
\tag{4.15}
$$

Figure 4.3 shows the weight functions for (4.14) and (4.15). Both show how the weights reduce as r_i increases beyond c. Indeed, for (4.15) the weights are zero whenever $r_i > c$. This implies that the corresponding points in the configurations are ignored so that one is fitting only to a subset of the total number of points in the configurations. This might be useful when there is reason to believe that two configurations may match well in some parts but are very different in other parts. The use of a common unweighted centroid would greatly reduce the possibility of finding partial matches, because of the adverse effects of points with large residuals (but zero weight). Consequently, one must use the weighted centroids, in which case points with zero weights would not contribute and the matching pairs of points would be referred to their own centroids.

An isotropic scaling parameter may be fitted in the usual way (see Sections 3.3 and 4.3). For weights given in a diagonal matrix $\mathbf{W}$, the estimate of s:

$$\hat{s} = \mathrm{trace}(\mathbf{X}_2'\mathbf{W}\mathbf{X}_1\mathbf{Q})/\mathrm{trace}(\mathbf{X}_2'\mathbf{W}\mathbf{X}_1).$$

Verboon and Heiser (1992) report that including a scaling factor with robust estimation may lead to 'severely distorted solutions'. It seems that scaling and downweighting the larger residuals interact in an unpredictable way.

Algorithm 4.2 is a simple implementation of the Verboon and Heiser majorisation algorithm. It is given for the weights (4.14), but if required, step 3 may be easily modified to give the weights (4.15). The possibility is included of estimating an isotropic scaling factor which is indicated by an input parameter $t = 0$ ($t \neq 0$ indicates that scaling is to be ignored). We have not included a step for updating translations according to the current weights. It will be seen that one scaling step is given for each rotation step. Numerous variations on this are possible with p rotations followed by q scaling steps for different settings of p and q. In the algorithm we have used the weights from the rotation step for estimating scaling but strictly every time $\mathbf{X}$ is updated, new weights should be calculated. We have not investigated the best strategy based on these possibilities.

For comparison with the robust analysis, users will often do an initial unweighted orthogonal Procrustes analysis delivering the rotation $\mathbf{Q}$. Then, it makes sense to initiate Algorithm 4.2 with $\mathbf{X}_1\mathbf{Q}$ rather than with $\mathbf{X}_1$.

At the end of this algorithm, $\mathbf{X} = s\mathbf{X}_1\mathbf{Q}$. To identify s and $\mathbf{Q}$ we note that: $\|\mathbf{X}\| = \|s\mathbf{X}_1\mathbf{Q}\| = s^2\|\mathbf{X}_1\|$, which identifies s. Then, if $\mathbf{Y} = \mathbf{X}_1\mathbf{Q} = \mathbf{X}/s$, we have that $\mathbf{Q} = (\mathbf{X}_1'\mathbf{X}_1)^{-1}\mathbf{X}_1'\mathbf{Y}$.

Algorithm 4.2

Step 1 Initialise: Set c, $lastres = -1, \mathbf{X} = \mathbf{X}_1$
Step 2 r_i^2 = sum-of-squares of ith row of $\mathbf{R} = \mathbf{X} - \mathbf{X}_2$, $i = 1, 2, \ldots, N$
Step 3 if $r_i < c$ then $w_i = 1$ else $w_i t = c/r_i$, $i = 1, 2, \ldots, N$
Step 4 $SVD(\mathbf{X}'\mathbf{W}\mathbf{X}_2) = \mathbf{U\Sigma V}'$, $\mathbf{Q} = \mathbf{VU}'$
Step 5 $\mathbf{X} = \mathbf{XQ}$ (update $\mathbf{X}_1$)
Step 6 if $t = 0$, go to *step 9*
Step 7 $s = \mathrm{trace}\,(\mathbf{X}'\mathbf{W}\mathbf{X}_2)/\mathrm{trace}(\mathbf{X}'\mathbf{W}\mathbf{X})$
Step 8 $\mathbf{X} = s\mathbf{X}$ (update $\mathbf{X}_1$)
Step 9 $res = \|\mathbf{X} - \mathbf{X}_2\|$
Step 10 if $|lastres - res| < \varepsilon$ exit
Step 11 $lastres = res$, go to *step* 2.

Above, we mentioned that the Mosteller and Tukey weights (4.15) give zero weight to residuals larger than the threshold c, and that this is useful when interested in the possibility of partial matches. Siegel and Benson (1982) offer an alternative way of handling such problems. Essentially Siegel and Benson are interested in finding a rotation, scaling, and translation that minimises the median residual. They point out that when more than half of the N points have a perfect match then their residuals are zero and hence so is the median residual. Thus, in this situation, minimising the median criterion must find the perfect partial match. When the partial match is not perfect, the criterion may be expected to perform well. For the case $P = 2$, Seigel and Benson (1982) offer an algorithm for minimising the median criterion. This operates by fixing on one point of $\mathbf{X}_2$ and finding the median angle between, and the median ratio associated with, all other points of $\mathbf{X}_1$ and $\mathbf{X}_2$. The median of these two quantities over all fixed points of $\mathbf{X}_2$ gives the rotation and scaling. Translation is a final step that takes the median residuals in the directions of the coordinate axes. This dependence on the arbitrary choice of reference axes perhaps could be eliminated by iterating the process. However, we have not pursued this possibility, as the Verboon and Heiser approach, together with the Mosteller and Tukey weights, seems to offer a more general algorithm and one with assured convergence properties.

Finally, the L_1-norm may be fitted into the Huber robust estimation and IRLS framework, basically by using weights $1/|r_i|$. We do not go into details because the algorithm is likely to be slow, with uncertain convergence properties (see Ekblom 1987 for a discussion of this approach in the context of multiple linear regression).

5
Projection Procrustes problems

Now, $\mathbf{T}$ is constrained to be a projection matrix, which we shall denote by $\mathbf{P}$, in line with the notation of Section 1.3.2. Appendix C gives the basic properties of projection matrices. Figure 5.1 illustrates the geometry for the case of three points and where $P_1 = 2$ and $P_2 = 1$. Section 5.7 gives further information on the geometrical interpretation of the variants of optimal projection discussed in this chapter.

The constrained version of (3.1) is to minimise:

$$\|\mathbf{X}_1\mathbf{P} - \mathbf{X}_2\| = \text{trace}(\mathbf{P}'\mathbf{X}_1'\mathbf{X}_1\mathbf{P} - 2\mathbf{X}_2'\mathbf{X}_1\mathbf{P} + \mathbf{X}_2'\mathbf{X}_2) \tag{5.1}$$

where ${}_{P_1}\mathbf{P}_{P_2}$ with $P_2 < P_1$, satisfies $\mathbf{P}'\mathbf{P} = \mathbf{I}$. With $P_2 > P_1$, the columns of $\mathbf{P}$ cannot be independent and so $\mathbf{P}'\mathbf{P} = \mathbf{I}$ is impossible. It is, however, possible to replace $\mathbf{P}$ by $\mathbf{P}'$ and this case is considered in Section 5.7.

We may introduce an arbitrary orthogonal matrix ${}_{P_1}\mathbf{Q}_{P_1}$ and write $\mathbf{X}_1\mathbf{P} = (\mathbf{X}_1\mathbf{Q})\mathbf{Q}'\mathbf{P})$ where $\mathbf{X}_1\mathbf{Q}$ gives a rotation of $\mathbf{X}_1$ and $\mathbf{QP}$ remains orthonormal. Thus, the optimal projection includes a rotation in the higher dimensional space, as was implied by the caption to Figure 5.1. In (5.1) the term trace$(\mathbf{P}'\mathbf{X}_1'\mathbf{X}_1\mathbf{P})$ does not simplify so, unlike the orthogonal Procrustes problem, now the least-squares criterion differs from the inner-product criterion. Consequently the two criteria have to be treated separately. It turns out that the more simple is the inner-product criterion, so we begin with that.

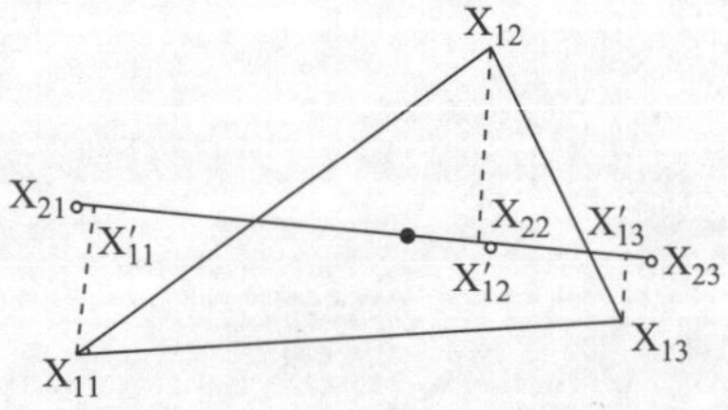

Fig. 5.1. The triangle $X_{11}X_{12}X_{13}$ is shown in a rotated position where its orthogonal projections $X'_{11}X'_{12}X'_{13}$ onto the line $X_{21}X_{22}X_{23}$ minimise the sum-of-squares of the residuals $X'_{11}X_{21}$, $X'_{12}X_{22}$, $X'_{13}X_{23}$. The projected positions are at the feet of the dashed lines.

5.1 Projection Procrustes by inner-product

The inner-product version of projection Procrustes has an analytical solution (Cliff 1966). Now we have to maximise trace($\mathbf{X}_2'\mathbf{X}_1\mathbf{P}$), the orthonormal version of (3.3). We start by writing $\mathbf{X}_2'\mathbf{X}_1 = {}_{P_2}\mathbf{U}_{P_2}\mathbf{\Sigma}_{P_1}\mathbf{V}_{P_1}'$ in its singular value decomposition form with full orthogonal matrices. Thus, $\mathbf{\Sigma}$ is rectangular with the P_2 non-zero singular values on the 'diagonal' in non-increasing order. Permuting within the trace operation shows that we have to maximise trace($\mathbf{\Sigma H}$) where ${}_{P_1}\mathbf{H}_{P_2} = \mathbf{V}'\mathbf{PU}$ an orthonormal matrix with $\mathbf{H}'\mathbf{H} = \mathbf{I}$. The maximum must occur when $\mathbf{H}$ has units down its 'diagonal' and is otherwise zero (i.e. $h_{ii} = 1$, and $h_{ij} = 0$ when $i \neq j$). Then $\mathbf{P} = \mathbf{VHU}' = \mathbf{V}_{P_2}\mathbf{U}'$ where $\mathbf{V}_{P_2}$ represents the matrix formed from the first P_2 columns of $\mathbf{V}$, agreeing with Cliff's result.

As noted by ten Berge and Knol (1984), using the inner-product criterion to estimate projections, turns out to be the same as using the least-squares criterion to estimate an orthogonal matrix $\mathbf{Q}$ after padding out $\mathbf{X}_2$ with $P_1 - P_2$ columns of zeros (see Section 4.4). This follows from writing the criterion as trace($\mathbf{X}_1'\mathbf{X}_2\mathbf{P}'$) and appealing to the rotational interpretation of $\mathbf{P}'$ (Appendix C). Alternatively, we may proceed directly, noting that min $\|\mathbf{X}_1\mathbf{Q} - (\mathbf{X}_2, \mathbf{0})\|$ is attained for the same $\mathbf{Q}$ as that which satisfies max(trace($\mathbf{Q}'\mathbf{X}_1'(\mathbf{X}_2, \mathbf{0})$)) which is the same as maximising trace($\mathbf{X}_2'\mathbf{X}_1\mathbf{P}$) where $\mathbf{P}$ is obtained as the first P_2 columns of $\mathbf{Q}$. It is also true that $\mathbf{X}_1\mathbf{P}$ and $\mathbf{X}_2$ have optimal orthogonal orientation in the P_2-dimensional space. This follows from noting that $\mathbf{X}_2'\mathbf{X}_1\mathbf{P} = \mathbf{U\Sigma}_{P_2}\mathbf{V}'\mathbf{V}_{P_2}\mathbf{U}' = \mathbf{U\Sigma}_{P_2}\mathbf{U}'$, which is symmetric and positive semi-definite (p.s.d). The result then follows immediately from Section 4.2.

5.2 Necessary but not sufficient conditions

With projection Procrustes we have just established above that $\mathbf{X}_2'\mathbf{X}_1\mathbf{P}$ being p.s.d. and symmetric is a necessary condition but we shall now show that, unlike with orthogonal transformations, it is not sufficient to show that the trace is maximised.

It will be convenient to write the singular value decomposition in the form $\mathbf{X}_2'\mathbf{X}_1 = {}_{P_2}\mathbf{U}_{P_2}\mathbf{\Sigma}_{P_2}\mathbf{V}_{P_1}'$ so that $\mathbf{V}$ is now orthonormal with $\mathbf{V}'\mathbf{V} = \mathbf{I}$ and $\mathbf{\Sigma}$ is (square) diagonal. As shown in Section 5.1, the maximal trace of $\mathbf{X}_2'\mathbf{X}_1\mathbf{P}$ occurs for $\mathbf{P} = \mathbf{VU}'$. Consider the P_1 by P_2 matrix $\mathbf{P}^* = (\mathbf{V}_\top, \mathbf{V}_\perp)\mathbf{U}'$ where $\mathbf{V}_\top$ consists of the first R, say, columns of $\mathbf{V}$ and $\mathbf{V}_\perp$ consists of any $P_2 - R$ columns from the null-space of $\mathbf{V}$ (i.e. columns that are orthogonal to the columns of $\mathbf{V}$). This is always possible when $0 < P_2 - R \leq P_1 - P_2$; only when $P_1 = P_2$ can no suitable value of R be found. By construction, the matrix $\mathbf{P}^*$ is orthonormal. We have that:

$$\mathbf{X}_2'\mathbf{X}_1\mathbf{P}^* = ({}_{P_2}\mathbf{U}_{P_2}\mathbf{\Sigma}_{P_2}\mathbf{V}_{P_1}')(\mathbf{V}_\top, \mathbf{V}_\perp)\mathbf{U}' = {}_{P_2}\mathbf{U}_{P_2}\mathbf{\Sigma}_{P_2}\mathbf{J}_R\mathbf{U}'$$

where $\mathbf{J}_R$ is a diagonal matrix with units in its first R positions, zero elsewhere. Thus:

$$\mathbf{X}_2'\mathbf{X}_1\mathbf{P}^* = \mathbf{U\Sigma}_R\mathbf{U}'$$

where $\mathbf{\Sigma}_R$ consists of only the first R singular values of $\mathbf{\Sigma}$. Clearly, $\mathbf{U\Sigma}_R\mathbf{U}'$ is symmetric and, because trace($\mathbf{\Sigma}_R$) < trace($\mathbf{\Sigma}$), it has smaller trace than $\mathbf{X}_2'\mathbf{X}_1\mathbf{P}$.

This shows that even when an orthonormal matrix $\mathbf{P}^*$ symmetricises $\mathbf{X}_2'\mathbf{X}_1$, this is not sufficient to guarantee that trace$(\mathbf{X}_2'\mathbf{X}_1\mathbf{P}^*)$ is maximised. In this construction we have chosen to preserve the first R columns of $\mathbf{V}$; different choices of columns will generate different symmetric matrices and different orthonormal matrices with different suboptimal traces.

We note that when $\mathbf{P}$ is orthogonal, so that $P_1 = P_2$, the construction fails. Then, $\mathbf{V}_\perp$ is null and $\mathbf{V}_\top = \mathbf{V}$ so $\mathbf{P}^* = \mathbf{VU}'$, giving the optimal trace, consistent with the necessary and sufficient conditions of Section 4.2.

5.3 Two-sided projection Procrustes by inner-product

The inner-product criterion (3.4) takes the form trace$(\mathbf{P}_2'\mathbf{X}_2'\mathbf{X}_1\mathbf{P}_1)$. Now we may project onto an R-dimensional subspace, so that the dimensions of $\mathbf{P}_1$ and $\mathbf{P}_2$ are $P_1 \times R$ and $P_2 \times R$, respectively. We continue to assume, without loss of generality, that $P_2 \leq P_1$ and we must also have that $R \leq P_2$ because we cannot project from a lower dimensional subspace into a containing higher dimensional space. Using the previous singular value decomposition gives

$$\text{trace}(\mathbf{P}_2'\mathbf{X}_2'\mathbf{X}_1\mathbf{P}_1) = \text{trace}(\mathbf{P}_2'\mathbf{U}_S\mathbf{\Sigma}_S\mathbf{V}_S'\mathbf{P}_1) = \text{trace}(\mathbf{H\Sigma K}')$$

where ${}_R\mathbf{H}_S = \mathbf{P}_2'\mathbf{U}$ and ${}_R\mathbf{K}_S = \mathbf{P}_1'\mathbf{V}$ are orthonormal matrices such that $\mathbf{HH}' = \mathbf{I}_S$ and $\mathbf{KK}' = \mathbf{I}_S$. Here, we have assumed that rank $(\mathbf{X}_2'\mathbf{X}_1) = S > R$, the geometrical meaning of which is discussed in Section 5.7. We have that trace$(\mathbf{H\Sigma K}')$ = trace$(\mathbf{K}'\mathbf{H\Sigma})$ = trace$(\sum_{i=1}^S \sigma_i \mathbf{h}_i \mathbf{k}_i') = \sum_{i=1}^S \sigma_i \mathbf{k}_i'\mathbf{h}_i$. Each term is maximised when $\mathbf{k}_i'\mathbf{h}_i = 1$, which requires that $\mathbf{h}_i = \mathbf{k}_i$. If we arrange this for $i = 1, 2, 3, \ldots$, after step R we arrive at:

$${}_S\mathbf{K}_R'\mathbf{H}_S = \begin{pmatrix} \mathbf{I}_R & \\ & 0 \end{pmatrix}$$

with

$$\mathbf{H} = \mathbf{K} = (\mathbf{H}_R' \quad 0)$$

where $\mathbf{H}_R$ is an arbitrary $R \times R$ orthogonal matrix, so that $\mathbf{HH}' = \mathbf{KK}' = \mathbf{I}_R$ satisfying the required orthonormality condition. It is not possible to add further non-zero columns to $\mathbf{H}$ and $\mathbf{K}$ without violating their orthonormality. It follows that:

$$\mathbf{K}' = \mathbf{V}'\mathbf{P}_1 = \begin{pmatrix} \mathbf{H}_R \\ 0 \end{pmatrix}, \quad \mathbf{H}' = \mathbf{U}'\mathbf{P}_2 = \begin{pmatrix} \mathbf{H}_R \\ 0 \end{pmatrix},$$

and hence:

$$\begin{aligned} (\mathbf{V}_R)'\mathbf{P}_1 &= \mathbf{H}_R, \quad \mathbf{V}_\perp'\mathbf{P}_1 = 0, \\ (\mathbf{U}_R)'\mathbf{P}_2 &= \mathbf{H}_R, \quad \mathbf{U}_\perp'\mathbf{P}_2 = 0. \end{aligned} \tag{5.2}$$

where $\mathbf{V}_R$ and $\mathbf{U}_R$ are the vectors corresponding to the R largest singular values of $\mathbf{\Sigma}$ and $\mathbf{V}_\perp$ and $\mathbf{U}_\perp$ are vectors corresponding to *non-null* singular values in $\mathbf{\Sigma}$, excluded because $R < S$. There are also the $P_1 - S$ null vectors $\mathbf{V}_{0\perp}$ and the $P_2 - S$ null vectors $\mathbf{U}_{0\perp}$ to be considered. We now note that a solution to the above equations is given by:

$$\mathbf{P}_1 = \mathbf{V}_R\mathbf{H}_R, \quad \mathbf{P}_2 = \mathbf{U}_R\mathbf{H}_R. \tag{5.3}$$

However, it is possible that $\mathbf{P}_1$ may also include vectors orthogonal to $\mathbf{V}_R$, so we consider the possibility of a more general solution:

$$\mathbf{P}_1 = \mathbf{V}_R\mathbf{H}_R + \mathbf{V}_\perp\mathbf{L} + \mathbf{V}_{0\perp}\mathbf{M},$$

where $\mathbf{L}$ and $\mathbf{M}$ are arbitrary matrices each with R columns. The orthonormality constraint now gives:

$$\mathbf{P}_1'\mathbf{P}_1 = \mathbf{I} = \mathbf{I} + \mathbf{L}'\mathbf{L} + \mathbf{M}'\mathbf{M}.$$

Thus, $\mathbf{L}'\mathbf{L} + \mathbf{M}'\mathbf{M} = \mathbf{0}$ whose zero diagonal, being formed from sums of squares, implies that $\mathbf{L} = \mathbf{0}$ and $\mathbf{M} = \mathbf{0}$. A similar argument applies to $\mathbf{P}_2$. Thus, the particular solution (5.3)

$$\mathbf{P}_1 = \mathbf{V}_R\mathbf{H}_R, \quad \mathbf{P}_2 = \mathbf{U}_R\mathbf{H}_R,$$

is seen to be the general solution required. Note that if $P_2 = R$ then (5.2) immediately gives $\mathbf{P}_2 = \mathbf{U}_R\mathbf{H}_R$; then $\mathbf{P}_1$ follows as above.

Thus (5.3) gives the projected configurations $\mathbf{X}_1\mathbf{V}_R\mathbf{H}_R$ and $\mathbf{X}_2\mathbf{U}_R\mathbf{H}_R$ with $\sigma_1 + \sigma_2 + \cdots + \sigma_R$ as the maximal trace of the inner-product criterion. Since $\mathbf{H}_R$ is an arbitrary orthogonal matrix, it may as well be taken to be $\mathbf{I}_R$. Because $\mathbf{P}_2'\mathbf{X}_2'\mathbf{X}_1\mathbf{P}_1 = \mathbf{H}_R'\mathbf{U}_R'(\mathbf{U}_s\mathbf{\Sigma}_S\mathbf{V}_S')\mathbf{V}_R'\mathbf{H}_R = \mathbf{H}_R'\mathbf{\Sigma}_R\mathbf{H}_R$ which is symmetric and p.s.d., Section 4.2 shows that $\mathbf{X}_1\mathbf{P}_1$ and $\mathbf{X}_2\mathbf{P}_2$ are optimally orthogonally oriented in the s-dimensional space.

Projection operators do not preserve size, so with these inner-product criteria correlational interpretations are not directly available. We could define:

$$r(\mathbf{X}_1\mathbf{P}, \mathbf{X}_2) = \frac{\text{trace}(\mathbf{X}_2'\mathbf{X}_1\mathbf{P})}{\sqrt{\|\mathbf{X}_1\mathbf{P}\|\|\mathbf{X}_2\|}} \quad \text{and} \quad r(\mathbf{X}_1\mathbf{P}_1, \mathbf{X}_2\mathbf{P}_2) = \frac{\text{trace}(\mathbf{P}_2'\mathbf{X}_2'\mathbf{X}_1\mathbf{P}_1)}{\sqrt{\|\mathbf{X}_1\mathbf{P}_1\|\|\mathbf{X}_2\mathbf{P}_2\|}}$$

but these are difficult functions to optimise. In practice, the use of the inner product may be acceptable when $\mathbf{X}_1$ and $\mathbf{X}_2$ are first centred and normalised, but the correlational interpretation remains questionable.

5.3.1 TUCKER'S PROBLEM

ten Berge (1979) gives an interesting application where the inner-product criterion is fully justified. This relates to a problem of Tucker (1951), who requires the maximisation of trace$(\mathbf{T}_2'\mathbf{X}_2'\mathbf{X}_1\mathbf{T}_1)$ subject to the constraints $\mathbf{T}_i'\mathbf{X}_i'\mathbf{X}_i\mathbf{T}_i = \mathbf{I}$, for $i = 1, 2$. Here the transformations are not themselves projections but the transformed matrices $\mathbf{X}_i\mathbf{T}_i$ are constrained to be orthonormal. With these constraints, the inner-product and correlational-criteria coincide; so, indeed does the two-sided least-squares criterion trace$\|\mathbf{X}_1\mathbf{T}_1 - \mathbf{X}_2\mathbf{T}_2\|$.

To make progress, ten Berge (1979) reparameterises $\mathbf{X}_i$ in the form:

$$\mathbf{X}_i = \mathbf{Y}_i\mathbf{L}_i, \quad \text{where } i = 1, 2$$

where $\mathbf{Y}_i$ is orthonormal, so that $\mathbf{Y}_i'\mathbf{Y}_i = \mathbf{I}_{P_i}$. This implies that $\mathbf{X}_i'\mathbf{X}_i = \mathbf{L}_i'\mathbf{L}_i$; any non-singular $\mathbf{L}_i$ satisfying this equation will suffice. ten Berge (1979) chooses the lower triangular (Choleski) decomposition but we could also, for example, use the S.V.D., $\mathbf{X}_i = \mathbf{U}_i\boldsymbol{\Sigma}_i\mathbf{V}_i'$ partitioned either as $\mathbf{X}_i = \mathbf{U}_i(\boldsymbol{\Sigma}_i\mathbf{V}_i')$ or $\mathbf{X}_i = \mathbf{U}_i\mathbf{V}_i'(\mathbf{V}_i\boldsymbol{\Sigma}_i\mathbf{V}_i')$, where valid choices of $\mathbf{L}_i$ are the terms in parenthesis and the corresponding $\mathbf{Y}_i$ are $\mathbf{U}_i$, and $\mathbf{U}_i\mathbf{V}_i'$, respectively.

Thus, if we define $\mathbf{P}_i = \mathbf{L}_i\mathbf{T}_i$, then $\mathbf{X}_i\mathbf{T}_i = \mathbf{Y}_i\mathbf{P}_i$, for $i = 1, 2$ and the constraints $\mathbf{T}_i'\mathbf{X}_i'\mathbf{X}_i\mathbf{T}_i = \mathbf{I}$ imply that $\mathbf{P}_1$ and $\mathbf{P}_2$ are orthonormal. The reparameterised form of Tucker's criterion is to maximise:

$$\text{trace}(\mathbf{P}_2'\mathbf{Y}_2'\mathbf{Y}_1\mathbf{P}_1)$$

subject to $\mathbf{P}_i'\mathbf{P}_i = \mathbf{I}_r$ for $i = 1, 2$. We have reduced Tucker's problem to that discussed at the beginning of this section, whose solution is given by (5.3) where, now, we use the singular value decomposition $\mathbf{Y}_2'\mathbf{Y}_1 = \mathbf{U}\boldsymbol{\Sigma}\mathbf{V}'$ giving $\mathbf{X}_1\mathbf{T}_1 = \mathbf{Y}_1\mathbf{V}$, $\mathbf{X}_2\mathbf{T}_2 = \mathbf{Y}_2\mathbf{U}$, ignoring the arbitrary orthogonal matrix $\mathbf{H}_R$.

5.3.2 MEREDITH'S PROBLEM

ten Berge (1979) also draws attention to a link between the Tucker problem, outlined above, and a problem due to Meredith (1964). The background is in factor analysis and its relevance to our main interest in matching configurations is slender. Furthermore, in some ways the link is closer to the one-sided inner-product criterion than to the two-sided. However, we think it worth a mention both for its intrinsic interest and to give something of the flavour of a significant part of the factor analysis literature. Meredith (1964) required the minimum of:

$$\sum_{i=1}^{2} \|\mathbf{X}_i\mathbf{T}_i^* - \mathbf{Z}\|$$

subject to the constraint $\mathbf{Z}_R'\mathbf{Z}_R = \mathbf{I}_R$. Thus, the transformations $\mathbf{T}_i^*$ are not specified to have any particular form but they have to approximate an orthonormal matrix $\mathbf{Z}$. In factor analysis, this would be regarded as finding a pair of 'oblique rotations' (meaning 'not orthogonal') to match $\mathbf{X}_1$ and $\mathbf{X}_2$ to a common matrix $\mathbf{Z}$. In our terminology, the transformations $\mathbf{T}_i^*$ neither represent rotations nor oblique axes; Chapter 6 discusses oblique axes transformation as defined in Appendix D.

We may make the same reparameterisations $\mathbf{X}_i = \mathbf{Y}_i\mathbf{L}_i$ with $\mathbf{Y}_i'\mathbf{Y}_i = \mathbf{I}_{p_i}$ as for the Tucker problem, and write $\mathbf{L}_i\mathbf{T}_i^* = \mathbf{S}_i$. Then, Meredith's criterion is to minimise:

$$\sum_{i=1}^{2} \|\mathbf{Y}_i\mathbf{S}_i - \mathbf{Z}\|,$$

over $\mathbf{S}_1$, $\mathbf{S}_2$, and $\mathbf{Z}$. Either by differentiation with respect to $\mathbf{S}_1$ and $\mathbf{S}_2$, or directly from linear least-squares theory, for *given* $\mathbf{Z}$, we have that:

$$S_i = (\mathbf{Y}_i'\mathbf{Y}_i)^{-1}\mathbf{Y}_i'\mathbf{Z} = \mathbf{Y}_i'\mathbf{Z} \quad i = 1, 2 \tag{5.4}$$

Thus

$$\begin{aligned}\sum_{i=1}^{2} \|\mathbf{Y}_i\mathbf{S}_i - \mathbf{Z}\| &= \sum_{i=1}^{2} \|(\mathbf{Y}_i\mathbf{Y}_i' - \mathbf{I})\mathbf{Z}\| \\ &= \text{trace} \sum_{i=1}^{2} \mathbf{Z}'(\mathbf{I} - \mathbf{Y}_i\mathbf{Y}_i')\mathbf{Z} \\ &= 2R - \text{trace}\mathbf{Z}'(\mathbf{Y}_1\mathbf{Y}_1' + \mathbf{Y}_2\mathbf{Y}_2')\mathbf{Z}.\end{aligned}$$

The minimum occurs when the trace term is maximised, which, as is well-known, is given by choosing $\mathbf{Z}$ to be given by the first R eigenvectors satisfying:

$$(\mathbf{Y}_1\mathbf{Y}_1' + \mathbf{Y}_2\mathbf{Y}_2')\mathbf{Z} = \mathbf{Z}\boldsymbol{\Lambda} \tag{5.5}$$

where $\boldsymbol{\Lambda}$ is the diagonal matrix of the R biggest eigenvalues. Thus (5.5) gives $\mathbf{Z}$ and then (5.4) gives $\mathbf{S}_1$ and $\mathbf{S}_2$. Finally, $\mathbf{X}_i\mathbf{T}_i^* = \mathbf{Y}_i\mathbf{S}_i$, gives the fitted values and if the transformations themselves are required, then we have $\mathbf{T}_i^* = \mathbf{L}_i^{-1}\mathbf{S}_i$, $i = 1, 2$.

This solves Meredith's problem but it remains to be shown how this result links with Tucker's problem. We follow a simplified form of the argument given by ten Berge (1979) and write (5.5) as:

$$(\mathbf{Y}_1 \quad \mathbf{Y}_2)\begin{pmatrix}\mathbf{Y}_1' \\ \mathbf{Y}_2'\end{pmatrix}\mathbf{Z} = \mathbf{Z}\boldsymbol{\Lambda} \tag{5.6}$$

which we know has the same eigenvalues and related eigenvectors as:

$$\begin{pmatrix}\mathbf{Y}_1' \\ \mathbf{Y}_2'\end{pmatrix}(\mathbf{Y}_1 \quad \mathbf{Y}_2) = \begin{pmatrix}\mathbf{I} & \mathbf{Y}_1'\mathbf{Y}_2 \\ \mathbf{Y}_2'\mathbf{Y}_1 & \mathbf{I}\end{pmatrix} = \begin{pmatrix}\mathbf{I} & \mathbf{V}\boldsymbol{\Sigma}\mathbf{U}' \\ \mathbf{U}\boldsymbol{\Sigma}\mathbf{V}' & \mathbf{I}\end{pmatrix}.$$

Now

$$\left.\begin{aligned}\begin{pmatrix}\mathbf{I} & \mathbf{V}\boldsymbol{\Sigma}\mathbf{U}' \\ \mathbf{U}\boldsymbol{\Sigma}\mathbf{V}' & \mathbf{I}\end{pmatrix}\begin{pmatrix}\mathbf{V} \\ \mathbf{U}\end{pmatrix} &= \begin{pmatrix}\mathbf{V} \\ \mathbf{U}\end{pmatrix}(\mathbf{I} + \boldsymbol{\Sigma}) \\ \begin{pmatrix}\mathbf{I} & \mathbf{V}\boldsymbol{\Sigma}\mathbf{U}' \\ \mathbf{U}\boldsymbol{\Sigma}\mathbf{V}' & \mathbf{I}\end{pmatrix}\begin{pmatrix}\mathbf{V} \\ -\mathbf{U}\end{pmatrix} &= \begin{pmatrix}\mathbf{V} \\ -\mathbf{U}\end{pmatrix}(\mathbf{I} - \boldsymbol{\Sigma})\end{aligned}\right\}. \tag{5.7}$$

The relationship between the eigenvectors of (5.6) and (5.7) follows from the identity $(\mathbf{A}\mathbf{A}')\mathbf{Z} = \mathbf{Z}\boldsymbol{\Lambda} \Rightarrow (\mathbf{A}'\mathbf{A})(\mathbf{A}'\mathbf{Z}\boldsymbol{\Lambda}^{-1/2}) = (\mathbf{A}'\mathbf{Z}\boldsymbol{\Lambda}^{-1/2})\boldsymbol{\Lambda}$. Thus, if $\mathbf{Z}$ are the normalised eigenvectors of $\mathbf{AA}'$ then the eigenvectors of $\mathbf{A}'\mathbf{A}$ are $\mathbf{A}'\mathbf{Z}\mathbf{\Lambda}^{-1/2}$ and these eigenvectors are properly normalised because $(\mathbf{\Lambda}^{-1/2}\mathbf{Z}'\mathbf{A})(\mathbf{A}'\mathbf{Z}\mathbf{\Lambda}^{-1/2}) = \mathbf{\Lambda}^{-1/2}(\mathbf{\Lambda})\mathbf{\Lambda}^{-1/2} = \mathbf{I}$. Setting $\mathbf{A} = (\mathbf{Y}_1, \mathbf{Y}_2)$, comparing the eigenvectors of (5.6) and (5.7) gives:

$$\frac{1}{\sqrt{2}}\begin{pmatrix}\mathbf{V}\\ \mathbf{U}\end{pmatrix} = \begin{pmatrix}\mathbf{Y}_1'\\ \mathbf{Y}_2'\end{pmatrix}\mathbf{Z}\mathbf{\Lambda}^{-1/2} \quad \text{and} \quad \mathbf{\Lambda} = \mathbf{I} + \mathbf{\Sigma}.$$

We have taken the eigenvalues $\mathbf{I} + \mathbf{\Sigma}$ rather than $\mathbf{I} - \mathbf{\Sigma}$ because the former must contain the r biggest eigenvalues. The factor $1/\sqrt{2}$ is needed to give the correct normalisation of the eigenvectors on the left-hand-side. Separating the upper and lower parts of the previous matrix equation and taking the eigenvalues to the left-hand-side gives:

$$\tfrac{1}{\sqrt{2}}\mathbf{V}(\mathbf{I}+\mathbf{\Sigma})^{1/2} = \mathbf{Y}_1'\mathbf{Z} \quad \tfrac{1}{\sqrt{2}}\mathbf{U}(\mathbf{I}+\mathbf{\Sigma})^{1/2} = \mathbf{Y}_2'\mathbf{Z}. \tag{5.8}$$

The fitted values are given by $\mathbf{X}_i\mathbf{T}_i^* = \mathbf{Y}_i\mathbf{S}_i$ which from (5.4) gives:

$$\mathbf{X}_i\mathbf{T}_i^* = \mathbf{Y}_i\mathbf{Y}_i'\mathbf{Z}, \quad i = 1, 2.$$

Now, using (5.8) we have:

$$\left.\begin{aligned}\mathbf{X}_1\mathbf{T}_1^* &= \tfrac{1}{\sqrt{2}}\mathbf{Y}_1\mathbf{V}(\mathbf{I}+\mathbf{\Sigma})^{1/2}\\ \mathbf{X}_2\mathbf{T}_2^* &= \tfrac{1}{\sqrt{2}}\mathbf{Y}_1\mathbf{U}(\mathbf{I}+\mathbf{\Sigma})^{1/2}\end{aligned}\right\}. \tag{5.9}$$

The fitted values in (5.9) are columnwise proportional to the fitted values found for the Tucker problem. It follows that the pairwise correlations (corrected for the mean or not) between columns of the fitted values are the same for both methods. It follows therefore that the Tucker congruence coefficient r_{T} (3.9) between the fitted values are the same for each method. This is not true for the RV coefficient (3.5). In fact we can establish the exact relationship.

For Meridith's problem we have that

$$\|\mathbf{Y}_i\mathbf{S}_i\| = \|\mathbf{S}_i\|, \quad i = 1, 2$$

Now, $\|\mathbf{S}_i\| = \|\mathbf{Y}_i'\mathbf{Z}\|$ which from (5.8) gives:

$$\|\mathbf{S}_1\| = \|\mathbf{S}_2\| = \tfrac{1}{2}\mathrm{trace}(\mathbf{I}+\mathbf{\Sigma}),$$

incidently showing that the two sets of fitted values are of equal *size*. Further,

$$\begin{aligned}\mathrm{trace}(\mathbf{S}_2'\mathbf{Y}_2'\mathbf{Y}_1\mathbf{S}_1) &= \mathrm{trace}(\mathbf{Z}'\mathbf{Y}_2\mathbf{Y}_2'\mathbf{Y}_1\mathbf{Y}_1'\mathbf{Z})\\ &= \tfrac{1}{2}\mathrm{trace}\{(\mathbf{I}+\mathbf{\Sigma})^{1/2}\mathbf{U}'(\mathbf{Y}_2'\mathbf{Y}_1)\mathbf{V}(\mathbf{I}+\mathbf{\Sigma})^{1/2}\}\\ &= \tfrac{1}{2}\mathrm{trace}((\mathbf{I}+\mathbf{\Sigma})\mathbf{\Sigma}).\end{aligned}$$

Putting these results together gives:

$$r_V(\text{MEREDITH}) = \frac{\text{trace}(\mathbf{\Sigma} + \mathbf{\Sigma}^2)}{\text{trace}(\mathbf{I} + \mathbf{\Sigma})}$$

The corresponding result for Tucker is:

$$r_V(\text{TUCKER}) = \frac{1}{R}\text{trace}(\mathbf{\Sigma}).$$

Comparing the two forms we see that when $\sigma_1 = \sigma_2 = \sigma_3 = \cdots = \sigma_R$ the two RV coefficients agree. More generally, if the variance of the singular values is small compared with their mean, then the two coefficients will be approximately equal.

5.3.3 GREEN'S PROBLEM

Yet another problem arising from oblique, meaning non-orthogonal, factor matching is discussed by Green (1969) with improvements by ten Berge (1983a). The problem is to minimise:

$$\|\mathbf{X}_1\mathbf{T} - \mathbf{X}_2\|$$

subject to:

$$\mathbf{T}'\mathbf{R}\mathbf{T} = \mathbf{S}$$

where

$$\mathbf{R} = \mathbf{X}_1'\mathbf{X}_1 = {}_{P_1}\mathbf{F}_R\mathbf{F}'_{P_1} \quad \text{and} \quad \mathbf{S} = {}_{P_2}\mathbf{G}_S\mathbf{G}'_{P_2}. \tag{5.10}$$

Here, it is recognised that rank $\mathbf{R} = R \leq P_1$ and rank $\mathbf{S} = S \leq P_2$. The constraint $\mathbf{T}'\mathbf{R}\mathbf{T} = \mathbf{S}$ implies that we must have $S \leq R$. Expanding the least-squares criterion shows that it is equivalent to maximising:

$$\text{trace}\,\mathbf{Z}\mathbf{T}$$

where $\mathbf{Z} = \mathbf{X}_2'\mathbf{X}_1$. Green (1969) first established a general form for all $\mathbf{T}$ that satisfy the constraints and then optimised the trace criterion over this general form. Expanding the constraint gives:

$$\mathbf{T}'\mathbf{F}\mathbf{F}'\mathbf{T} = \mathbf{G}\mathbf{G}'$$

from which it follows that:

$$\mathbf{F}'\mathbf{T} = \mathbf{P}\mathbf{G}' \tag{5.11}$$

for any orthonormal $\mathbf{P}$, satisfying ${}_S\mathbf{P}'_R\mathbf{P}_S = \mathbf{I}_S$. We may use the columns of $\mathbf{F}$ as a basis for expressing $\mathbf{T}$, to give:

$$\mathbf{T} = \mathbf{F}_R\mathbf{K} + \mathbf{F}_\perp\mathbf{L} \quad \text{for arbitrary } \mathbf{K} \text{ and } \mathbf{L}$$

where $\mathbf{F}_\perp$ has $P_1 - R$ independent columns, orthogonal to the R columns of $\mathbf{F}$. Pre-multiplying both sides by $\mathbf{F}'$ gives $\mathbf{F}'\mathbf{T} = \mathbf{F}'\mathbf{F}\mathbf{K}$, so that:

$$\mathbf{K} = (\mathbf{F}'\mathbf{F})^{-1}\mathbf{F}'\mathbf{T}$$

and from (5.11)

$$\mathbf{K} = (\mathbf{F}'\mathbf{F})^{-1}\mathbf{P}\mathbf{G}'$$

giving:

$$\mathbf{T} = \mathbf{F}(\mathbf{F}'\mathbf{F})^{-1}\mathbf{P}\mathbf{G}' + \mathbf{F}_{\perp}\mathbf{L}. \tag{5.12}$$

ten Berge (1983a) introduced the term in $\mathbf{F}_{\perp}$ and showed that it had no effect on the trace criterion. This follows from noting that (5.12) gives:

$$\text{trace}\,\mathbf{ZT} = \text{trace}(\mathbf{ZF}(\mathbf{F}'\mathbf{F})^{-1}\mathbf{PG}' + \mathbf{ZF}_{\perp}\mathbf{L})$$

and then considering the term $\mathbf{ZF}_{\perp}\mathbf{L}$. From the definition of $\mathbf{R}$ in (5.10) we have:

$$(\mathbf{F}'_{\perp}\mathbf{X}'_1)(\mathbf{X}_1\mathbf{F}_{\perp}) = \mathbf{0} \Rightarrow \mathbf{X}_1\mathbf{F}_{\perp} = \mathbf{0}.$$

Thus

$$\mathbf{ZF}_{\perp}\mathbf{L} = \mathbf{X}'_2(\mathbf{X}_1\mathbf{F}_{\perp})\mathbf{L} = \mathbf{0},$$

so that, finally

$$\text{trace}\,\mathbf{ZT} = \text{trace}\,\mathbf{ZF}(\mathbf{F}'\mathbf{F})^{-1}\mathbf{PG}. \tag{5.13}$$

We may now move to the second phase of maximising the trace. The only free term in (5.14) is the orthonormal matrix $\mathbf{P}$. Rearranging (5.13) gives:

$$\text{trace}\,\mathbf{ZT} = \text{trace}\,\mathbf{GZF}(\mathbf{F}'\mathbf{F})^{-1}\mathbf{P}.$$

This is the problem of Section 5.1 now defining $\mathbf{GZF}(\mathbf{F}'\mathbf{F})^{-1} = \mathbf{U\Sigma V}'$, giving $\mathbf{P} = \mathbf{V}_{P_2}\mathbf{U}'$. Thus, the solution to Green's problem is to set:

$$\mathbf{T} = \mathbf{F}(\mathbf{F}'\mathbf{F})^{-1}\mathbf{V}_{P_2}\mathbf{U}'\mathbf{G}' + \mathbf{F}_{\perp}\mathbf{L},$$

where the term $\mathbf{F}_{\perp}\mathbf{L}$ may be ignored without loss.

ten Berge (1983a) shows that other oblique axis problems reduce to special cases of Green's problem and hence may be subsumed by Green's solution.

5.4 Projection Procrustes by least squares

The least squares normal equations for $\mathbf{P}$ derived from (5.1) are:

$$\mathbf{X}'_1\mathbf{X}_1\mathbf{P} - \mathbf{P\Lambda} = \mathbf{X}'_1\mathbf{X}_2 \tag{5.14}$$

where $\mathbf{\Lambda}$ is a symmetric matrix of Lagrange multipliers associated with the $\frac{1}{2}p_2(p_2+1)$ constraints $\mathbf{P}'\mathbf{P} = \mathbf{I}$. Pre-multiplying by $\mathbf{P}'$ gives:

$$\mathbf{P}'\mathbf{X}'_1\mathbf{X}_1\mathbf{P} - \mathbf{\Lambda} = \mathbf{P}'\mathbf{X}'_1\mathbf{X}_2$$

showing that $\mathbf{P}'\mathbf{X}'_1\mathbf{X}_2$ is symmetric and hence by the result of Section 4.2, $\mathbf{X}_1\mathbf{P}$ and $\mathbf{X}_2$ must be optimally oriented as was also found for the inner-product criterion. Of course, the two approaches give different estimates of $\mathbf{P}$ and different fitted values.

Equation (5.14) has no closed form solution but Green and Gower (1981) (see also Gower 1984) propose the following algorithmic solution (see Section 3.5):

Algorithm 5.1

Step 1 Initialisation: add $P_1 - P_2$ zero columns to $\mathbf{X}_2$.
Step 2 Use the orthogonal Procrustes procedure (Section 4.2) to fit $\mathbf{X}_1$ to the new $\mathbf{X}_2$, replacing $\mathbf{X}_1$ by its rotated form.
Step 3 If the sum-of-squares of the differences between the final $P_1 - P_2$ columns of $\mathbf{X}_1$ and $\mathbf{X}_2$ is less than some threshold, exit.
Step 4 Replace the final $P_1 - P_2$ columns of $\mathbf{X}_2$ by the corresponding columns of the rotated $\mathbf{X}_1$ and return to step 2.

When an isotropic scaling factor s is to be included, step 2 should be included in the estimation of the scaling in the orthogonal Procrustes procedure and $\mathbf{X}_1$ replaced by its scaled and rotated form. At the conclusion of the algorithm the final versions of s and $\mathbf{Q}$ and, hence $\mathbf{P}$, may be derived from $s\mathbf{X}_{1(\text{initial})}\mathbf{Q} = \mathbf{X}_{1(\text{final})}$. Thus, $s^2\|\mathbf{X}_1\|_{\text{initial}} = \|\mathbf{X}_1\|_{\text{final}}$ and $\mathbf{Q} = 1/s(\mathbf{X}_1'\mathbf{X}_1)^{-1}_{\text{initial}}(\mathbf{X}_1'\mathbf{X}_1)_{\text{final}}$.

Algorithm 5.1 is illustrated here using data from Harman (1976) quoted by Gruvaeus (1970, table 1, p. 498). Gruvaeus provides a three-column target matrix in his table 3, p. 500 from which we have extracted the first two columns.

$$\mathbf{X}_1 = \begin{pmatrix} 0.90 & -0.09 & -0.03 \\ 0.83 & 0.09 & -0.04 \\ 0.87 & -0.01 & 0.07 \\ 0.55 & 0.79 & -0.07 \\ 0.56 & 0.65 & 0.04 \\ 0.63 & 0.60 & 0.03 \\ 0.28 & 0.27 & 0.45 \\ 0.38 & 0.20 & 0.63 \\ 0.38 & 0.19 & 0.77 \end{pmatrix}, \quad \text{The target } \mathbf{X}_2 = \begin{pmatrix} 0.98 & 0 \\ 0.76 & 0 \\ 0.86 & 0 \\ 0 & 1.00 \\ 0 & 0.84 \\ 0 & 0.77 \\ 0 & 0 \\ 0 & 0 \\ 0 & 0 \end{pmatrix}.$$

The stopping criterion $\varepsilon = 10^{-6}$ resulted in convergence after 12 iterations. The

$$\mathbf{P} \text{ obtained is } \begin{pmatrix} 0.8255 & 0.2308 \\ -0.4176 & 0.8637 \\ -0.3797 & -0.4481 \end{pmatrix}, \quad \mathbf{X}_1\mathbf{P} \text{ was } \begin{pmatrix} 0.7919 & 0.1434 \\ 0.6628 & 0.2872 \\ 0.6958 & 0.1608 \\ 0.1507 & 0.8406 \\ 0.1757 & 0.6727 \\ 0.2582 & 0.6502 \\ -0.0524 & 0.0962 \\ -0.0090 & -0.0219 \\ -0.0580 & -0.0932 \end{pmatrix}$$

with $\|\mathbf{X}_1\mathbf{P} - \mathbf{X}_2\| = 0.4133$

The orthogonal Procrustes procedure must reduce the residual sum-of-squares at each step and is bounded below by zero, so convergence is assured. At convergence the final $P_1 - P_2$ columns of $\mathbf{X}_1$ and $\mathbf{X}_2$ will agree and finally $\mathbf{Q} = \mathbf{I}$. Also, defining $\bar{\mathbf{P}}$ to be an orthogonal complement of $\mathbf{P}$, so that $(\mathbf{P}, \bar{\mathbf{P}})$ is an orthogonal matrix, we have that $\mathbf{X}_1(\text{final}) = \mathbf{X}_1(\mathbf{P}, \bar{\mathbf{P}})$, so if we use the orthogonal Procrustes procedure to rotate the given $\mathbf{X}_1$ to fit the final $\mathbf{X}_1$ we shall obtain the orthogonal matrix $(\mathbf{P}, \bar{\mathbf{P}})$ and hence the required orthonormal matrix $\mathbf{P}$. Alternatively, $(\mathbf{P}, \bar{\mathbf{P}}) = (\mathbf{X}_1'\mathbf{X}_1)^{-1}(\mathbf{X}_1'\mathbf{X}_1(\text{final}))$.

We note that when $P_2 = 1$ the above algorithm may be used in place of the algorithm for oblique Procrustes analysis of Browne (1967) described in Section 6.1. However, Browne's is a special purpose algorithm and should be the more efficient.

5.4.1 THE KOCHAT AND SWAYNE APPROACH

An alternative approach, based on an algorithm developed by Koschat and Swayne (1991) in a different context (see Section 8.1), is as follows. The key observation of Koschat and Swayne is that if $\mathbf{X}_1'\mathbf{X}_1 = \rho^2\mathbf{I}$, then (5.1) simplifies to $\|\mathbf{X}_1\mathbf{P} - \mathbf{X}_2\| = \rho^2 P + \|\mathbf{X}_2\| - 2\text{trace}(\mathbf{X}_2'\mathbf{X}_1\mathbf{P})$ so we have only to maximise the inner-product $\text{trace}(\mathbf{X}_2'\mathbf{X}_1\mathbf{P})$ which is solved by the singular value decomposition of $\mathbf{X}_2'\mathbf{X}_1$ as shown by Cliff (1966) (see Section 5.1). We now consider writing (5.1) in the form:

$$\left\| \begin{pmatrix} \mathbf{X}_1 \\ \mathbf{X}_0 \end{pmatrix} \mathbf{P} - \begin{pmatrix} \mathbf{X}_2 \\ \mathbf{X}_0\mathbf{P} \end{pmatrix} \right\|.$$

If we choose $\mathbf{X}_0$ so that $\mathbf{X}_1'\mathbf{X}_1 + \mathbf{X}_0'\mathbf{X}_0 = \rho^2\mathbf{I}$, then we have satisfied Koschat and Swayne's condition but have left the criterion unchanged, so may maximise the inner product to achieve the least-squares solution. We discuss possible choices of $\mathbf{X}_0$ and ρ^2 below. The snag is that $\mathbf{P}$ appears in the term $\mathbf{X}_0\mathbf{P}$. We therefore consider the sequence defined by:

$$T_{i+1}(\mathbf{P}_{i+1}) = \left\| \begin{pmatrix} \mathbf{X}_1 \\ \mathbf{X}_0 \end{pmatrix} \mathbf{P}_{i+1} - \begin{pmatrix} \mathbf{X}_2 \\ \mathbf{X}_0\mathbf{P}_i \end{pmatrix} \right\|.$$

where $\mathbf{P}_i$ is assumed to be known and, therefore, $\mathbf{P}_{i+1}$ may be found by maximising the trace of the inner product. Thus, $\mathbf{P}_i$ is taken as fixed and then $T_{i+1}(\mathbf{P}_{i+1})$ is to be considered as a function only of $\mathbf{P}_{i+1}$. This suggests an algorithm which, starting from some initial setting $\mathbf{P}_1$, will hopefully converge to give the desired value of $\mathbf{P}$. Convergence of such an algorithm is easily established, as follows. We first note that because $\mathbf{P}_{i+1}$ minimises T_{i+1}, then replacing it by $\mathbf{P}_i$ cannot decrease the residual sum-of-square. Thus:

$$T_{i+1}(\mathbf{P}_{i+1}) \le T_{i+1}(\mathbf{P}_i) = \|\mathbf{X}_1\mathbf{P}_i - \mathbf{X}_2\|.$$

Also, separating the top and bottom halves of T_{i+1} gives:

$$T_{i+1}(\mathbf{P}_{i+1}) = \|\mathbf{X}_1\mathbf{P}_{i+1} - \mathbf{X}_2\| + \|\mathbf{X}_0(\mathbf{P}_{i+1} - \mathbf{P}_i)\| \ge \|\mathbf{X}_1\mathbf{P}_{i+1} - \mathbf{X}_2\|.$$

Putting together these two inequalities gives:

$$\|\mathbf{X}_1\mathbf{P}_{i+1} - \mathbf{X}_2\| \leq T_{i+1}(\mathbf{P}_{i+1}) \leq \|\mathbf{X}_1\mathbf{P}_i - \mathbf{X}_2\|,$$

showing that the basic least-squares criterion decreases at each step of the algorithm. Since the criterion is bounded below by zero, this shows that the algorithm converges. As with all algorithms of this kind, it does not show that it necessarily converges to a global minimum or even to a minimum—perhaps a saddle-point is reached. Koschat and Swayne (1991) have done simulation studies to investigate performance. They caution against using a systematic start for $\mathbf{P}$ and recommend using random starts. The number of random starts depends on P_1 and P_2 and when $P_1 = P_2 = P$ they give $(1 - 1/P)^m$ as an upper bound on the probability of missing a global minimum with m random starts.

Koschat and Swayne derive $\mathbf{X}_0$ from the Choleski triangular decomposition of $\rho^2\mathbf{I} - \mathbf{X}_1'\mathbf{X}_1$. For this to give a real solution, ρ^2 must be sufficiently large to ensure that the matrix be p.s.d. Thus we must have $\rho^2 \geq \lambda$, the maximal eigenvalue of the matrix $\mathbf{X}_1'\mathbf{X}_1$; in fact Koschat and Swayne choose $\rho^2 = (1.1)\lambda$, presumably to avoid the possibility of round-off generating the square-roots of small negative quantities that the choice $\rho^2 = \lambda$ might induce. It seems to us that these considerations can be bypassed. This follows from expanding the inner product $\mathbf{Z}_{i+1}$ derived from $\mathbf{T}_{i+1}(\mathbf{P}_{i+1})$, to give:

$$\mathbf{Z}_{i+1} = \mathbf{X}_2'\mathbf{X}_1 + \mathbf{P}_i'\mathbf{X}_0'\mathbf{X}_0 = \mathbf{X}_2'\mathbf{X}_1 + \mathbf{P}_i'(\rho^2\mathbf{I} - \mathbf{X}_1'\mathbf{X}_1),$$

which does not require an explicit form of $\mathbf{X}_0$ and would allow us to set ρ^2 to any non-zero quantity. Whether some settings of ρ^2 improve convergence, or might ameliorate the local minimum problem, is an open question; a scaling of $\rho^2 = \|\mathbf{X}_1\|$ would ensure commensurability. So our variant of the algorithm is to form:

$$\mathbf{Z}_{i+1} = \mathbf{X}_2'\mathbf{X}_1 + \mathbf{P}_i'(\rho^2\mathbf{I} - \mathbf{X}_1'\mathbf{X}_1),$$

and then from the SVD $\mathbf{Z}_{i+1} = \mathbf{U\Sigma V}'$ derive $\mathbf{P}_{i+1} = \mathbf{VU}'$ from $\mathbf{U}$ and the first P_2 columns of $\mathbf{V}$, which correspond to the P_2 non-zero singular values of the $P_2 \times P_1$ matrix $\mathbf{Z}_{i+1}$. More fully:

Algorithm 5.2

Step 1 Initialise $\mathbf{P}$, either from the inner-product solution or by a random choice. Set $\rho^2 = \|\mathbf{X}_1\|$ or λ.
Step 2 Form $\mathbf{P}_1 = \mathbf{P}$ and $\mathbf{Z} = \mathbf{X}_2'\mathbf{X}_1 + \mathbf{P}_1'(\rho^2\mathbf{I} - \mathbf{X}_1'\mathbf{X}_1)$.
Step 3 Form SVD $\mathbf{Z} = \mathbf{U\Sigma V}'$ and set $\mathbf{P} = \mathbf{VU}'$.
Step 4 If $\|\mathbf{P} - \mathbf{P}_1\|$ is greater than threshold go to step 2.
Step 5 Evaluate $\mathbf{X}_1\mathbf{P}$ and $\|\mathbf{X}_1\mathbf{P} - \mathbf{X}_2\|$.

The Gower-modified Koschat and Swayne (1991) algorithm was run with the input based on data from Harman (1976) and Gruvaeus (1970, see also Algorithm 5.1), where

$$\mathbf{X}_1 = \begin{pmatrix} 0.90 & -0.09 & -0.03 \\ 0.83 & 0.09 & -0.04 \\ 0.87 & -0.01 & 0.07 \\ 0.55 & 0.79 & -0.07 \\ 0.56 & 0.65 & 0.04 \\ 0.63 & 0.60 & 0.03 \\ 0.28 & 0.27 & 0.45 \\ 0.38 & 0.20 & 0.63 \\ 0.38 & 0.19 & 0.77 \end{pmatrix}, \quad \text{and} \quad \mathbf{X}_2 = \begin{pmatrix} 0.98 & 0 \\ 0.76 & 0 \\ 0.86 & 0 \\ 0 & 1.00 \\ 0 & 0.84 \\ 0 & 0.77 \\ 0 & 0 \\ 0 & 0 \\ 0 & 0 \end{pmatrix}.$$

The stopping criterion was set to 10^{-6} and convergence was reached in 23 iterations with a sum-of-squares $\|\mathbf{X}_1\mathbf{P} - \mathbf{X}_2\| = 0.4133$. The projection matrix obtained was

$$\mathbf{P} = \begin{pmatrix} 0.8252 & 0.2292 \\ -0.4158 & 0.8650 \\ -0.3823 & -0.4463 \end{pmatrix}, \text{ giving an } \mathbf{X}_1\mathbf{P} = \begin{pmatrix} 0.7916 & 0.1418 \\ 0.6628 & 0.2859 \\ 0.6953 & 0.1595 \\ 0.1521 & 0.8407 \\ 0.1765 & 0.6728 \\ 0.2589 & 0.6500 \\ -0.0532 & 0.0969 \\ -0.0104 & -0.0211 \\ -0.0598 & -0.0922 \end{pmatrix}.$$

Agreeing with the result obtained by Algorithm 5.1.

At convergence we have:

$$\mathbf{Z} = \mathbf{U\Sigma V}' = \mathbf{X}_2'\mathbf{X}_1 + \mathbf{P}'(\rho^2\mathbf{I} - \mathbf{X}_1'\mathbf{X}_1) \quad \text{where } \mathbf{P} = \mathbf{VU}'.$$

Therefore, $\mathbf{ZP} = \mathbf{U\Sigma U}'$ is symmetric and so is $\mathbf{X}_2'\mathbf{X}_1\mathbf{P}$. As we saw in Section 5.2 convergence to symmetricity is necessary but not sufficient to show that a global maximum of trace($\mathbf{X}_2'\mathbf{X}_1\mathbf{P}$) has been reached; nor does it show that a global minimum of the least-squares criterion has been reached.

Kiers and ten Berge (1992) have shown that the Koschat and Swayne (1991) algorithm is a special case of an algorithm that uses majorisation to minimise the trace of a very general matrix quadratic function.

5.4.2 THE WIND MODEL

This section gives an example of how the Procrustes projection problem can arise in practice. Constantine and Gower (1982) introduced what they termed the *wind model*. In this model one envisages N cities situated in a continent over which there is a prevailing jet-stream with velocity v. The cities are connected by aeroplanes that fly with velocity V relative to the jet-stream. Thus if d_{ij} is the distance between cities i and j and θ_{ij} is the angle that the jet stream makes with the flight path, then the flight times t_{ij} and t_{ji} in the two directions are:

$$t_{ij} = \frac{d_{ij}}{V + v\cos\theta_{ij}}, \quad t_{ji} = \frac{d_{ij}}{V - v\cos\theta_{ij}}.$$

The average flight time between the two cities is:

$$m_{ij} = \tfrac{1}{2}(t_{ij} + t_{ji}) = \frac{V}{V^2 - v^2\cos^2\theta_{ij}} d_{ij},$$

which, when v is small compared with V, gives:

$$m_{ij} \approx \frac{d_{ij}}{V}.$$

Thus, the average flight times are proportional to the distances between the cities and an MDS of the symmetric matrix $\mathbf{M} = \{m_{ij}\}$ will recover the map of the continent.

Next, we note that the average difference in the flight times is:

$$n_{ij} = \frac{1}{2}(t_{ij} - t_{ji}) = \frac{v\cos\theta_{ij}}{V^2 - v^2\cos^2\theta_{ij}} d_{ij} \approx \frac{v}{V}\cos\theta_{ij}\frac{d_{ij}}{V}.$$

The matrix $\mathbf{N} = \{n_{ij}\}$ is skew-symmetric and its elements are approximately proportional to $d_{ij}\cos\theta_{ij}$. From Figure 5.2 we see that $d_{ij}\cos\theta_{ij}$ is the length of the projection of the distance between cities i and j (denoted by points P_i and P_j) onto the direction of the jet-stream. It follows that $n_{ij} = n_i - n_j$ where $n_z(i = 1, \ldots, N)$ are coordinates of the projections along the jet-stream. The

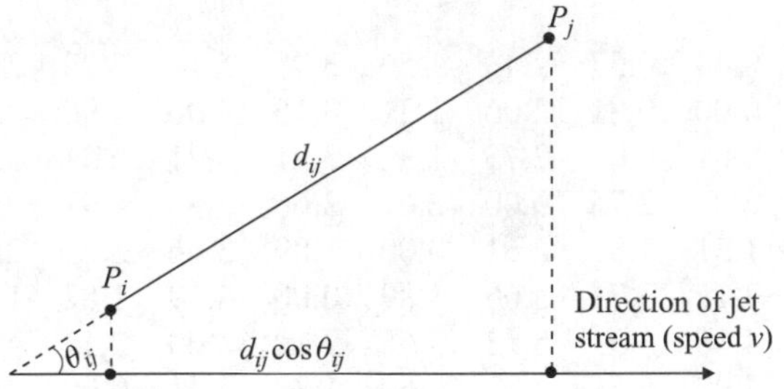

Fig. 5.2. The wind model. Showing cities i and j denoted by points P_i and P_j and their projection $d_{ij}\cos\theta_{ij}$ onto the direction of the jet stream.

Table 5.1. Flying times, in hours, between ten cities in the USA.

		1	2	3	4	5	6	7	8	9	10
1	New York	0.00	0.48	0.33	2.57	1.70	3.77	5.53	5.73	3.44	2.07
2	Washington	0.40	0.00	0.73	2.16	1.41	3.42	5.31	5.36	2.97	1.67
3	Boston	0.40	0.88	0.00	2.93	2.00	4.06	5.73	6.04	3.83	2.41
4	Miami	2.28	1.95	2.57	0.00	2.61	3.83	6.07	5.32	2.26	2.35
5	Chicago	1.42	1.21	1.65	2.62	0.00	2.07	3.90	4.05	2.26	0.60
6	Denver	3.13	2.87	3.36	3.49	1.71	0.00	2.26	1.99	1.87	1.51
7	Seattle	4.76	4.62	4.89	5.60	3.41	2.12	0.00	2.22	3.97	3.45
8	Los Angeles	4.73	4.44	4.98	4.62	3.33	1.64	2.00	0.00	2.70	3.06
9	Houston	2.82	2.43	3.15	1.94	1.93	1.90	4.12	3.08	0.00	1.42
10	St Louis	1.70	1.39	1.98	2.27	0.51	1.78	3.85	3.68	1.66	0.00

interesting thing is that this one-dimensional structure can be identified by an SVD of **N**, the first two dimensions giving an approximately linear plot. When this happens the coordinates n_i may be easily estimated as the row means of **N** (see, for example, Constantine and Gower 1978).

Thus we have (i) a map of the cities obtained from the MDS of **M**, or indeed provided independently, and (ii) a set of points on a line obtained from the row-means of **N**. Also, we know that these points are proportional to the coordinates of the projections of the cities onto the jet-stream. Hence, the direction of the jet-stream relative to the map may be obtained by a projection Procrustes analysis with $\mathbf{X}_1$ being the map coordinates and $\mathbf{X}_2$ the linear coordinates obtained from **N**. Because the scales are different, a scaling factor s should be included. The projection matrix **P** provides direction cosines for the estimated orientation and s gives an estimate of the relative velocities v/V.

For example, Table 5.1 shows flight times between ten cities of the USA. These have not been derived from any airline schedule but were calculated assuming a flight speed of 750 Kmph and a wind speed of 75 Kmph of a jet stream bearing 30° to the west–east direction. The derived matrices **M** and **N** are as follows:

$$\mathbf{M} = \begin{pmatrix} 0.00 & 0.44 & 0.37 & 2.43 & 1.56 & 3.45 & 5.14 & 5.23 & 3.13 & 1.88 \\ 0.44 & 0.00 & 0.81 & 2.06 & 1.31 & 3.15 & 4.96 & 4.90 & 2.70 & 1.53 \\ 0.37 & 0.81 & 0.00 & 2.75 & 1.82 & 3.71 & 5.31 & 5.51 & 3.49 & 2.19 \\ 2.43 & 2.06 & 2.75 & 0.00 & 2.61 & 3.66 & 5.83 & 4.97 & 2.10 & 2.31 \\ 1.56 & 1.31 & 1.82 & 2.61 & 0.00 & 1.89 & 3.66 & 3.69 & 2.09 & 0.55 \\ 3.45 & 3.15 & 3.71 & 3.66 & 1.89 & 0.00 & 2.19 & 1.82 & 1.89 & 1.65 \\ 5.14 & 4.96 & 5.31 & 5.83 & 3.66 & 2.19 & 0.00 & 2.11 & 4.05 & 3.65 \\ 5.23 & 4.90 & 5.51 & 4.97 & 3.69 & 1.82 & 2.11 & 0.00 & 2.89 & 3.37 \\ 3.13 & 2.70 & 3.49 & 2.10 & 2.09 & 1.89 & 4.05 & 2.89 & 0.00 & 1.54 \\ 1.88 & 1.53 & 2.19 & 2.31 & 0.55 & 1.65 & 3.65 & 3.37 & 1.54 & 0.00 \end{pmatrix},$$

$$
\mathbf{N} = \begin{pmatrix}
0.00 & 0.04 & -0.03 & 0.15 & 0.14 & 0.32 & 0.39 & 0.50 & 0.31 & 0.18 \\
-0.04 & 0.00 & -0.07 & 0.11 & 0.10 & 0.28 & 0.34 & 0.46 & 0.27 & 0.14 \\
0.03 & 0.07 & 0.00 & 0.18 & 0.17 & 0.35 & 0.42 & 0.53 & 0.34 & 0.22 \\
-0.15 & -0.11 & -0.18 & 0.00 & -0.01 & 0.17 & 0.24 & 0.35 & 0.16 & 0.04 \\
-0.14 & -0.10 & -0.17 & 0.01 & 0.00 & 0.18 & 0.25 & 0.36 & 0.17 & 0.04 \\
-0.32 & -0.28 & -0.35 & -0.17 & -0.18 & 0.00 & 0.07 & 0.18 & -0.01 & -0.13 \\
-0.39 & -0.34 & -0.42 & -0.24 & -0.25 & -0.07 & 0.00 & 0.11 & -0.08 & -0.20 \\
-0.50 & -0.46 & -0.53 & -0.35 & -0.36 & -0.18 & -0.11 & 0.00 & -0.19 & -0.31 \\
-0.31 & -0.27 & -0.34 & -0.16 & -0.17 & 0.01 & 0.08 & 0.19 & 0.00 & -0.12 \\
-0.18 & -0.14 & -0.22 & -0.04 & -0.04 & 0.13 & 0.20 & 0.31 & 0.12 & 0.00
\end{pmatrix}.
$$

A principal coordinate analysis of **M** had two large eigenvalues of 36.27, 9.46 and the remainder being close to zero. Evidently, we have a good match to **M**. The coordinates $\mathbf{X}_1$ were found to be:

New York	1.91	0.74
Washington	1.67	0.37
Boston	2.10	1.05
Miami	1.80	−1.69
Chicago	0.37	0.50
Denver	−1.45	0.00
Seattle	−3.20	1.33
Los Angeles	−3.09	−0.78
Houston	−0.29	−1.49
St Louis	0.19	−0.02

and the map is shown in Figure 5.3.

Next, we show in Figure 5.4 the pair of singular vectors corresponding to the dominant singular value of the SVD of **N**. As expected, this gives a nearly linear scatter of points, confirming the uni-dimensionality of the vector **n**. The orientation of the figure is entirely arbitrary. We could obtain **n**, which is to be our matrix $\mathbf{X}_2$,

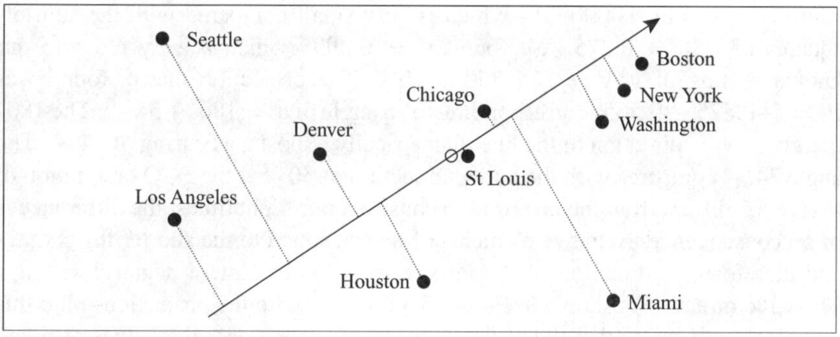

Fig. 5.3. The dominant dimensions of a principal coordinates analysis of **M**. Also shown is the estimated wind direction and the projections of the cities onto this direction. O is the origin.

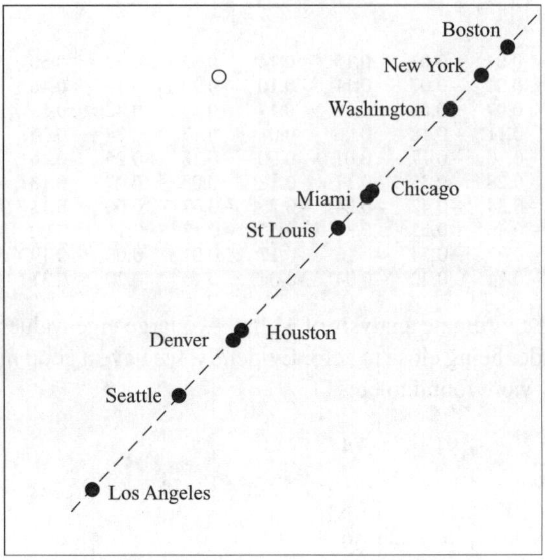

Fig. 5.4. The first plane (hedron) of an SVD of **N**, showing strong linear structure. O is the origin.

by calculating coordinates along the line shown in Figure 5.4 and multiplied by half the distance of the origin from the line. However, it is easier to take the row means of **N**, which give the least squares best fitting matrix of the form $\mathbf{n1}' - \mathbf{1n}'$. The vector **n**, so found, is:

$$\mathbf{n}' = \mathbf{X}_2' = -0.199, -0.158, -0.232, -0.052, 0.120, 0.187, 0.299, 0.108, \\ -0.014.$$

Finally we have to solve the projection Procrustes problem of minimising $\|s\mathbf{X}_2\mathbf{P} - \mathbf{X}_1\|$, using Algorithm 5.1. The projection Procrustes fit had a residual sum-of-squares of 0.00000314 which is very small compared with the sum-of-squares of $\mathbf{n}'\mathbf{n} = 0.275$. We found $s = 0.0995$ which is very close to our known scaling of $v/V = 75/750 = 0.1$. The projection matrix found was $\mathbf{P}' = (-0.825, -0.565)$ corresponding to an angle of $\alpha = 180^\circ + 34.4^\circ$. The 180° merely sets the direction of the line diametrically opposite a bearing of 34.4°. The angle 34.4° compares with the known direction of 30° but the PCO orientation of Figure 5.3 differs from the one used to construct our flight times; the difference is of no consequence. Negative elements of **n** correspond to the shorter flight times and therefore the direction of the jet stream is towards these negative settings. The wind direction is shown in Figure 5.3 together with the projections onto this direction, confirming that the projections are consistent with the values of **n** and the line shown in Figure 5.4.

Thus, given the flight times, we have reconstructed the map of the USA, found the direction of the jet stream and estimated the relative velocities of the wind and

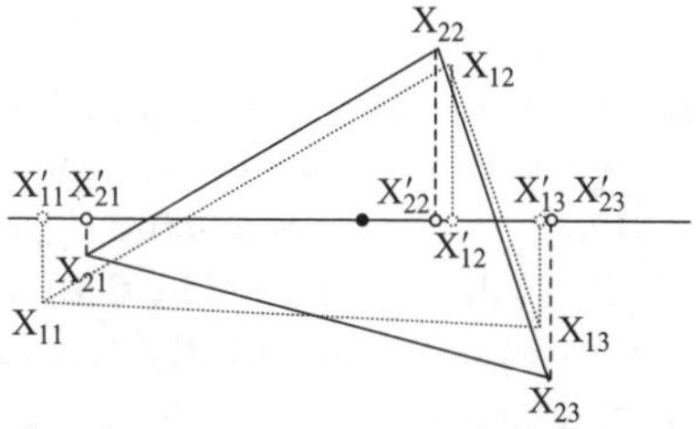

Fig. 5.5. Two triangles ($X_{11}X_{12}X_{13}$ and $X_{21}X_{22}X_{23}$) projected onto a line, such that their projections on this line ($X'_{11}X'_{12}X'_{13}$ and $X'_{21}X'_{22}X'_{23}$) match as close as possible. Note that we have shown the projected line as passing through the centroid; however, any parallel line would give the same fit.

flight. The example is artificial but we would expect the method to work well with real scheduled flight times.

5.5 Two-sided projection Procrustes by least-squares

There seems to have been no direct attack on the two-sided orthonormal version of (3.2) reported in the literature (but see Chapter 9 for the K-sets version). An obvious algorithm is to use alternating least-squares, first fixing $\mathbf{P}_2$ and estimating $\mathbf{P}_1$ by either of the methods of the previous section and then fixing $\mathbf{P}_1$ and estimating $\mathbf{P}_2$; continue the process until convergence.

The geometrical meaning of two-sided projection is that projections of $\mathbf{X}_1$ and $\mathbf{X}_2$ are to be matched. Normally, these projected versions would both be in some specified number R of dimensions so that $\mathbf{P}_1$ and $\mathbf{P}_2$ would both have $R < P_1$, P_2 columns. Figure 5.5 shows an example for the case $P_1 = P_2 = 2$ and $R = 1$.

In principle, the projections could be in different numbers of dimensions and matched by the orthogonal Procrustes procedure for configurations of differing dimensions (see Section 4.2). In this respect, it is worth noting that projected configurations given by the orthonormal version of (3.2) must be oriented in the R-dimensional space to be optimal orthogonal Procrustes fits. This is because if $\mathbf{X}_1\mathbf{P}_1$ and $\mathbf{X}_2\mathbf{P}_2$ can be rotated to a better fit in the r-dimensional space, then there exists an orthogonal matrix ${}_R\mathbf{Q}_R$ such that from the orthogonal Procrustes point of view $\mathbf{X}_1\mathbf{P}_1\mathbf{Q}$ is better than $\mathbf{X}_1\mathbf{P}_1$. This would imply that $\|\mathbf{X}_1\mathbf{P}_1\mathbf{Q} - \mathbf{X}_2\mathbf{P}_2\| < \|\mathbf{X}_1\mathbf{P}_{11} - \mathbf{X}_2\mathbf{P}_2\|$. But $\mathbf{P}_1\mathbf{Q}$ is itself an orthonormal matrix, which contradicts that $\mathbf{P}_1$ is optimal. It follows that $\mathbf{Q} = \mathbf{I}$ and that the projected configurations are optimally oriented.

5.6 Maximising the projected average configuration

In Section 4.4 we saw that Orthogonal Procrustes Analysis both minimises $\|\mathbf{X}_1\mathbf{Q} - \mathbf{X}_2\|$ and maximises $\|\mathbf{X}_1\mathbf{Q} + \mathbf{X}_2\|$, the latter of which may be expressed as $4\|\mathbf{G}\|$ where $\mathbf{G}$ is the average configuration $\frac{1}{2}(\mathbf{X}_1\mathbf{Q} + \mathbf{X}_2)$. If we define $\mathbf{G} = \frac{1}{2}(\mathbf{X}_1\mathbf{P} + \mathbf{X}_2)$ then maximising $\|\mathbf{G}\|$ is not attained for the same setting of $\mathbf{P}$ as for minimising $\|\mathbf{X}_1\mathbf{P} - \mathbf{X}_2\|$.

5.6.1 METHOD 1

An algorithm for maximising the average projected configuration may be developed as follows. Suppose $\mathbf{P}_i$ is a current estimate of $\mathbf{P}$, giving a current projected average $\mathbf{G}_i = \frac{1}{2}(\mathbf{X}_1\mathbf{P}_i + \mathbf{X}_2)$. Now, choose $\mathbf{P}_{i+1}$ to maximise the trace of the inner-product $\mathbf{G}_i'\mathbf{X}_1\mathbf{P}_{i+1}$; this is achieved by Cliff's (1966) singular value decomposition method discussed in Section 5.1. By definition:

$$\text{trace}(\mathbf{X}_1\mathbf{P}_i + \mathbf{X}_2)'\mathbf{X}_1\mathbf{P}_{i+1} \geq \text{trace}(\mathbf{X}_1\mathbf{P}_{i+}\mathbf{X}_2)'\mathbf{X}_1\mathbf{P}_i$$

and so:

$$\text{trace}(\mathbf{X}_1\mathbf{P}_{i+1}\mathbf{X}_2)'(\mathbf{X}_1\mathbf{P}_{i+1} + \mathbf{X}_2) \geq \text{trace}(\mathbf{X}_1\mathbf{P}_i + \mathbf{X}_2)'(\mathbf{X}_1\mathbf{P}_{i+}\mathbf{X}_2)$$

that is,

$$\text{trace}(\mathbf{G}_i'\mathbf{G}_{i+1}) \geq \|\mathbf{G}_i\|$$

By Cauchy's inequality $1/2(\|\mathbf{G}_i\| + \|\mathbf{G}_{i+1}\|) \geq \|\mathbf{G}_i\|$, so that $\|\mathbf{G}_{i+1}\| \geq \|\mathbf{G}_i\|$.

Thus, by finding the projection that maximises the trace of the inner-product of $\mathbf{X}_1$ with $\mathbf{G}_i$, we can increase $\|\mathbf{G}_i\|$ to $\|\mathbf{G}_{i+1}\|$. This gives the following algorithm:

Algorithm 5.3

Step 1 Evaluate the SVD $\mathbf{G}'\mathbf{X}_1 = \mathbf{U\Sigma V}'$ and set $\mathbf{P} = \mathbf{VU}'$.
Step 2 Set $\mathbf{G}_0 = 1/2(\mathbf{X}_1\mathbf{P} + \mathbf{X}_2)$
Step 3 If $\|\mathbf{G}_0 - \mathbf{G}\| < \varepsilon$ then exit
Step 4 Set $\mathbf{G} = \mathbf{G}_0$ and return to *step 1.*

On exit, $\mathbf{P}$ will be the required projection matrix and $\mathbf{G}$ will give the optimal average configuration. At convergence we must have:

$$2\mathbf{G}'\mathbf{X}_1 = (\mathbf{X}_1\mathbf{P} + \mathbf{X}_2)'\mathbf{X}_1 = \mathbf{U\Sigma V}' \quad \text{and} \quad \mathbf{P} = \mathbf{VU}'.$$

Post-multiplying by $\mathbf{P}$ gives:

$$\mathbf{P}'\mathbf{X}_1'\mathbf{X}_1\mathbf{P} + \mathbf{X}_2'\mathbf{X}_1\mathbf{P} = (\mathbf{U\Sigma V}')\mathbf{VU}' = \mathbf{U\Sigma U}'.$$

The first matrix on the left-hand-side and $\mathbf{U\Sigma U}'$ are both symmetric and so therefore is $\mathbf{X}_2'\mathbf{X}_1\mathbf{P}$. As we have already found in similar problems of this type, the inner-product of the final matrices must be symmetric. Incidentally, this gives an alternative proof of the result of Section 5.2 that the symmetry of $\mathbf{X}_2'\mathbf{X}_1\mathbf{P}$ is not sufficient to guarantee the maximum of $\text{trace}(\mathbf{X}_2'\mathbf{X}_1\mathbf{P})$ to which we may now add, nor to the maximum of $\text{trace}(\mathbf{G}'\mathbf{X}_1\mathbf{P})$ or, equivalently, $\|\mathbf{X}_1\mathbf{P} + \mathbf{X}_2\|$.

5.6.2 METHOD 2

Method 1 maximises an inner product. Now, we examine the possibility of maximising $\|\mathbf{X}_1\mathbf{P}+\mathbf{X}_2\|$ through a least-squares algorithm. Defining $\bar{\mathbf{P}}$ as in Section 5.4, we note that:

$$\text{Max}\|\mathbf{X}_1\mathbf{P}+\mathbf{X}_2\| = \text{Max}\|\mathbf{X}_1(\mathbf{P}, \bar{\mathbf{P}})+(\mathbf{X}_2, -\mathbf{X}_1\bar{\mathbf{P}})\| = \text{Max}\|\mathbf{X}_1\mathbf{Q}+(\mathbf{X}_2, -\mathbf{X}_1\bar{\mathbf{P}})\|.$$

We have already seen that the orthogonal matrix $\mathbf{Q}$ that maximises the latter function is the same $\mathbf{Q}$ that satisfies $\text{Min}\|\mathbf{X}_1\mathbf{Q} - (\mathbf{X}_2, -\mathbf{X}_1\bar{\mathbf{P}})\|$ which is an ordinary orthogonal Procrustes problem as discussed in Chapter 4. These results suggest an algorithm (see Gower 1995) for maximising the average projected configuration by iteratively updating $(\mathbf{X}_2, -\mathbf{X}_1\bar{\mathbf{P}})$ by replacing the final $P_1 - P_2$ columns by the negative of the corresponding columns of $\mathbf{X}_1\mathbf{Q}$. This procedure is similar to that used in Algorithm 5.1. There, the updating of the final columns zeroises their contribution and so reduces the value of the least-squares criterion, as is appropriate for minimisation. However, the current proposal does not zeroise the contribution and may even increase it. Hence, we must seek an alternative approach.

Consider the maximisation of:

$$S = \|\mathbf{X}_1(\mathbf{P}_{i+1}, \mathbf{P}_{i+1}^{\perp}) - (\tfrac{1}{2}(\mathbf{X}_2 + \mathbf{X}_1\mathbf{P}_i), \mathbf{0}\|$$

where $\mathbf{P}_{i+1}^{\perp}$ is the orthogonal complement of $\mathbf{P}_{i+1}$. Thus, $\mathbf{Q}_{i+1} = (\mathbf{P}_{i+1}, \mathbf{P}_{i+1}^{\perp})$ is an orthogonal matrix and for given $\mathbf{P}_i$ the minimisation of S is an ordinary orthogonal Procrustes problem. Indeed, we know (Section 4) that this minimisation is achieved by maximising the inner product:

$$\text{trace}[(\tfrac{1}{2}(\mathbf{X}_2 + \mathbf{X}_1\mathbf{P}_i), \mathbf{0})'\mathbf{X}_1(\mathbf{P}_{i+1}, \mathbf{P}_{i+1}^{\perp})] = \text{trace}[\tfrac{1}{2}(\mathbf{X}_2 + \mathbf{X}_1\mathbf{P}_i)'\mathbf{X}_1(\mathbf{P}_{i+1})].$$

Thus, again we have the inner-product criterion of Method 1, above, which may therefore be expressed as the following least-squares algorithm:

Algorithm 5.4

Step 1 Initialise $\mathbf{Y} = \mathbf{X}_2$, say, appended by $P_1 - P_2$ zero columns.

Step 2 Derive the orthogonal Procrustes solution for $\mathbf{X}_1$ and $\mathbf{Y}$, to give $\mathbf{P}_i$.

Step 3 Replace $\mathbf{X}_1$ by its rotated form and $\mathbf{Y}$ by $1/2(\mathbf{X}_2 + \mathbf{X}_1\mathbf{P}_i)$ appended by $P_1 - P_2$ zero columns.

Step 4 Repeat from *step 2* with the current versions of $\mathbf{X}_1$ and $\mathbf{Y}$, until the residual sum-of-squares stabilises.

The Algorithm 5.3 was run with the data from Harman (1976) and Gruveaus (1970, see also Algorithm 5.1). $\mathbf{X}_1$ (Harman) is

$$\begin{pmatrix} 0.90 & -0.09 & -0.03 \\ 0.83 & 0.09 & -0.04 \\ 0.87 & -0.01 & 0.07 \\ 0.55 & 0.79 & -0.07 \\ 0.56 & 0.65 & 0.04 \\ 0.63 & 0.60 & 0.03 \\ 0.28 & 0.27 & 0.45 \\ 0.38 & 0.20 & 0.63 \\ 0.38 & 0.19 & 0.77 \end{pmatrix}$$

with $\mathbf{X}_2$ (Gruvaeus)

$$\begin{pmatrix} 0.98 & 0 \\ 0.76 & 0 \\ 0.86 & 0 \\ 1.00 & 0 \\ 0 & 0.84 \\ 0 & 0.77 \\ 0 & 0 \\ 0 & 0 \\ 0 & 0 \end{pmatrix}.$$

The stopping criterion was set at 0.001, and the algorithm converged in 2 iterations producing

$$\mathbf{P} = \begin{pmatrix} 0.9243 & 0.3398 & 0.1737 \\ -0.3622 & 0.9245 & 0.1189 \\ -0.1202 & -0.1729 & 0.9776 \end{pmatrix}$$

and

$$\mathbf{X}_1\mathbf{P} = \begin{pmatrix} 0.8681 & 0.2278 & 0.1163 \\ 0.7394 & 0.3721 & 0.1158 \\ 0.7994 & 0.2743 & 0.2184 \\ 0.2307 & 0.9293 & 0.1211 \\ 0.2774 & 0.7843 & 0.2137 \\ 0.3614 & 0.7636 & 0.2101 \\ 0.1069 & 0.2670 & 0.5207 \\ 0.2031 & 0.2051 & 0.7057 \\ 0.1899 & 0.1717 & 0.8414 \end{pmatrix}$$

and hence a group average configuration

$$\frac{1}{2}(\mathbf{X}_1\mathbf{P} + \mathbf{X}_2) = \begin{pmatrix} 0.9241 & 0.1139 & 0.0582 \\ 0.7497 & 0.1861 & 0.0579 \\ 0.8297 & 0.1372 & 0.1092 \\ 0.1154 & 0.9647 & 0.0606 \\ 0.1387 & 0.8122 & 0.1069 \\ 0.1807 & 0.7668 & 0.1051 \\ 0.0535 & 0.1335 & 0.5354 \\ 0.1016 & 0.1026 & 0.7329 \\ 0.0950 & 0.0859 & 0.8857 \end{pmatrix}.$$

As with minimising the residual sum-of–squares given by projection, we have that $\mathbf{X}_1(\text{final}) = \mathbf{X}_1(\mathbf{P}, \bar{\mathbf{P}})$ so if we use the orthogonal Procrustes procedure to rotate the given $\mathbf{X}_1$ to fit the final $\mathbf{X}_1$, we shall obtain the orthogonal matrix $(\mathbf{P}, \bar{\mathbf{P}})$ and hence the required orthonormal matrix $\mathbf{P}$.

5.7 Rotation into higher dimensions

We now consider the minimisation of:

$$\|\mathbf{X}_2\mathbf{P}' - \mathbf{X}_1\|, \tag{5.15}$$

where $\mathbf{P}$ remains orthonormal with $P_1 > P_2$. Note that the transformation must now be attached to $\mathbf{X}_2$ rather than the usual $\mathbf{X}_1$. As explained in Appendix C.1, transformation by $\mathbf{P}'$ can be interpreted as a rotation of $\mathbf{X}_2$ into the higher dimensional space, here occupied by $\mathbf{X}_1$. Unlike with projection, being a rotation, the least-squares and inner-product criteria should be consistent. This is indeed so, because:

$$\|\mathbf{X}_2\mathbf{P}' - \mathbf{X}_1\| = \|\mathbf{X}_2\| + \|\mathbf{X}_1\| - 2\,\mathrm{trace}(\mathbf{X}_1'\mathbf{X}_2\mathbf{P}')$$

so that minimising $\|\mathbf{X}_2\mathbf{P}' - \mathbf{X}_1\|$ is the same as maximising $2\mathrm{trace}(\mathbf{X}_1'\mathbf{X}_2\mathbf{P}')$.

Thus, equation (5.15) does not represent a new type of Procrustes problem. It is merely an alternative way of expressing Cliff's inner-product criterion (Section 5.1) and is equivalent to the device of padding $\mathbf{X}_2$ by $P_1 - P_2$ zero columns and using orthogonal Procrustes methods (Section 4.5).

5.8 Some geometrical considerations

In the above, we have considered several Procrustean problems involving orthonormal projection matrices. In this section, we review the different geometrical interpretations for these together with some additional problems. Appendix C outlines the basic properties of projection matrices. Of particular importance for understanding the different criteria to be considered, is the distinction between a double matrix ${}_P\mathbf{P}_R\mathbf{P}'_P$ and the single form ${}_P\mathbf{P}_R$. In the former case, ${}_N\mathbf{X}_P\mathbf{P}_R\mathbf{P}'_P$ gives the coordinates relative to the P axes to which the configuration $\mathbf{X}$ is referred. In the latter case, ${}_N\mathbf{X}_P\mathbf{P}_R$ gives the coordinates of projected points relative to R axes in the sub-space spanned by the columns of $\mathbf{P}$. Indeed, the columns of $\mathbf{P}$ give the directions of these new axes, relative to the P original axes. This distinction is vital when we are concerned with two configurations $\mathbf{X}_1$ and $\mathbf{X}_2$—are they to be considered, at least initially, as being embedded in a common space (in which case the form $\mathbf{PP}'$ is relevant) or are they to be considered as occupying disjoint spaces (in which case the form $\mathbf{P}$ is relevant)? To give a clearer understanding of what is involved we give some simple diagrams in Figure 5.6.

Initially, we consider $\mathbf{X}_1$ to be a two-dimensional configuration, of three points, and $\mathbf{X}_2$ to be one-dimensional, but embedded in the two dimensions. Thus, $N = 3$, $P_1 = 2$, and $P_2 = 1$. Thus, in Figure 5.6(a) we see that the residuals $\mathbf{X}_1\mathbf{P} - \mathbf{X}_2$ are evaluated in a common space of $P_2 = 1$ dimension but in Figure 5.6(b) the residuals $\mathbf{X}_1\mathbf{PP}' - \mathbf{X}_2$ are evaluated in the $P_1 = 2$ dimensional space which contains the configuration $\mathbf{X}_2$ in a $P_2 = 1$-dimensional subspace. This implies that $\mathbf{X}_2$ exists in P_1-dimensions, perhaps evidenced by padding out its coordinate representation by $P_1 - P_2$ columns of zeros—though this is not the only way. In

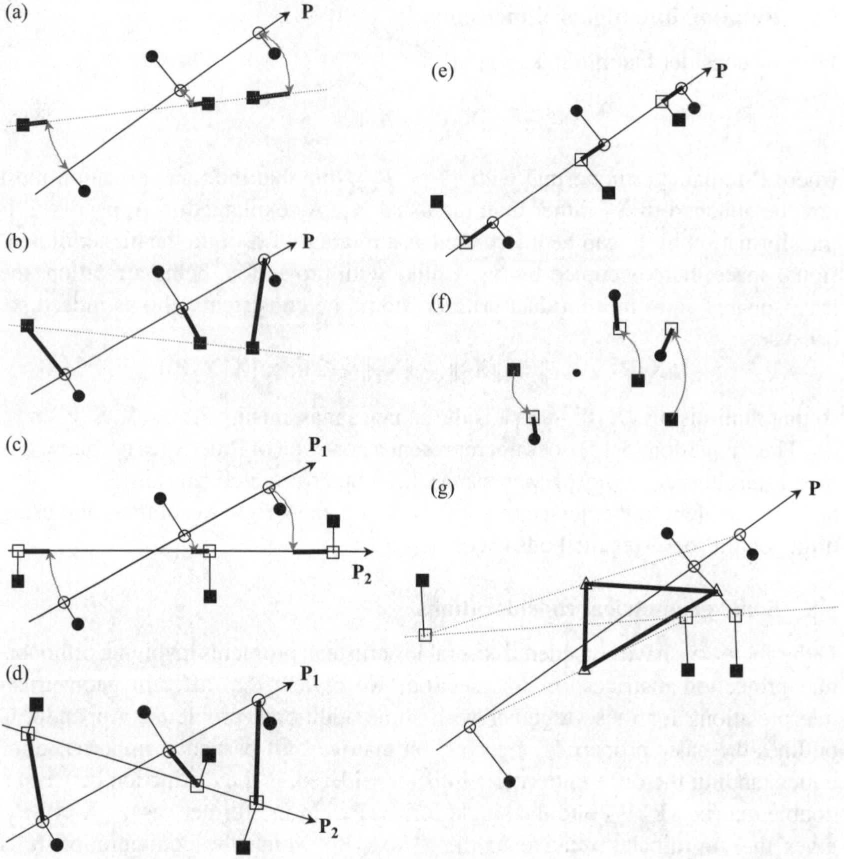

Fig. 5.6. Illustrative examples of some projection (rotation in f) variants. Legend: closed symbols (•, ■) represent the data, where • is $\mathbf{X}_1$ and ■ is $\mathbf{X}_2$, and open symbols (○, □) represent projected data (rotated in Figure 5.6(f)). Thick connecting lines stand for residuals (they denote the group average configuration in 5.6(g)), thin lines show a projection onto directions indicated by $\mathbf{P}$, $\mathbf{P}_1$, and $\mathbf{P}_2$. Double arrows (5.6(a)) denotes a superposition of the two data sets, one-sided arrows (5.6(f)) denotes a proper rotation. N.B. These diagrams are purely illustrative; we have made no attempt to show the true optimal positions of the configurations.

problems concerned with matching configurations, there is usually no reason to believe that corresponding columns of $\mathbf{X}_1$ and $\mathbf{X}_2$ refer to the same things; indeed, the main thrust is to find transformations that do embed the two configurations in a common space, in such a way that their relative positioning is in some way optimal. Then, it would be wise to avoid using the $\mathbf{PP}'$ form.

In Figure 5.6(a) we illustrate the minimisation of $\|\mathbf{X}_1\mathbf{P} - \mathbf{X}_2\|$ (Section 5.4). The best projection $\mathbf{P}$, here given by a single vector, because $P_2 = 1$, is shown together with the projections of $\mathbf{X}_1$ onto it. The effect is two-dimensional;

however, the criterion is one-dimensional and $\mathbf{X}_1\mathbf{P}$, the fitted configuration, and $\mathbf{X}_2$ have to be embedded in the same one-dimensional space. This process is one of superimposition where $\mathbf{p}_i$ (the ith column of $\mathbf{P}$) is made to line up with $\mathbf{e}_i$ (the ith of the original coordinate axes). This can be achieved by an orthogonal rotation which, for convenience, is shown as a rotation of $\mathbf{X}_1\mathbf{P}$ onto $\mathbf{X}_2$ but could equally well have been rotated in the opposite direction. The residuals in the common space are highlighted.

In contrast, Figure 5.6(b) illustrates the minimisation of $\|\mathbf{X}_1\mathbf{P}\mathbf{P}' - \mathbf{X}_2\|$. Again the best projection $\mathbf{P}$, is given by a (different) single vector, and is shown together with the projections of $\mathbf{X}_1$ onto it. Now, the criterion residuals are two-dimensional as is shown in the highlighting. Up to now, we have not discussed how to minimise this criterion but it is easy to see that we require the minimum of:

$$\text{trace}\mathbf{P}'(\mathbf{X}_1'\mathbf{X}_1' - \mathbf{X}_1'\mathbf{X}_2 - \mathbf{X}_2'\mathbf{X}_1)\mathbf{P}$$

which, as is well-known, is an eigenvalue problem setting $\mathbf{P}$ to be the P_2 leading eigenvectors of the symmetric matrix $\mathbf{X}_1'\mathbf{X}_1' - \mathbf{X}_1'\mathbf{X}_2 - \mathbf{X}_2'\mathbf{X}_1$.

Next we move on to the two-sided problems and consider $\mathbf{X}_1$ and $\mathbf{X}_2$ each to be two-dimensional configurations of three points which are to be projected onto $R = 1$ dimension. Thus, $N = 3$, $R = 1$, $P_1 = 2$, and $P_2 = 2$.

In Figure 5.6(c) we illustrate the minimisation of the two-sided criterion $\|\mathbf{X}_1\mathbf{P}_1 - \mathbf{X}_2\mathbf{P}_2\|$ (Section 5.5). Now we show two projection matrices $\mathbf{P}_1$ and $\mathbf{P}_2$ each given by a single vector, because $R = 1$. The criterion operates on residuals in the R-dimensional space, so one of the spaces spanned by the columns of $\mathbf{P}_1$ and $\mathbf{P}_2$ has to be superimposed on the other. In this simple example, the two directions have to be superimposed as is indicated by the rotation arrows.

Figure 5.6(d) illustrates the minimisation of the two-sided criterion $\|\mathbf{X}_1\mathbf{P}_1\mathbf{P}_1' - \mathbf{X}_2\mathbf{P}_2\mathbf{P}_2'\|$. Now the criterion operates on residuals in the P_1-dimensional space. In the example, $\mathbf{X}_1$ and $\mathbf{X}_2$ are both two-dimensional and occupy the same space but either configuration could be in a subspace of the other. We could have made them both occupy *different* two-dimensional parts of a three- or four-dimensional space, in which case different optimal solutions would exist. This is not the case with those criteria that are expressed in terms of $\mathbf{P}$ rather than $\mathbf{P}\mathbf{P}'$. We have not examined the process of fitting this model but it is unlikely to have a closed-form solution. Presumably, an alternating least-squares algorithm using the one-sided eigenvalue solution, given above in the discussion of the one-sided problem (Figure 5.6(b)), would suffice. Another possibility with two-sided projections is that $\mathbf{P}_1\mathbf{P}_1'$ and $\mathbf{P}_2\mathbf{P}_2'$ may project onto subspaces of different numbers of dimensions both embedded in the biggest space—we do not pursue this further.

Figure 5.6(e) shows another problem, not previously discussed—that of minimising $\|(\mathbf{X}_1 - \mathbf{X}_2)\mathbf{P}\|$. That is we seek the same optimal projection for both configurations. The criterion has no meaning unless both configurations occupy the same number of dimensions, that is, $P_1 = P_2$, taken in the example to be two, with $R = 1$. The solution is given by the leading eigenvectors of $(\mathbf{X}_1 - \mathbf{X}_2)'(\mathbf{X}_1 - \mathbf{X}_2)$. In this case, precisely the same solution occurs if we seek to minimise $\|(\mathbf{X}_1 - \mathbf{X}_2)\mathbf{PP}'\|$. This is because the initial relative position of the two configurations is critical and the above formulation regards the configurations of $\mathbf{X}_1$ and $\mathbf{X}_2$ to be given. However, even when $\mathbf{X}_1$ and $\mathbf{X}_2$ occupy the same number of dimensions, their relative orientations in this space may be arbitrary. Then, we may require the minimum of $\|(\mathbf{X}_1\mathbf{Q} - \mathbf{X}_2)\mathbf{P}\| = \|(\mathbf{X}_1 - \mathbf{X}_2\mathbf{Q}')\mathbf{QP}\|$ which is a more difficult algorithmic problem.

For comparison with all these projections, in Figure 5.6(f) we show the orthogonal rotation of two two-dimensional configurations.

Figure 5.6(g) shows an example of maximising the group-average (Section 5.6). We have chosen max $\|\mathbf{X}_1\mathbf{PP}' + \mathbf{X}_2\|$ but the reader will readily be able to extend to the other related criteria max $\|\mathbf{X}_1\mathbf{P} + \mathbf{X}_2\|$, max $\|\mathbf{X}_1\mathbf{P}_1 + \mathbf{X}_2\mathbf{P}_2\|$, and max $\|\mathbf{X}_1\mathbf{P}_1\mathbf{P}_1' + \mathbf{X}_2\mathbf{P}_2\mathbf{P}_2'\|$, not to mention max $\|(\mathbf{X}_1 + \mathbf{X}_2)\mathbf{P}\|$, etc. The geometry is similar to that of Fig. 5.6(b) but now, rather than minimise the sum-of-squares of the residuals we have to maximise the size of the group-average triangle that is high-lighted in the figure. This is equivalent to maximising the sum-of-squares of the distances of the triangle vertices from the origin. Both configurations are shown as two-dimensional and projected onto one dimension.

What we have not shown in this series of figures are the parallel figures for maximising the inner-product criteria (Sections 5.1 and 5.3) or the corresponding Tucker's congruence coefficient or Escoufier's RV coefficient or any other goodness-of-fit criteria that might be considered. The same basic geometry applies but rather than minimising the sums of squares of residuals (the thick lines in Figure 5.6), these criteria involve angles as well as distances and are rather difficult to illustrate simply.

6
Oblique Procrustes problems

Matching configurations defined relative to oblique axes is a labyrinthine topic; we shall try to plot a course through the labyrinth. Although our outlook in this book is based on matching configurations, oblique Procrustes methods are so strongly rooted in factor analysis that it seems appropriate to begin with a note on this aspect here. The basic model for factor analysis is:

$$_P\mathbf{y}_1 = {}_P\mathbf{X}_F\mathbf{f}_1 + {}_P\mathbf{e}_1$$

where **y** represents P observable variables, **f** represents F ($<P$) unobserved *factors*, assumed to be independent with unit variances. **X** is termed the matrix of *factor loadings*, the kth column of which gives the loadings of the P variables on the kth factor. The vector **e** represents an error term of P independent *specific factors*, so that cov(**e**) is a diagonal matrix, of variances. The model is not fully determined, as it may be rewritten:

$$\mathbf{y} = (\mathbf{XT})(\mathbf{T}^{-1}\mathbf{f}) + \mathbf{e}$$

where **T** is an arbitrary non-singular matrix, defining a new loading matrix **XT** and new factors $\mathbf{T}^{-1}\mathbf{f}$ which are no longer independent because:

$$\text{cov}(\mathbf{T}^{-1}\mathbf{f}) = \mathbf{T}^{-1}\text{cov}(\mathbf{f})(\mathbf{T}')^{-1} = (\mathbf{T}'\mathbf{T})^{-1}$$

If we retain the requirement of unit variances for the factors, it follows that $(\mathbf{T}')^{-1}$ may be interpreted as a matrix **C** of direction cosines and then $\mathbf{XT} = \mathbf{X}(\mathbf{C}')^{-1}$ gives the parallel axes coordinates of **X** relative to a set of oblique axes (equation (D.2) in Appendix D). Alternatively, one might regard **XT** in a context where the elements of **T** are interpreted as regression coefficients. If these are standardised, we shall require $\text{diag}(\mathbf{TT}') = \mathbf{I}$ so we arrive at $\mathbf{T} = \mathbf{C}'$. Thus, we see how different constraints on **T**, that may be interpreted in terms of a matrix of direction cosines, arise naturally in a factor analysis context. The factor analysis literature on oblique axes is very extensive and we do not attempt to cover it here. Furthermore, the term *oblique* is often used to mean *not orthogonal* and may have no interpretation in terms of geometrically interpretable oblique axes. We have already encountered some problems of this kind in Section 5.3. We now return to more direct geometrical interpretations that are the concern of the remainder of this chapter.

Appendix D gives the basic geometry and algebra for oblique axes. Now **T** will be the transformation that takes coordinates $\mathbf{X}_1$ measured in one coordinate system into $\mathbf{X}_2$ measured in another coordinate system. Usually, $\mathbf{X}_1$ will refer to coordinates relative to classical orthogonal Cartesian axes and $\mathbf{X}_2$ will have values (coordinates) relative to one of other of the two oblique axes representations mentioned in Appendix D—(i) the projection method or (ii) the parallel axes method based on vector-sums. The direction cosines of the oblique axes relative to some set of Cartesian axes are given as the columns of a square matrix **C**. Then, the elements of $\mathbf{C}'\mathbf{C}$ give the cosines of the angles between all pairs of oblique axes and hence $\text{diag}(\mathbf{C}'\mathbf{C}) = \mathbf{I}$, so that $\mathbf{C}'\mathbf{C}$ is very like a correlation matrix and sometimes may be so interpreted.

We may match $\mathbf{X}_1$ to $\mathbf{X}_2$ in several ways but shall discuss only four of the possibilities. The four methods depend on the two oblique axes representations (projection or parallel axes) and on whether or not we take into account the associated distance metrics given in the Appendix (D.4) and (D.5). Much depends on how we interpret $\mathbf{X}_2$. If $\mathbf{X}_2$ is merely a target matrix, as we have already seen in the Mosier (1939) and Hurley and Cattell (1962) approaches, then it is legitimate to choose **C** to minimise:

$$\|\mathbf{X}_1\mathbf{C} - \mathbf{X}_2\| \tag{6.1}$$

for the projection method and to minimise:

$$\|\mathbf{X}_1(\mathbf{C}')^{-1} - \mathbf{X}_2\| \tag{6.2}$$

for the parallel axes method. In (6.1) and (6.2) $\mathbf{X}_1$ is referred to rectangular Cartesian axes and we are seeking oblique axes, with directions defined by the columns of **C**, in such a way that its projection, respectively parallel axes, coordinates match as well as possible the target matrix $\mathbf{X}_2$. No explicit assumption is made about any coordinate representation of $\mathbf{X}_2$; the match is purely algebraic. However, if $\mathbf{X}_2$ is regarded as giving coordinates relative to the oblique axes, then both $\mathbf{X}_2$ and the transformed version of $\mathbf{X}_1$ are in the same space, where distance is defined by (D.4) and (D.5). Now, the natural least-squares criteria should be in the appropriate metric, leading to the minimisation of:

$$\text{trace}(\mathbf{X}_1\mathbf{C} - \mathbf{X}_2)(\mathbf{C}'\mathbf{C})^{-1}(\mathbf{X}_1\mathbf{C} - \mathbf{X}_2) = \|\mathbf{X}_1 - \mathbf{X}_2\mathbf{C}^{-1}\| \tag{6.3}$$

for the projection method, and:

$$\text{trace}(\mathbf{X}_1(\mathbf{C}')^{-1} - \mathbf{X}_2)\mathbf{C}'\mathbf{C}(\mathbf{X}_1(\mathbf{C}')^{-1} - \mathbf{X}_2)' = \|\mathbf{X}_1 - \mathbf{X}_2\mathbf{C}'\| \tag{6.4}$$

for the parallel axes method. An interesting thing about using the oblique axes metrics is that the effect is to turn the problem into an ordinary least-squares criterion but with the transformation attached to $\mathbf{X}_2$ rather than to $\mathbf{X}_1$. Therefore, *for consistency of the notation of attaching the transformation to* $\mathbf{X}_1$, we shall rewrite (6.3) and (6.4) as:

$$\|\mathbf{X}_1\mathbf{C}^{-1} - \mathbf{X}_2\| \tag{6.5}$$

and

$$\|\mathbf{X}_1\mathbf{C}' - \mathbf{X}_2\| \tag{6.6}$$

but it should be remembered that the roles of $\mathbf{X}_1$ and $\mathbf{X}_2$ have been interchanged; in particular, $\mathbf{X}_1$ is now the target matrix.

Before studying these four criteria further we indicate the many other oblique axes Procrustes problems that could be developed. First, there is no reason why one of the configurations must have Cartesian coordinates. Perhaps $\mathbf{X}_1$ is given relative to projection oblique axes and $\mathbf{X}_2$ relative to parallel oblique axes with the same direction cosines. Then, by referring both configurations to their common Cartesian system we would minimise $\|\mathbf{X}_1\mathbf{C}^{-1} - \mathbf{X}_2\mathbf{C}'\|$ or, equivalently, working in terms of the metric of $\mathbf{X}_2$ regarded as a target matrix, we would minimise $\text{trace}(\mathbf{X}_1(\mathbf{C}'\mathbf{C})^{-1} - \mathbf{X}_2)(\mathbf{C}'\mathbf{C})(\mathbf{X}_1(\mathbf{C}'\mathbf{C})^{-1} - \mathbf{X}_2)'$ with the metric (D.5). In deriving the latter, $\mathbf{X}_1\mathbf{C}^{-1}$ expresses the configuration in terms of Cartesian axes and the further multiplication $(\mathbf{X}_1\mathbf{C}^{-1})(\mathbf{C}')^{-1} = \mathbf{X}_1(\mathbf{C}'\mathbf{C})^{-1}$ is the usual transformation to the parallel axes system of $\mathbf{X}_2$. Second, perhaps we may wish to allow $\mathbf{X}_2$ to be given an orthogonal rotation $\mathbf{Q}$ before being matched to $\mathbf{X}_1$. Note that, even though it is expressed relative to oblique axes, rotation of $\mathbf{X}_2$ does not change the configuration though it *does* change the target coordinates. This leads to the minimisation of $\|\mathbf{X}_1\mathbf{C} - \mathbf{X}_2\mathbf{Q}\|$ and its many variants. Rotation of $\mathbf{X}_1$ would have no effect, because $\mathbf{QC}$ continues to satisfy $\text{diag}(\mathbf{C}'\mathbf{Q}'\mathbf{QC}) = \text{diag}(\mathbf{C}'\mathbf{C}) = \mathbf{I}$ and therefore remains a direction cosine matrix for axes with the same mutual directions as before. Third, $\mathbf{X}_1$ and $\mathbf{X}_2$ may be referred to different sets of oblique axes, leading to the minimisation of $\|\mathbf{X}_1\mathbf{C}_1 - \mathbf{X}_2\mathbf{C}_2\|$ and its 16 variants obtained by replacing $\mathbf{C}_1$ by $\mathbf{C}_1'$, $\mathbf{C}_1^{-1}$, or $(\mathbf{C}_1')^{-1}$ and similarly for $\mathbf{C}_2$. Thus, although our following discussion may seem lengthy, it is by no means exhaustive. We now return to some of the details of the four oblique axes problems introduced above.

Thus, we are concerned with $\mathbf{T} = \mathbf{C}$, $(\mathbf{C}')^{-1}$, $\mathbf{C}^{-1}$, and $\mathbf{C}'$, all subject to the constraint $\text{diag}(\mathbf{C}'\mathbf{C}) = \mathbf{I}$. For these forms of $\mathbf{T}$, we need to minimise:

$$\|\mathbf{X}_1\mathbf{T} - \mathbf{X}_2\| + \text{trace}\{\mathbf{\Lambda}(\text{diag}(\mathbf{C}'\mathbf{C}) - \mathbf{I})\} \tag{6.7}$$

where $\mathbf{\Lambda}$ is a diagonal matrix of Lagrange multipliers associated with the p constraints. We may differentiate (6.7) with respect to $\mathbf{T}$, giving:

$$(\mathbf{X}_1'\mathbf{X}_1)\mathbf{T} - \mathbf{X}_1'\mathbf{X}_2 = \frac{\partial \text{trace}\{\mathbf{\Lambda}(\text{diag}(\mathbf{C}'\mathbf{C})\}}{\partial \mathbf{T}}. \tag{6.8}$$

The only complication is the partial differential coefficient, which defines a different function of $\mathbf{T}$ for each of the four criteria. Proceeding generally, we have:

$$\text{trace}\{\mathbf{\Lambda}(\text{diag}(\mathbf{C}'\mathbf{C})\} = \sum_{i=1}^{P}\sum_{j=1}^{P}\lambda_j c_{ij}^2$$

so that:

$$\frac{1}{2}\frac{\partial \text{trace}\{\mathbf{\Lambda}\text{diag}(\mathbf{C}'\mathbf{C})\}}{\partial t_{ij}} = \text{trace}\left[\mathbf{C\Lambda}\left(\frac{\partial \mathbf{C}}{\partial t_{ij}}\right)'\right] \tag{6.9}$$

which may be used to simplify (6.8) in the special cases to be considered below. We could have differentiated (6.8) with respect to $\mathbf{C}$, giving the simple form $\mathbf{C\Lambda}$ on the right-hand-side and transferring the complications to the left. The final result is unaltered, and the one form can be obtained from the other by simple matrix operations, as we shall see.

Applying (6.9) to the four cases we are considering, gives the following results:

(i) When $\mathbf{T} = \mathbf{C}$ then $\partial\mathbf{C}/\partial t_{ij} = \partial\mathbf{T}/\partial t_{ij} = \mathbf{e}_i\mathbf{e}_j'$ where $\mathbf{e}_i$ is a unit column-vector with a unit in the ith position and zero elsewhere. Thus (6.9) becomes $\text{trace}(\mathbf{C\Lambda e}_j\mathbf{e}_i') = \mathbf{e}_i'\mathbf{C\Lambda e}_j$ which is the (i, j)th term of $\mathbf{C\Lambda}$.

(ii) When $\mathbf{T} = (\mathbf{C}')^{-1}$, then $\partial\mathbf{C}/\partial t_{ij} = \partial(\mathbf{T}')^{-1}/\partial t_{ij} = -(\mathbf{T}')^{-1}\mathbf{e}_j\mathbf{e}_i'(\mathbf{T}')^{-1} = -\mathbf{Ce}_j\mathbf{e}_i'\mathbf{C}$. Thus (6.9) becomes $\text{trace}(-\mathbf{C\Lambda C}'\mathbf{e}_i\mathbf{e}_j'\mathbf{C}') = -\mathbf{e}_i'\mathbf{C\Lambda C}'\mathbf{Ce}_j$ which is the (i, j)th term of $-\mathbf{C\Lambda C}'\mathbf{C}$.

(iii) When $\mathbf{T} = \mathbf{C}^{-1}$, then $\partial\mathbf{C}/\partial t_{ij} = \partial\mathbf{T}^{-1}/\partial t_{ij} = -\mathbf{T}^{-1}\mathbf{e}_i\mathbf{e}_j'\mathbf{T}^{-1} = -\mathbf{Ce}_i\mathbf{e}_j'\mathbf{C}$. Thus (6.9) becomes $\text{trace}(-\mathbf{C\Lambda C}'\mathbf{e}_j\mathbf{e}_i'\mathbf{C}') = -\mathbf{e}_i'\mathbf{C}'\mathbf{C\Lambda C}'\mathbf{e}_j$ which is the (i, j)th term of $-\mathbf{C}'\mathbf{C\Lambda C}'$.

(iv) When $\mathbf{T} = \mathbf{C}'$, then $\partial\mathbf{C}/\partial t_{ij} = \partial(\mathbf{T}')/\partial t_{ij} = \mathbf{e}_j\mathbf{e}_i'$. Thus (6.9) becomes $\text{trace}(\mathbf{C\Lambda e}_i\mathbf{e}_j') = \mathbf{e}_i'\mathbf{\Lambda C}'\mathbf{e}_j$ which is the (i, j)th term of $\mathbf{\Lambda C}'$.

Gathering these results together, we find the normal equations for the four cases are:

$$\begin{aligned}
(\mathbf{X}_1'\mathbf{X}_1)\mathbf{C} - \mathbf{X}_1'\mathbf{X}_2 &= \mathbf{C\Lambda} \\
(\mathbf{X}_1'\mathbf{X}_1)(\mathbf{C}')^{-1} - \mathbf{X}_1'\mathbf{X}_2 &= -\mathbf{C\Lambda C}'\mathbf{C} \\
(\mathbf{X}_1'\mathbf{X}_1)\mathbf{C}^{-1} - \mathbf{X}_1'\mathbf{X}_2 &= -\mathbf{C}'\mathbf{C\Lambda C}' \\
(\mathbf{X}_1'\mathbf{X}_1)\mathbf{C}' - \mathbf{X}_1'\mathbf{X}_2 &= \mathbf{\Lambda C}'
\end{aligned} \tag{6.10}$$

If we had differentiated with respect to $\mathbf{C}$ rather than $\mathbf{T}$, we would always have $\mathbf{C\Lambda}$ on the right-hand-side. Thus, the last equation of (6.10) would be replaced by its transpose, and the second would be replaced by pre- and post-multiplying by $\mathbf{C}^{-1}$, transposing the result and changing the sign; similarly for the third equation. Particular solutions to some of the equations (6.10) have been suggested in the literature and are briefly discussed below. In the following, we suggest and recommend a uniform approach applicable to all equations.

6.1 The projection method

This is the problem first posed by Mosier (1939) and, indirectly, by Hurley and Cattell (1962), in the context of seeking oblique rotations of factor-loadings $\mathbf{X}_1$ that match $\mathbf{X}_2$. When Cartesian coordinates are projected onto oblique axes, then equation (D.1) of Appendix D shows that they become $\mathbf{X}_1\mathbf{C}$ and we are led to minimising (6.1). We note that

$$\|\mathbf{X}_1\mathbf{C} - \mathbf{X}_2\| = \sum_{j=1}^{P_2} \|\mathbf{X}_1\mathbf{c}_j - \mathbf{y}_j\|$$

where $\mathbf{c}_j$ is the jth column of $\mathbf{C}$ and $\mathbf{y}_j$ is the jth column of $\mathbf{X}_2$. Thus the minimum of (6.1) occurs when $\|\mathbf{X}_1\mathbf{c}_j - \mathbf{y}_j\|$ is minimum for $j = 1, 2, \ldots, P_2$ and we have only to solve the problem for any vector $\mathbf{c}_j$ where $\mathbf{c}_j'\mathbf{c}_j = 1$.* For simplicity, we drop the suffix j and consider the minimisation of $\|\mathbf{X}_1\mathbf{c} - \mathbf{y}\|$ where $\mathbf{c}'\mathbf{c} = 1$. The solution to this problem is given in Appendix E. In the notation of the Appendix we set $\mathbf{X} = \mathbf{X}_1$, $\mathbf{x} = \mathbf{c}$, $\mathbf{y} = \mathbf{y}$ with the constraints given by the matrix $\mathbf{B} = \mathbf{I}$ and $k = 1$ and use Algorithm E2.1. The Appendix is inspired by the solution given by Browne (1967) to the problem discussed in this section but it differs in the following respects: (i) it deals with a more general problem, that includes other oblique axes Procrustes problems (ii) it achieves simplifications in the algebra by using the two-sided eigen-decomposition rather than Browne's use of the classical spectral decomposition, available to him because of the simple form $\mathbf{B} = \mathbf{I}$ (iii) it identifies the minimum as being unique, showing that there are no local minima, leading to (iv) a more simple algorithm for computing the minimum and hence $\mathbf{c}$.

Note that the minimisation of $\|\mathbf{Xc}-\mathbf{y}\|$ is the same problem as that of Projection Procrustes Analysis (Chapter 5) with $P_2 = 1$ and therefore Algorithm 5.1 may be used. However, that algorithm is designed to handle general values of P_2 and will be less efficient than the special-purpose algorithm of Browne and its modifications described here.

Example 6.1 The method is illustrated using the data from Browne (1967). We use Algorithm E2.1. with $\mathbf{X}$ = Browne's $\mathbf{F}$, $\mathbf{y}$ = Browne's $\boldsymbol{\phi}$, $\mathbf{B} = \mathbf{I}\,(4\times 4)$ and $k = 1$.

$$\text{Thus, } \mathbf{X} = \begin{pmatrix} -0.5280 & 0.6280 & 0.2680 & -0.0370 \\ -0.1920 & 0.7810 & -0.3060 & -0.0210 \\ 0.7510 & 0.3680 & 0.0420 & 0 \\ -0.4240 & 0.2240 & 0.4520 & -0.0350 \\ 0.5970 & 0.3100 & 0.3710 & -0.1100 \\ 0.6520 & 0.1330 & 0.3600 & 0.0200 \\ 0.1230 & 0.4100 & -0.5470 & -0.1090 \\ 0.5160 & 0.4100 & -0.1070 & 0.1370 \\ -0.1970 & 0.3410 & 0.1550 & 0.2840 \\ -0.1360 & 0.1110 & 0.2890 & -0.1030 \\ -0.2300 & -0.1000 & 0.2500 & -0.2920 \\ -0.2700 & 0.0990 & 0.2480 & 0.3350 \end{pmatrix} \text{ and } \mathbf{y} = \begin{pmatrix} 0.30 \\ 0.30 \\ 0 \\ 0.15 \\ 0 \\ 0 \\ 0 \\ 0.25 \\ 0 \\ 0 \\ 0 \\ 0.30 \end{pmatrix}.$$

* For most oblique axes problems we shall require that $p_1 = p_2 = p$ but, because in this section the solutions for each axis are independent, there is no need to impose this restriction here. In principle, when $p_1 > p_2$ the algorithmic strategy of Section 3.5 is available for all the oblique axes procedures discussed in this chapter.

The stopping criterion was set to $\varepsilon = 10^{-5}$ with which convergence was achieved in 21 steps. Browne (1967) with $\varepsilon = 10^{-6}$ achieved convergence in only 3 steps but he was using the more computationally demanding Newton–Raphson method for finding the roots of $f(\lambda)$ rather than the simple bisection process of Algorithm E2.1. See Appendix E for comments on the relative merits of Newton–Raphson and bisection. To start the bisection process we find that $\lambda_- = 1/\gamma_1 = 2.3521$ and $\lambda_+ = 1/\gamma_P = 0.3298$. Because the latter is positive, we have one of the special cases handled by Algorithm E2.1 which, in this case, replaces λ_- by -0.9844. The final root found was $\lambda_0 = 0.2482$ corresponding to a residual sum-of-squares 0.26.

The projected $\mathbf{X}$, $\mathbf{Xx}$ ($\mathbf{Ft}$ in Browne) written as a row-vector is:

0.267, 0.251, 0.037, 0.130, −0.042, 0.010, −0.015, 0.197, 0.417, −0.020, −0.259, 0.396.

The transformation vector $\mathbf{c}$ ($\mathbf{t}$ in Browne, $\mathbf{x}$ in appendix) written as a row-vector is:

−0.1237, 0.3444, 0.0722, 0.9278

6.2 The parallel axes or vector-sums method

When Cartesian coordinates are expressed as vector-sums of directions parallel to oblique axes, then equation (D.2) of Appendix D shows that they become $\mathbf{X}_1(\mathbf{C}')^{-1}$. Thus, if we are to match these to a given matrix of coordinates $\mathbf{X}_2$ then we are led to minimise (6.2) leading to the second of the normal equations (6.10). We note that the approximate method of Mosier and of Hurley and Cattell remains available for this problem. This would replace the regression estimate $\mathbf{T}$ given by (3.10) with an appropriate normalisation. In this case, normalisation requires that $\text{diag}(\mathbf{T}'\mathbf{T})^{-1} = \mathbf{I}$ so we have to normalise the rows of $\mathbf{T}^{-1}$. We have the same reservations about this approach, as we do with the similar approach when used with projection oblique axes, but it might be useful to provide an initial setting for some iterative algorithm for determining $\mathbf{C}$. A direct attack on this problem has been given by Gruvaeus (1970). Gruvaeus proposes an algorithm which handles constraints without appealing to Lagrange multipliers but which requires a general function minimiser. He applies his method to the oblique Procrustes problem (6.2) using the Fletcher and Powell (1963) function optimisation method (which does not require derivatives) in a variant described by Jöreskog (1967). We propose a different solution in Section 6.4.

Browne and Kristof (1969) consider minimisation in the parallel axes metric (D.5), where $\mathbf{T} = \mathbf{C}'$, seeking to solve the final of the normal equations given in (6.10); recall that $\mathbf{X}_1$ is now the target matrix. From (6.10) we immediately have:

$$(\mathbf{X}_1'\mathbf{X}_1 - \mathbf{\Lambda})\mathbf{C}' = \mathbf{X}_1'\mathbf{X}_2 \tag{6.11}$$

whence the constraint on direction cosines gives:

$$\text{diag}[(\mathbf{X}_1'\mathbf{X}_1 - \mathbf{\Lambda})^{-1}\mathbf{X}_1'\mathbf{X}_2\mathbf{X}_2'\mathbf{X}_1(\mathbf{X}_1'\mathbf{X}_1 - \mathbf{\Lambda})^{-1}] = \mathbf{I}.$$

This gives P polynomial equations, each of degree $2P$, for the roots $\mathbf{\Lambda}$, which Kristof and Browne (1969) solve by using the Newton–Raphson technique. Having found $\mathbf{\Lambda}$, $\mathbf{C}$ is obtained directly from (6.11); we refer the reader to Kristof and Browne's paper for details. Here we propose an alternative strategy that may be used for all three remaining methods. Our approach is inspired by the results of the previous section where the direction of each axis is found independently of the others. Independence no longer holds, but given the direction of $P-1$ of the axes, we may seek the best direction of the remaining axis. Assuming that we can find this direction, we may proceed iteratively by repeating the procedure for the other axes, each step reducing the value of the criterion (6.10).

First, we show how this works in the simpler case already considered in Section 6.1 of minimising

$$\|\mathbf{X}_1\mathbf{C} - \mathbf{X}_2\| = \text{trace}(\mathbf{C}'\mathbf{A}\mathbf{C}) - 2\text{trace}(\mathbf{C}'\mathbf{Z}) + \text{Constant}$$

where $\mathbf{A} = \mathbf{X}_1'\mathbf{X}_1$ and $\mathbf{Z} = \mathbf{X}_1'\mathbf{X}_2$. We write $\mathbf{C} = (\mathbf{C}_*, \mathbf{c})$ where $\mathbf{C}_*$ represents the first $(P-1)$ columns of $\mathbf{C}$ and $\mathbf{c}$ represents the final column giving the preferred direction to be determined conditional on a given matrix $\mathbf{C}_*$. Similarly we partition $\mathbf{Z}$ conformally to give $\mathbf{Z} = (\mathbf{Z}_*, \mathbf{z})$. We adopt this partition for simplicity and when other preferred directions are adopted we imagine them as being permuted into the final column.* Substituting the partitioned form of $\mathbf{C}$ and $\mathbf{Z}$ into the criterion, shows that we have to minimise $\text{trace}(\mathbf{C}_*'\mathbf{A}\mathbf{C}_*) + \mathbf{c}'\mathbf{A}\mathbf{c} - 2\text{trace}(\mathbf{C}_*'\mathbf{Z}_*) - 2\mathbf{c}'\mathbf{z}$. Differentiating with respect to $\mathbf{c}$ with the constraint $\mathbf{c}'\mathbf{c} = 1$, gives $\mathbf{A}\mathbf{c} - \mathbf{z} = \lambda\mathbf{c}$, the normal equations for the projection method. We then proceed as in the remainder of Section 6.1, using the results of Appendix E. Because $\mathbf{C}_*$ is not involved in this solution, as we have already seen, each column of $\mathbf{C}$ may be handled independently and P steps will be sufficient to estimate $\mathbf{C}$.

Turning to the criterion:

$$\|\mathbf{X}_1\mathbf{C}' - \mathbf{X}_2\| = \text{trace}(\mathbf{C}\mathbf{A}\mathbf{C}') - 2\text{trace}(\mathbf{C}\mathbf{Z}) + \text{Constant}$$

we adopt the partition $\mathbf{Z} = \binom{\mathbf{Z}_*}{\mathbf{z}'}$ [N.B. partition is by rows rather than by columns used above] and partition $\mathbf{A} = \begin{pmatrix} \mathbf{A}_* & \mathbf{a} \\ \mathbf{a}' & \alpha \end{pmatrix}$. After a little algebraic manipulation we find that the terms in $\mathbf{c}$ in the criterion may be written: $2\mathbf{c}'\mathbf{C}_*\mathbf{a} - 2\mathbf{c}'\mathbf{z}$, which after differentiation with the constraint $\mathbf{c}'\mathbf{c} = 1$ yields $\lambda\mathbf{c} = \mathbf{C}_*\mathbf{a} - \mathbf{z}$ and therefore $\lambda^2 = \|\mathbf{C}_*\mathbf{a} - \mathbf{z}\|$. With this setting of $\mathbf{c}$ we have that $2\mathbf{c}'\mathbf{C}_*\mathbf{a} - 2\mathbf{c}'\mathbf{z} = 2\lambda$, so the biggest reduction in sums of squares is obtained by taking the *negative* square root of λ^2. Thus, finally:

$$\mathbf{c} = \frac{\mathbf{z} - \mathbf{C}_*\mathbf{a}}{\{(\mathbf{C}_*\mathbf{a} - \mathbf{z})'(\mathbf{C}_*\mathbf{a} - \mathbf{z})\}^{1/2}} \tag{6.12}$$

* Of course, computational algorithms will not physically permute columns but will use pointer techniques. We may imagine the ith column of $\mathbf{C}$ as being interchanged with the final column. This implies similar changes in the columns of $\mathbf{X}_1$ or $\mathbf{X}_2$. When $\mathbf{T} = \mathbf{C}$ or $(\mathbf{C}')^{-1}$ the ith and final columns of $\mathbf{X}_2$ should be interchanged; when $\mathbf{T} = \mathbf{C}'$ or $\mathbf{C}^{-1}$ the ith and final columns of $\mathbf{X}_1$ should be interchanged.

a solution that is simpler than finding the roots of a polynomial but which has the disadvantage of involving the other columns $\mathbf{C}_*$ of $\mathbf{C}$. That the new vector $\mathbf{c}$ is a function of the settings in the columns of $\mathbf{C}_*$ implies that previously computed columns of $\mathbf{C}$ may need to be recomputed. Nevertheless, the use of (6.12) reduces the value of the criterion so its iterative use ensures convergence. There is no guarantee that the global optimum is reached but this is a problem with all algorithms in this area. Much depends on the initial setting of $\mathbf{C}$. We may compute an initial setting of $\mathbf{C}$ by using the Hurley and Cattell normalisation of the unconstrained solution, or merely set $\mathbf{C} = \mathbf{I}$.

Algorithm 6.1

Step 1 Initialise $\mathbf{C}$, compute $S_1 = \|\mathbf{X}_1\mathbf{C}' - \mathbf{X}_2\|$, set $i = 0$.
Step 2 Set $i = i \mod (P) + 1$, $S = S_1$.
Step 3 Replace the ith column of $\mathbf{C}$ by $\mathbf{c}_i$, using (6.12). Compute $S_1 = \|\mathbf{X}_1\mathbf{C}' - \mathbf{X}_2\|$.
Step 4 If $S - S_1 < \varepsilon$ (threshold), exit, else go to *step 2*.

The same basic strategy will be used for the other oblique criteria discussed in Section 6.3 but with appropriate changes to the calculation of S_1 and replacements for (6.12).

6.3 The cases $\mathbf{T} = \mathbf{C}^{-1}$ and $\mathbf{T} = (\mathbf{C}')^{-1}$

We deal with these remaining cases together and rather than attempting to solve the normal equations given in (6.10) we pursue the iterative procedure discussed in Section 6.2. The direct approach requires the inverse of the partitioned matrix $\mathbf{C} = (\mathbf{C}_*, \mathbf{c})$; we shall write this inverse as $\binom{\mathbf{G}}{\mathbf{g}'}$. Then direct multiplication $\binom{\mathbf{G}}{\mathbf{g}'}(\mathbf{C}_*, \mathbf{c})$ verifies that:

$$\left.\begin{aligned} \mathbf{G} &= (\mathbf{C}_*'\mathbf{C}_*)^{-1}\mathbf{C}_*'\left\{\mathbf{I} - \frac{\mathbf{c}'\mathbf{c}\mathbf{K}}{\mathbf{c}'\mathbf{K}\mathbf{c}}\right\} \\ \text{and} \quad \mathbf{g}' &= \frac{\mathbf{c}'\mathbf{K}}{\mathbf{c}'\mathbf{K}\mathbf{C}} \end{aligned}\right\} \tag{6.13}$$

where $\mathbf{K} = \mathbf{I} - \mathbf{C}_*(\mathbf{C}_*'\mathbf{C}_*)^{-1}\mathbf{C}_*'$ with $\mathbf{g}'\mathbf{g} = 1/(\mathbf{c}'\mathbf{K}\mathbf{c})$.

A problem with (6.13) is that $\mathbf{G}$ is a function of $\mathbf{c}$ and this greatly complicates the estimation of $\mathbf{c}$, or equivalently $\mathbf{g}$. This complication can be avoided by writing $\mathbf{C} = \mathbf{Q}\mathbf{D}$ where $\mathbf{Q}$ is orthogonal and $\mathbf{D}$ is an upper triangular matrix. Triangularisation is easily accomplished by pre-multiplying $\mathbf{C}$ by a series of P Householder transforms (Appendix (F)), chosen to annihilate to zero the parts of the successive columns of $\mathbf{C}$ below the diagonal, yielding $\mathbf{Q}'\mathbf{C} = \mathbf{D}$. These Householder transforms are functions of the successive columns of $\mathbf{C}$ and triangularisation is achieved in $P - 1$ steps, before the final column is used; it follows that $\mathbf{Q}$ is independent of $\mathbf{c}$. It is easy to invert a triangular matrix.

In the following we continue to use the symbol **G** for an inverse although this differs from the **G** of (6.13) which is not used again. Partitioning **D** we have:

$$\mathbf{D} = \begin{pmatrix} \mathbf{D}_* & \mathbf{d}_* \\ & d \end{pmatrix} \quad \text{with inverse} \quad \mathbf{G} = \mathbf{D}^{-1} = \begin{pmatrix} \mathbf{G}_* & \mathbf{g}_* \\ & g \end{pmatrix}$$

where $\mathbf{D}_*$ and $\mathbf{G}_*$ are both upper triangular matrices. Because $\mathbf{D} = \mathbf{Q}'\mathbf{C}$, $\mathbf{D}_*$ is independent of **c** and only the final column $\binom{\mathbf{d}_*}{d} = \mathbf{Q}'\mathbf{c}$ is a function of **c**. Next we show that $\mathbf{D}^{-1}$ has similar properties.

From: $$\mathbf{I} = \mathbf{D}\mathbf{D}^{-1} = \begin{pmatrix} \mathbf{D}_* & \mathbf{d}_* \\ & d \end{pmatrix} \begin{pmatrix} \mathbf{G}_* & \mathbf{g}_* \\ & g \end{pmatrix} = \begin{pmatrix} \mathbf{D}_*\mathbf{G}_* & \mathbf{D}_*\mathbf{g}_* + g\mathbf{d}_* \\ & gd \end{pmatrix},$$

we have that $g = 1/d$, $\mathbf{D}_*\mathbf{g}_* = -g\mathbf{d}_* = -\mathbf{d}_*/d$ and, inversely from $\mathbf{D}^{-1}\mathbf{D} = \mathbf{I}$, $d = 1/g$, $\mathbf{G}_*\mathbf{d}_* = -d\mathbf{g}_* = -\mathbf{g}_*/g$. Also $\mathbf{D}_*\mathbf{G}_* = \mathbf{G}_*\mathbf{D}_* = \mathbf{I}$, so that $\mathbf{G}_*$ is independent of **c**. Because $\text{diag}(\mathbf{D}'\mathbf{D}) = \text{diag}(\mathbf{C}'\mathbf{Q}\mathbf{Q}'\mathbf{C}) = \text{diag}(\mathbf{C}'\mathbf{C}) = \mathbf{I}$, we see that **D** is itself a direction cosine matrix. In particular, its final column is normalised so that:

$$\mathbf{d}_*'\mathbf{d}_* + d^2 = 1$$

which, through the relationships just established, may be expressed in terms of the final column of the inverse as:

$$\mathbf{g}_*'\mathbf{D}_*'\mathbf{D}_*\mathbf{g}_* - g^2 = -1$$

with matrix form:

$$(\mathbf{g}_*'\mathbf{D}_*' \quad g) \begin{pmatrix} \mathbf{I} & \\ & -1 \end{pmatrix} \begin{pmatrix} \mathbf{D}_*\mathbf{g}_* \\ g \end{pmatrix} = -1$$

or

$$\mathbf{s}'\mathbf{J}\mathbf{s} = -1, \tag{6.14}$$

where $\mathbf{J} = \mathbf{I} - 2\mathbf{e}\mathbf{e}'$, **e** being a unit vector with unity in its final position (and so zero, elsewhere) and

$$\mathbf{s} = \begin{pmatrix} \mathbf{D}_*\mathbf{g}_* \\ g \end{pmatrix} = \begin{pmatrix} \mathbf{D}_* & \\ & 1 \end{pmatrix} \mathbf{g} = \mathbf{E}\mathbf{g} \tag{6.15}$$

so defining **E**.

With these preliminaries we may recast the criteria as follows. First, we deal with $\mathbf{T} = \mathbf{C}^{-1}$

$$\|\mathbf{X}_1\mathbf{C}^{-1} - \mathbf{X}_2\| = \|\mathbf{X}_1\mathbf{G}\mathbf{Q}' - \mathbf{X}_2\| = \|\mathbf{X}_1\mathbf{G} - \mathbf{X}_2\mathbf{Q}\|$$

and in this expression only the final column, $\mathbf{X}_1\mathbf{g}$, of $\mathbf{X}_1\mathbf{G}$ involves $\mathbf{g}_*$ and g and hence **c**. Thus, denoting the final column of $\mathbf{X}_2\mathbf{Q}$ by **y**, and recalling from (6.15) that $\mathbf{E}\mathbf{g} = \mathbf{s}$, we shall write:

$$\mathbf{X}_1\mathbf{g} = \mathbf{X}_1\mathbf{E}^{-1}\mathbf{s}, \text{ say.}$$

So we have to minimise:

$$\|\mathbf{Xs} - \mathbf{y}\| \tag{6.16}$$

subject to the constraint (6.14) where $\mathbf{X} = \mathbf{X}_1\mathbf{E}^{-1}$. The solution is given in Appendix E with $\mathbf{X} = \mathbf{X}_1\mathbf{E}^{-1}$, $\mathbf{x} = \mathbf{s}$, and $\mathbf{y} = \mathbf{y}$ and the constraint matrix $\mathbf{B} = \mathbf{J}$ with $k = -1$.

The above can be summarised in algorithmic form as in Algorithm 6.2. Recall that we have been concerned only in updating one column of $\mathbf{C}$, which is assumed to be permuted into the final column of $\mathbf{C}$. The whole process must be repeated iteratively as in Algorithm 6.1 but with $\mathbf{C}$ there replaced by $\mathbf{C}^{-1}$.

Summary Algorithm 6.2

Step 1 Triangularise, that is, Form $\mathbf{C} = \mathbf{QD}$, for $\mathbf{Q}$ orthogonal and $\mathbf{D}$ upper triangular.

Step 2 Form $\mathbf{D}^{-1}$.

Step 3 Form $\mathbf{X} = \mathbf{X}_1\mathbf{E}^{-1}$, Set $\mathbf{y}$ to the final column of $\mathbf{X}_2\mathbf{Q}$.

Step 4 Use the algorithm of Appendix E, with $\mathbf{B} = \mathbf{J}$ and $k = -1$, to minimise $\|\mathbf{Xs} - \mathbf{y}\|$.

Step 5 From $\mathbf{s} = \frac{1}{\mathrm{d}}\binom{-\mathbf{d}_*}{1}$, derive $\mathbf{d}$, whence $\mathbf{c} = \mathbf{Qd}$.

A similar method is available for the remaining problem of minimising $\|\mathbf{X}_1(\mathbf{C}')^{-1} - \mathbf{X}_2\|$. As before, this begins with the triangularisation of $\mathbf{C}$ to give a direction cosine matrix $\mathbf{D}$ which may be partitioned to give the inverse $\mathbf{G}$ with elements $\mathbf{G}_*$, $\mathbf{g}_*$, and g. Thus, the criterion becomes:

$$\begin{aligned}\|\mathbf{X}_1(\mathbf{C}')^{-1} - \mathbf{X}_2\| &= \|\mathbf{C}^{-1}\mathbf{X}_1' - \mathbf{X}_2'\| = \|\mathbf{GQ}'\mathbf{X}_1' - \mathbf{X}_2'\| \\ &= \left\|\begin{pmatrix}\mathbf{G}_* & \mathbf{g}_* \\ & g\end{pmatrix}\mathbf{Q}'\mathbf{X}_1' - \mathbf{X}_2'\right\| = \|\mathbf{gx}_1' - \mathbf{Y}'\|\end{aligned}$$

where $\mathbf{Q} = (\mathbf{Q}_*, \mathbf{q})$, $\mathbf{x}_1 = \mathbf{X}_1\mathbf{q}$, and $\mathbf{Y} = \mathbf{X}_2 - \mathbf{X}_1\mathbf{Q}_*(\mathbf{G}_*', 0)$.
$\|\mathbf{gx}_1' - \mathbf{Y}'\|$ is not quite in the form required by Appendix E. We may put it in the required form by the following series of algebraic manipulations:

$$\|\mathbf{gx}_1' - \mathbf{Y}'\| = (\mathbf{x}_1'\mathbf{x}_1)(\mathbf{g}'\mathbf{g}) - 2\mathbf{x}_1'\mathbf{Yg} + \|\mathbf{Y}\| = (\mathbf{x}_1'\mathbf{x}_1)\|\mathbf{g} - \mathbf{x}_1'\mathbf{Y}/(\mathbf{x}_1'\mathbf{x}_1)\| + \text{constant}.$$

Thus, minmising $\|\mathbf{gx}_1' - \mathbf{Y}'\|$ w.r.t. $\mathbf{g}$ is the same as minimising $\|\mathbf{g} - \mathbf{x}_1'\mathbf{Y}/(\mathbf{x}_1'\mathbf{x}_1)\|$ w.r.t. $\mathbf{g}$. The latter is now in the correct form but we need an explicit expression for the constraint on $\mathbf{g}$. This derives immediately from $\mathbf{s} = \mathbf{Eg}$ of (6.15) so that $\mathbf{s}'\mathbf{Js} + 1 = 0$ becomes $\mathbf{g}'\mathbf{E}'\mathbf{JEg} + 1 = 0$. Thus the constraint matrix of Appendix E is given by $\mathbf{B} = \mathbf{E}'\mathbf{JE} = \mathbf{E}'(\mathbf{I} - 2\mathbf{ee}')\mathbf{E} = \begin{pmatrix}\mathbf{D}_*'\mathbf{D}_* & \\ & 1\end{pmatrix} - 2\begin{pmatrix}0 & \\ & 1\end{pmatrix} = \begin{pmatrix}\mathbf{D}_*'\mathbf{D}_* & \\ & -1\end{pmatrix}$. The latter matrix is the same as $\mathbf{C}'\mathbf{C}$ with its last row and column replaced by a diagonal value of -1, otherwise zero. Although this setting of $\mathbf{B}$ is more complicated than $\mathbf{J}$, it remains easy to compute.

Thus, we may now apply Appendix E in either of two ways:

(i) $\mathbf{X} = \mathbf{E}^{-1}, \mathbf{x} = \mathbf{s}, \mathbf{y} = \mathbf{Y}'\mathbf{x}_1/(\mathbf{x}_1'\mathbf{x}_1), \mathbf{B} = \mathbf{J}$, and $k = -1$.

(ii) $\mathbf{X} = \mathbf{I}, \mathbf{x} = \mathbf{g}, \mathbf{y} = \mathbf{Y}'\mathbf{x}_1/(\mathbf{x}_1'\mathbf{x}_1), \mathbf{B} = \mathbf{E}'\mathbf{JE}$ (calculated from $\mathbf{C}'\mathbf{C}$ as described above), $k = -1$.

With these settings substituted for *step 4*, Algorithm 6.2 remains available. The required direction $\mathbf{c}$ derives from $\mathbf{s}$ or $\mathbf{g}$ through the relationships $\mathbf{s} = \mathbf{Eg} = \frac{1}{\mathrm{d}}\binom{-\mathbf{d}_*}{1}$, and $\mathbf{c} = \mathbf{Qd}$.

6.4 Summary of results

For convenience, we summarise the settings of $\mathbf{X}$, $\mathbf{y}$, $\mathbf{B}$, and k used for the various settings of $\mathbf{T}$. We include the case $\mathbf{T} = \mathbf{C}'$ although the closed form solution (6.12) is more convenient; it may be readily verified that the settings given in Table 6.1 are consistent with (6.12).

Table 6.1. Settings of $\mathbf{X}$, $\mathbf{y}$, $\mathbf{B}$, and k needed to determine $\mathbf{x}$ that minimises $\|\mathbf{Xx} - \mathbf{y}\|$ under the constraint $\mathbf{x}'\mathbf{Bx} = k$ for different forms of oblique transformations $\mathbf{T}$. The returned setting of $\mathbf{x}$ will have to be changed into a direction cosine $\mathbf{c}$ except when $\mathbf{T} = \mathbf{C}$, when it is already in the correct form.

$\mathbf{T}$	$\mathbf{B}$	k	$\mathbf{X}$	$\mathbf{y}$	$\mathbf{x}$
$\mathbf{C}'$	$\mathbf{I}$	1	$\mathbf{I}$	$(\mathbf{z} - \mathbf{C}_*\mathbf{a})/\|\mathbf{z} - \mathbf{C}_*\mathbf{a}\|^{1/2}$	$\mathbf{c}$
$\mathbf{C}$	$\mathbf{I}$	1	$\mathbf{X}_1$	$fc(\mathbf{X}_1'\mathbf{X}_2)$	$\mathbf{c}$
$\mathbf{C}^{-1}$	$\mathbf{J}$	-1	$\mathbf{X}_1\mathbf{E}^{-1}$	$fc(\mathbf{X}_2\mathbf{Q})$	$\mathbf{s}$
$(\mathbf{C}')^{-1}$	$\mathbf{J}$	-1	$\mathbf{E}^{-1}$	$\mathbf{Yx}_1/\|\mathbf{x}_1\|$	$\mathbf{s}$
$(\mathbf{C}')^{-1}$	$\mathbf{E}'\mathbf{JE}$	-1	$\mathbf{I}$	$\mathbf{Yx}_1/\|\mathbf{x}_1\|$	$\mathbf{G}$

Notes:
fc means *final column of*
$\mathbf{a}$ is the vector of the first $P-1$ elements of $fc(\mathbf{X}_1'\mathbf{X}_1)$, $\mathbf{z} = fc(\mathbf{X}_2'\mathbf{X}_1)$

$\mathbf{C} = \mathbf{QD}, \mathbf{D} = \begin{pmatrix} \mathbf{D}_* & \mathbf{d}_* \\ & \mathrm{d} \end{pmatrix}, \mathbf{D}^{-1} = \begin{pmatrix} \mathbf{G}_* & \mathbf{g}_* \\ & \mathrm{g} \end{pmatrix}$ (upper triangular).

$\mathbf{Q} = (\mathbf{Q}_*, \mathbf{q}), \mathbf{x}_1 = \mathbf{X}_1\mathbf{q}, \mathbf{Y} = \mathbf{X}_1\mathbf{Q}_*(\mathbf{G}_*, \mathbf{0}) - \mathbf{X}_2$

$\mathbf{E} = \begin{pmatrix} \mathbf{D}_* & \\ & 1 \end{pmatrix}, \mathbf{E}^{-1} = \begin{pmatrix} \mathbf{G}_* & \\ & 1 \end{pmatrix}, \mathbf{E}'\mathbf{JE} = \begin{pmatrix} \mathbf{D}_*'\mathbf{D}_* & \\ & -1 \end{pmatrix} = \mathbf{C}'\mathbf{C}$ (with final row/col $= -1$).

When $\mathbf{x}$ is given as $\mathbf{s}$ or $\mathbf{g}$, the corresponding direction cosine $\mathbf{c}$ may be derived from:

$$\mathbf{s} = \mathbf{Eg} = \frac{1}{\mathrm{d}}\begin{pmatrix} -\mathbf{d}_* \\ 1 \end{pmatrix}, \mathbf{c} = \mathbf{Qd}.$$

7
Other two-sets Procrustes problems

7.1 Permutations

Minimise: $\|\mathbf{X}_1\mathbf{T} - \mathbf{X}_2\|$ where ${}_{P_1}\mathbf{T}_{P_2}$ is a permutation matrix. Thus $\mathbf{X}_1\mathbf{T}$ permutes the columns of $\mathbf{X}_1$ and the problem is to find the permutation that best-fits $\mathbf{X}_2$. Because a permutation matrix is orthogonal it follows that we have to maximise trace $(\mathbf{X}_2'\mathbf{X}_1\mathbf{T})$. But a permutation matrix is special with all its elements zero or unity, so there is precisely one unit in every row and column of $\mathbf{T}$. A doubly stochastic matrix is one whose elements are non-negative and whose rows and columns all sum to unity. Hence permutation matrices are also special cases of doubly stochastic matrices. Thus, we consider the problem of maximising trace$(\mathbf{X}_2'\mathbf{X}_1\mathbf{T})$ over all doubly stochastic matrices. This criterion is linear in the elements of $\mathbf{T}$, subject to the linear constraints:

$$t_{ij} \geq 0, \qquad \text{for all } i, j$$

$$\sum_{i=1}^{P_1} t_{ij} = 1, \qquad \text{for all } j$$

$$\sum_{j=1}^{P_2} t_{ij} = 1, \qquad \text{for all } i.$$

This is a linear programming problem and hence its solution is at a vertex of the feasible region defined by the constraints. All vertices correspond to permutation matrices, so the solution to the linear programming problem for doubly stochastic matrices gives the best-fitting permutation matrix.

A similar problem, with a similar solution, is to minimise $\|\mathbf{T}\mathbf{X}_1 - \mathbf{X}_2\|$ where $\mathbf{T}$ is constrained to be a permutation matrix operating on the rows of $\mathbf{X}_1$. This allows us to tackle matching problems where the rows of $\mathbf{X}_1$ and $\mathbf{X}_2$ are known to agree but have become disordered. An alternating least-squares approach would allow matching with both row and column permutations. Similarly, permutation of rows may be combined with constrained transformations such as minimising $\|\mathbf{T}\mathbf{X}_1\mathbf{Q} - \mathbf{X}_2\|$ where here $\mathbf{T}$ represents a permutation and $\mathbf{Q}$ is orthogonal, as in our usual notation.

7.2 Reduced rank regression

Redundancy Analysis, also known as Reduced Rank Regression, is a variant of (3.1) in which the rank of $\mathbf{T}$ is specified in advance. Thus we wish to minimise $\|\mathbf{X}_1\mathbf{T} - \mathbf{X}_2\|$ subject to rank $\mathbf{T} = R$. Suppose $\mathbf{T}_0$ is the ordinary least-squares estimate of $\mathbf{T}$, then

$$\|\mathbf{X}_1\mathbf{T} - \mathbf{X}_2\| = \|\mathbf{X}_1\mathbf{T}_0 - \mathbf{X}_2\| + \|\mathbf{X}_1(\mathbf{T} - \mathbf{T}_0)\|$$

so it is necessary only to minimise $\|\mathbf{X}_1(\mathbf{T} - \mathbf{T}_0)\|$ subject to the constraint on rank. The solution to this minimisation problem for the rank R approximation to $\mathbf{T}_0$ may be regarded as being given by a weighted form of the Eckart–Young theorem. However, it is simple to proceed directly, noting that:

$$\|\mathbf{X}_1(\mathbf{T} - \mathbf{T}_0)\| = \text{trace}(\mathbf{T} - \mathbf{T}_0)'\mathbf{X}_1'\mathbf{X}_1(\mathbf{T} - \mathbf{T}_0) = \|(\mathbf{X}_1'\mathbf{X}_1)^{1/2}(\mathbf{T} - \mathbf{T}_0)\|. \tag{7.1}$$

Thus, if $(\mathbf{X}_1'\mathbf{X}_1)^{1/2}\mathbf{T}_0 = \mathbf{U\Sigma V}'$ is a conventional singular value decomposition then by the Eckart–Young theorem, we have that the rank r approximation to $(\mathbf{X}_1'\mathbf{X}_1)^{1/2}\mathbf{T}_0$ is given by $(\mathbf{X}_1'\mathbf{X}_1)^{1/2}\hat{\mathbf{T}}_0 = \mathbf{U}_R\mathbf{\Sigma}_R\mathbf{V}_R'$ and hence that:

$$\hat{\mathbf{T}}_0 = (\mathbf{X}_1'\mathbf{X}_1)^{-1/2}\mathbf{U}_R\mathbf{\Sigma}_R\mathbf{V}_R'. \tag{7.2}$$

By defining $\mathbf{W} = \mathbf{U}'(\mathbf{X}_1'\mathbf{X}_1)^{1/2}$ and $\mathbf{W}^R$ to represent the first R columns of $\mathbf{W}^{-1}$ we may rewrite these results as:

$$\mathbf{T}_0 = \mathbf{W}^{-1}\mathbf{\Sigma V}' \quad \text{and} \quad \hat{\mathbf{T}}_0 = \mathbf{W}^R\mathbf{\Sigma}_R\mathbf{V}_R'. \tag{7.3}$$

The solution (7.3) is in a similar form to the generalised singular value decomposition given by Gower and Hand (1995, p. 236) which suggests that $\mathbf{W}$ may be obtained as the solution to a two-sided eigenvalue problem. That this is indeed so, follows from noting that from the first equation of (7.1):

$$\mathbf{T}_0\mathbf{T}_0'\mathbf{W}' = (\mathbf{W}'\mathbf{W})^{-1}\mathbf{W}'\mathbf{\Sigma}^2 = (\mathbf{X}_1'\mathbf{X}_1)^{-1}\mathbf{W}'\mathbf{\Sigma}^2. \tag{7.4}$$

This is a two-sided eigenvalue problem in the matrices $\mathbf{T}_0\mathbf{T}_0'$ and $(\mathbf{X}_1'\mathbf{X}_1)^{-1}$ yielding the eigenvectors $\mathbf{W}'$. Note that the normalisation of the vectors is given by:

$$\mathbf{W}(\mathbf{X}_1'\mathbf{X}_1)^{-1}\mathbf{W}' = \mathbf{U}'(\mathbf{X}_1'\mathbf{X}_1)^{1/2}(\mathbf{X}_1'\mathbf{X}_1)^{-1}(\mathbf{X}_1'\mathbf{X}_1)^{1/2}\mathbf{U} = \mathbf{I}.$$

Because $\mathbf{WT}_0 = \mathbf{\Sigma V}'$ it follows that $\mathbf{W}_R\mathbf{T}_0 = \mathbf{\Sigma}_R\mathbf{V}_R'$ where $\mathbf{W}_R$ represents the first R rows of $\mathbf{W}$. From (7.3), this allows us to express $\hat{\mathbf{T}}_0$ in the elegant form:

$$\hat{\mathbf{T}}_0 = \mathbf{W}^R\mathbf{W}_R\mathbf{T}_0 \tag{7.5}$$

which may be compared with Appendix (D.2) and interpreted as a generalised projection in the metric $(\mathbf{X}_1'\mathbf{X}_1)^{-1}$.

7.3 Miscellaneous choices of T

A further possibility is to allow **T** to be a general matrix whose elements are constrained to be non-negative. The criterion function (3.1) remains quadratic in the elements t_{ij} but its minimisation is now subject to the linear constraints $t_{ij} \geq 0$ for all pairs i, j. Thus, the direct approach leads to a quadratic programming problem.

We might consider the Procrustes problem of minimising $\|\mathbf{X}_1\mathbf{T} - \mathbf{X}_2\|$ where **T** is a symmetric matrix. The constraints $t_{ij} = t_{ji}$ induces Lagrange multipliers that may be assembled into a skew-symmetric matrix $\mathbf{\Lambda}$. The normal equations are then:

$$\mathbf{X}_1'\mathbf{X}_1\mathbf{T} - \mathbf{X}_1'\mathbf{X}_2 = \mathbf{\Lambda}. \tag{7.6}$$

Because $\mathbf{\Lambda}$ is skew, adding the above to its transpose, gives:

$$(\mathbf{X}_1'\mathbf{X}_1\mathbf{T} + \mathbf{T}\mathbf{X}_1'\mathbf{X}_1) - (\mathbf{X}_1'\mathbf{X}_2 + \mathbf{X}_2'\mathbf{X}_1) = 0.$$

The spectral decomposition $\mathbf{X}_1'\mathbf{X}_1 = \mathbf{V}\mathbf{\Gamma}\mathbf{V}'$ (**V** orthogonal and $\mathbf{\Gamma}$ diagonal) allows this to be written:

$$(\mathbf{V}\mathbf{\Gamma}\mathbf{V}')\mathbf{T} + \mathbf{T}(\mathbf{V}\mathbf{\Gamma}\mathbf{V}') = \mathbf{X}_1'\mathbf{X}_2 + \mathbf{X}_2'\mathbf{X}_1 = \mathbf{C}, \text{ say},$$

which gives:

$$\mathbf{\Gamma}\mathbf{S} + \mathbf{S}\mathbf{\Gamma} = \mathbf{V}'\mathbf{C}\mathbf{V} \tag{7.7}$$

where $\mathbf{S} = \mathbf{V}'\mathbf{T}\mathbf{V}$ and hence $\mathbf{T} = \mathbf{V}\mathbf{S}\mathbf{V}'$. Equation (7.7) is linear in the elements of **S** and may be easily solved to give:

$$\begin{aligned} 2\gamma_i s_{ii} &= (\mathbf{V}'\mathbf{C}\mathbf{V})_{ii} \\ (\gamma_i + \gamma_j)s_{ij} &= (\mathbf{V}'\mathbf{C}\mathbf{V})_{ij}. \end{aligned} \tag{7.8}$$

Thus (7.8) gives **S** and hence **T**, as required.

We may derive the usual form of analysis of variance, as follows. First, we note that trace$\mathbf{T}\mathbf{\Lambda}$ = trace$\mathbf{\Lambda}\mathbf{T}$ (by permutation) = $-$trace$\mathbf{\Lambda}\mathbf{T}$ (by transposition); hence trace$\mathbf{T}\mathbf{\Lambda} = 0$. From (7.6) it follows that:

$$\mathbf{T}(\mathbf{X}_1'\mathbf{X}_1\mathbf{T} - \mathbf{X}_1'\mathbf{X}_2) = 0.$$

and thence

$$\|\mathbf{X}_1\mathbf{T} - \mathbf{X}_2\| = \|\mathbf{X}_2\| - \|\mathbf{X}_1\mathbf{T}\|$$

which is the required analysis of variance, expressing the total sum-of-squares $\|\mathbf{X}_2\|$ as the sum of orthogonal terms attributable to the fitted component $\mathbf{X}_1\mathbf{T}$ and the residuals $\mathbf{X}_1\mathbf{T} - \mathbf{X}_2$. A very similar argument applies to the case where **T** is constrained to be skew-symmetric, but now $\mathbf{\Lambda}$ is symmetric so we start from the difference between (7.6) and its transpose.

When $\mathbf{T}$ is to be symmetric and positive semi-definite (p.s.d.), we could find the nearest p.s.d. matrix to the general symmetric $\mathbf{T}$ found as above. This may be done by restricting the spectral decomposition of $\mathbf{T}$ to include only its positive eigenvalues. However, a more general approach is possible for all matrices $\mathbf{T}$ which form a *convex cone*. In this context, a cone is an algebraic construct such that if $\mathbf{T}_i$ and $\mathbf{T}_j$ are members of the cone then so is $\lambda\mathbf{T}_i + \mu\mathbf{T}_j$ for all $\lambda, \mu \geq 0$; normally, $\mathbf{T} = \mathbf{0}$ is included in the cone (its vertex). Geometrically, the cone property expresses that all points between any two points belonging to the cone are also members of the cone. In particular, the sum of two members of the cone must also be a member. The sum of two non-negative matrices is another non-negative matrix, the sum of two symmetric matrices is symmetric, and the sum of two p.s.d. matrices is another p.s.d. matrix. All these classes of matrix define convex cones. Orthogonal, projection and direction-cosine matrices do not form cones. Andersson and Elving (1997) have discussed an algorithm for computing $\mathbf{T}$ for classes of matrix that do form cones. Such problems have little to do with matching configurations and are beyond the scope of this book.

7.4 On simple structure rotations etc

Rather than handling the indeterminacy of factor loadings as above, a modern tendency is to use an internal criterion of simple structure such as Quartimax (Neuhaus and Wrigley 1954), Varimax (Kaiser 1958), and Orthomax (Harman 1960), each with its own algorithm. Jennrich (2001) has provided a general algorithmic approach that includes as special cases many such criteria based on orthogonal rotations. Simple structure criteria have no reference to an external target $\mathbf{X}_2$ and so are beyond the scope of this book.

An alternative approach to finding simple structure is to code $\mathbf{X}_2$ in binary form, where $x_{ij} = 1$ indicates a desired high loading of the ith variable on the jth factor and $x_{ij} = 0$ indicates a low loading. This latter form remains a type of Procrustes problem but least squares is no longer appropriate and is replaced by choosing $\mathbf{T}$ to maximise the ratio of the sum-of-squares of the squared high loadings to the sum-of-squares of the squared low loadings (see Lawley and Maxwell 1971).

7.5 Double Procrustes problems

Schönemann (1968) has considered estimating an orthogonal matrix $\mathbf{Q}$, which minimises $\|\mathbf{Q}'\mathbf{X}_1\mathbf{Q}-\mathbf{X}_2\|$. $\mathbf{Q}$ is orthogonal and therefore $\mathbf{X}_1$ and $\mathbf{X}_2$ must be square; it is further assumed that they are both symmetric. We refer to such transformations as *double* Procrustes problems, for the obvious reason that the transformation occurs on both sides of a single matrix $\mathbf{X}_1$ and also to distinguish from the cases, already discussed, of *two-sided* problems defined by (3.2) where transformations are applied to both matrices. That being said, because $\mathbf{Q}$ is an orthogonal matrix, Schönemann's criterion can be written $\|\mathbf{X}_1\mathbf{Q}-\mathbf{Q}\mathbf{X}_2\|$ which looks very much like a two-sided problem. Also, there are links with the problems discussed in Section 8.2 but there we are concerned with diagonal scaling matrices rather than orthogonal matrices. Schönemann introduced his criterion as a contribution towards solving a permutation problem. In Section 7.1 we showed how Permutation Procrustes problems could be solved by linear programming methods, so to some extent the original purpose of double Procrustes methods has been overtaken. Nevertheless we present the solution to minimising $\|\mathbf{Q}'\mathbf{X}_1\mathbf{Q} - \mathbf{X}_2\|$ for its interest and possible applications.

7.5.1 DOUBLE PROCRUSTES FOR SYMMETRIC MATRICES (ORTHOGONAL CASE)

In the usual way, the minimum of $\|\mathbf{Q}'\mathbf{X}_1\mathbf{Q} - \mathbf{X}_2\|$ occurs for the same value of $\mathbf{Q}$ that maximises trace$(\mathbf{X}_2\mathbf{Q}'\mathbf{X}_1\mathbf{Q})$. Regarding $\mathbf{X}_2\mathbf{Q}'\mathbf{X}_1$ as fixed, we know that the value of $\mathbf{Q}$ that maximises the trace is given by $\mathbf{Q} = \mathbf{V}\mathbf{U}'$ derived from the SVD, $\mathbf{X}_2\mathbf{Q}'\mathbf{X}_1 = \mathbf{U}\boldsymbol{\Sigma}\mathbf{V}'$. Thus

$$\mathbf{X}_2\mathbf{U}\mathbf{V}'\mathbf{X}_1 = \mathbf{U}\boldsymbol{\Sigma}\mathbf{V}'$$

and therefore

$$(\mathbf{U}'\mathbf{X}_2\mathbf{U})(\mathbf{V}'\mathbf{X}_1\mathbf{V}) = \boldsymbol{\Sigma}.$$

The left-hand side is a product of two symmetric matrices and the right-hand side is also symmetric, in fact, diagonal. It follows from Appendix F (Special Case) that the eigenvectors of the two symmetric matrices must be permutations of one another. Therefore, we need the spectral forms of $\mathbf{X}_1$ and $\mathbf{X}_2$. Because both are symmetric we may write the spectral forms $\mathbf{X}_1 = \mathbf{V}_1\boldsymbol{\Gamma}_1\mathbf{V}_1'$ and $\mathbf{X}_2 = \mathbf{V}_2\boldsymbol{\Gamma}_2\mathbf{V}_2'$, so the eigenvectors of $\mathbf{V}'\mathbf{X}_1\mathbf{V}$ are $\mathbf{V}'\mathbf{V}_1$ and the eigenvectors of $\mathbf{U}'\mathbf{X}_2\mathbf{U}$ are $\mathbf{U}'\mathbf{V}_2$. Thus from Appendix F there exists a permutation matrix $\boldsymbol{\Pi}$ such that:

$$\mathbf{V}'\mathbf{V}_1 = (\mathbf{U}'\mathbf{V}_2)\boldsymbol{\Pi}.$$

With this setting we find that our trace criterion becomes:

$$\text{trace}(\mathbf{X}_2\mathbf{Q}'\mathbf{X}_1\mathbf{Q}) = \text{trace}(\mathbf{V}_2\boldsymbol{\Gamma}_2\mathbf{V}_2'\mathbf{U}\mathbf{V}'\mathbf{V}_1\boldsymbol{\Gamma}_1\mathbf{V}_1'\mathbf{V}\mathbf{U}')$$

which, on substituting the previous result, simplifies to:

$$\text{trace}(\mathbf{X}_2\mathbf{Q}'\mathbf{X}_1\mathbf{Q}) = \text{trace}(\boldsymbol{\Gamma}_2\boldsymbol{\Pi}\boldsymbol{\Gamma}_1\boldsymbol{\Pi}').$$

What this means is that we have to permute the diagonal elements of $\boldsymbol{\Gamma}_1$ to maximise the sum of products with the corresponding elements of $\boldsymbol{\Gamma}_2$. The maximum occurs (See Hardy, Littlewood, and Polya 1952) when both sequences are in the same order. Thus, if we have used the usual form of spectral decompositions in which the eigenvalues are arranged in non-increasing order down the diagonal, the eigenvectors will already be in the required form. Then, $\boldsymbol{\Pi} = \mathbf{I}$ and $\mathbf{V}'\mathbf{V}_1 = \mathbf{U}'\mathbf{V}_2$. The fitted value $\mathbf{Q}'\mathbf{X}_1\mathbf{Q}$ is of special interest. It is given by:

$$\mathbf{Q}'\mathbf{X}_1\mathbf{Q} = \mathbf{U}\mathbf{V}'(\mathbf{V}_1\boldsymbol{\Gamma}_1\mathbf{V}_1')\mathbf{V}\mathbf{U}' = \mathbf{V}_2\boldsymbol{\Gamma}_1\mathbf{V}_2'.$$

Thus, the matrix fitted to $\mathbf{X}_2$ has the eigenvectors of $\mathbf{X}_2$ but the eigenvalues of $\mathbf{X}_1$. This makes sense, because orthogonal matrices are concerned only with orientation and not size. Size is represented by the eigenvalues which are unchanged, while the orientation is made to line up with the eigenvectors of $\mathbf{X}_2$.

7.5.2 DOUBLE PROCRUSTES FOR RECTANGULAR MATRICES (ORTHOGONAL CASE)

The general two-sided Procrustes problem requires orthogonal matrices $\mathbf{Q}_1$ and $\mathbf{Q}_2$, which minimise $\|\mathbf{Q}_2'\mathbf{X}_1\mathbf{Q}_1 - \mathbf{X}_2\|$. This is a more difficult problem than for symmetric matrices and was discussed by Schönemann (1968) but for the special case where $\mathbf{X}_2$ and $\mathbf{X}_1$ are both square but not symmetric. We shall treat the more general case where $\mathbf{X}_2$ and $\mathbf{X}_1$ are rectangular with P rows and Q columns. Thus $\mathbf{Q}_2$ is of order P and $\mathbf{Q}_1$ is of order Q. In the usual way, the least-squares solution requires the maximisation of trace$(\mathbf{X}_2'\mathbf{Q}_2'\mathbf{X}_1\mathbf{Q}_1)$. Regarding $\mathbf{Q}_2$ as fixed, we know that $\mathbf{Q}_1$ derives from the SVD of $\mathbf{X}_2'\mathbf{Q}_2'\mathbf{X}_1$ and, similarly, regarding $\mathbf{Q}_1$ as fixed, we know that $\mathbf{Q}_2$ derives from the SVD of $\mathbf{X}_2\mathbf{Q}_1'\mathbf{X}_1'$. Thus:

$$\begin{aligned} \mathbf{X}_2'\mathbf{Q}_2'\mathbf{X}_1 &= \mathbf{U}_1\boldsymbol{\Sigma}_1\mathbf{V}_1' \quad & \mathbf{Q}_1 &= \mathbf{V}_1\mathbf{U}_1' \\ \mathbf{X}_2\mathbf{Q}_1'\mathbf{X}_1' &= \mathbf{U}_2\boldsymbol{\Sigma}_2\mathbf{V}_2' \quad & \mathbf{Q}_2 &= \mathbf{V}_2\mathbf{U}_2'. \end{aligned} \tag{7.9}$$

Substituting for $\mathbf{Q}_1$ and $\mathbf{Q}_2$ and rearranging, gives:

$$\begin{aligned} \mathbf{U}_1'\mathbf{X}_2'\mathbf{U}_2\mathbf{V}_2'\mathbf{X}_1\mathbf{V}_1 &= \boldsymbol{\Sigma}_1 \\ \mathbf{U}_2'\mathbf{X}_2\mathbf{U}_1\mathbf{V}_1'\mathbf{X}_1'\mathbf{V}_2 &= \boldsymbol{\Sigma}_2. \end{aligned} \tag{7.10}$$

Writing $\mathbf{A} = \mathbf{V}_2'\mathbf{X}_1\mathbf{V}_1$ and $\mathbf{B} = \mathbf{U}_2'\mathbf{X}_2\mathbf{U}_1$, then (7.2) becomes:

$$\mathbf{B}'\mathbf{A} = \boldsymbol{\Sigma}_1, \quad \mathbf{B}\mathbf{A}' = \boldsymbol{\Sigma}_2 = \mathbf{A}\mathbf{B}'. \tag{7.11}$$

$\boldsymbol{\Sigma}_1$ and $\boldsymbol{\Sigma}_2$ are diagonal and so are certainly symmetric. It follows that the conditions of the result given in Appendix F for the two products of (7.11) both to be symmetric, are certainly satisfied. This result says that $\mathbf{A}$ and $\mathbf{B}$ share the same singular vectors. The SVDs of $\mathbf{A}$ and $\mathbf{B}$ derive from those of $\mathbf{X}_1$ and $\mathbf{X}_2$, which we shall write:

$$\mathbf{X}_1 = \mathbf{P}_1\boldsymbol{\Gamma}_1\mathbf{R}_1' \quad \mathbf{X}_2 = \mathbf{P}_2\boldsymbol{\Gamma}_2\mathbf{R}_2'$$

where we assume that $\boldsymbol{\Gamma}_1$ is of order R, the number of non-zero singular values of $\mathbf{A}$ and that $\boldsymbol{\Gamma}_2$ is of order $S \le R$, the number of non-zero singular values of $\mathbf{B}$. Then the SVDs of $\mathbf{A}$ and $\mathbf{B}$ are: $\mathbf{A} = (\mathbf{V}_2'\mathbf{P}_1)\boldsymbol{\Gamma}_1(\mathbf{V}_1'\mathbf{R}_1)'$, $\mathbf{B} = (\mathbf{U}_2'\mathbf{P}_2)\boldsymbol{\Gamma}_2(\mathbf{U}_1'\mathbf{R}_2)'$. Thus the S singular vectors $\mathbf{U}_2'\mathbf{P}_2$ must be a permutation of S of the R singular vectors $\mathbf{V}_2'\mathbf{P}_1$. Similarly, the S singular vectors $\mathbf{U}_1'\mathbf{R}_2$ must be the same permutation of S of the R singular vectors $\mathbf{V}_1'\mathbf{R}_1$. Writing $\boldsymbol{\Pi}$ for the $R \times S$ permutation matrix, we have that:

$$\mathbf{B}'\mathbf{A} = (\mathbf{U}_1'\mathbf{R}_2)\boldsymbol{\Gamma}_2\boldsymbol{\Pi}'\boldsymbol{\Gamma}_1\boldsymbol{\Pi}(\mathbf{V}_1'\mathbf{R}_1)'.$$

As previously, we choose $\boldsymbol{\Pi}$ to ensure that the product $\gamma_{11}\gamma_{21} + \gamma_{12}\gamma_{22} + \gamma_{13}\gamma_{23} + \cdots + \gamma_{1S}\gamma_{2S}$ is maximised. Thus, we have:

$$\mathbf{V}_2'\mathbf{P}_1 = \mathbf{U}_2'\mathbf{P}_2,$$

where now $\mathbf{P}_1$ has the first S columns, as determined by the optimal permutation, full R-column version of $\mathbf{P}_1$. It follows that:

$$\mathbf{P}_1 = \mathbf{Q}_2\mathbf{P}_2.$$

To determine $\mathbf{Q}_2$, we append $P-S$ extra columns $\mathbf{P}_{1\perp}$ to $\mathbf{P}_1$ to make an orthogonal matrix and then consider admissible forms of $\mathbf{C}$ that satisfy:

$$(\mathbf{P}_1, \mathbf{P}_{1\perp}) = \mathbf{Q}_2(\mathbf{P}_2, \mathbf{C}).$$

Because $(\mathbf{P}_1, \mathbf{P}_{1\perp})$ and $\mathbf{Q}_2$ are both orthogonal, then so is $(\mathbf{P}_2, \mathbf{C})$ and so $\mathbf{C}$ must have the form $\mathbf{P}_{2\perp}$. Inversion then gives:

$$\mathbf{Q}_2 = (\mathbf{P}_1, \mathbf{P}_{1\perp})(\mathbf{P}_2, \mathbf{P}_{2\perp})'$$

which may be written:

$$\mathbf{Q}_2 = \mathbf{P}_1\mathbf{P}_2' + \mathbf{P}_{1\perp}\mathbf{P}_{2\perp}'.$$

Similarly,

$$\mathbf{V}_1'\mathbf{R}_1 = \mathbf{U}_1'\mathbf{R}_2,$$

where now $\mathbf{R}_1$ has the first S columns, as determined by the optimal permutation, of the R-column version of $\mathbf{R}_1$. It follows that:

$$\mathbf{R}_1 = \mathbf{Q}_1\mathbf{R}_2,$$

leading to

$$\mathbf{Q}_1 = \mathbf{R}_1\mathbf{R}_2' + \mathbf{R}_{1\perp}\mathbf{R}_{2\perp}'.$$

With these values of $\mathbf{Q}_1$ and $\mathbf{Q}_2$ we find, paying special attention to the dimensionalities of all the matrices concerned, that:

$$\begin{aligned}\mathbf{Q}_2'\mathbf{X}_1\mathbf{Q}_1 &= \mathbf{P}_2\mathbf{\Gamma}_S\mathbf{R}_2' + \mathbf{P}_{2\perp}\mathbf{\Gamma}_{R-S}\mathbf{R}_{2\perp}' \\ &= (\mathbf{P}_2, \mathbf{P}_{2\perp})\begin{pmatrix}\mathbf{\Gamma}_S & \\ & \mathbf{\Gamma}_{R-S}\end{pmatrix}(\mathbf{R}_2, \mathbf{R}_{2\perp})'. \qquad (7.12)\end{aligned}$$

Here $\mathbf{\Gamma}_S$ contains the largest non-zero singular values of $\mathbf{\Gamma}_1$ and $\mathbf{\Gamma}_{R-S}$ contains the remaining $R-S$ non-zero singular values. Equation (7.12) is of similar form to the fitted values of previous double Procrustes problem (Section 7.5.1), with the non-zero singular values of $\mathbf{X}_1$ associated with the singular vectors of $\mathbf{X}_2$. Note that:

1. The 'best' S non-zero singular vectors of $\mathbf{X}_1$ and $\mathbf{X}_2$ line up. The remaining $R-S$ singular vectors associated with non-zero singular values may be arbitrarily permuted.
2. We have assumed that $R \geq S$, that is, $\mathbf{X}_2$ cannot have more non-zero singular values than $\mathbf{X}_1$. Presumably, when $S > R$ then: $\mathbf{Q}_2'\mathbf{X}_1 = \mathbf{P}_2\mathbf{\Gamma}_R\mathbf{P}_1'$ and the $S-R$ 'least important' singular vectors are irrelevant.
3. The residual sum-of-squares is:

$$\text{trace}(\mathbf{\Gamma}_1^2 + \mathbf{\Gamma}_2^2 - 2\mathbf{\Gamma}_S\mathbf{\Gamma}_2)$$

where $\mathbf{\Gamma}_S$ represents the S best singular values of $\mathbf{\Gamma}_1$.

8
Weighting, scaling, and missing values

Weighting and scaling are related concepts that are often difficult to disentangle. Broadly speaking, *weighting* (Section 8.1) refers to *given* quantities that themselves do not require estimation but need to be taken into account when estimating other parameters. By contrast, *scaling* refers to parameters that have to be estimated, as for the scalar parameter s for isotropic scaling that is estimated by (3.10). Anisotropic scaling (Section 8.3) involves estimating several parameters. Confusion arises when estimation of scaling is combined with estimating other parameters, say $\mathbf{T}$. Then, in alternating least-squares algorithms, the step for estimating $\mathbf{T}$ often involves regarding the current values of estimates of the scaling parameters as fixed, and hence, to be treated like weights. Many of the results given in this section may seem of esoteric interest only. This is sometimes so, but it turns out that results of little interest in the two sets case have serious applications with K sets (Chapters 9 and 13). Also included in this chapter is a discussion of how to handle missing values in two sets Procrustes problems (Section 8.2). The only connection this has with weighting or scaling is that indicator matrices, that have some relationship with weighting matrices, may be used to identify missing cells. K sets missing value estimation is discussed in Sections 9.1.3 and 9.1.7.

8.1 Weighting

In this section we treat several forms of weighting by a *given* matrix $\mathbf{W}$. The weights may refer to replication of the individual cells of $\mathbf{X}_1$ and/or $\mathbf{X}_2$, or they may refer to the rows of these matrices. Often, the weighting will have an effect on the estimation of translational effects, should these be of interest. Because $\mathbf{W}$ is given, it has no associated problems of normalisation that can occur when it is a scaling matrix as in Section 8.3. Although our results are expressed using the simple form of transformation $\mathbf{T}$, it is understood that this includes the general form $\mathbf{RTS}$ discussed in Section 8.3.

8.1.1 TRANSLATION WITH WEIGHTING

There is one important situation where simple centering does not suffice. This is when $\mathbf{X}_1$ is preceded by a matrix $\mathbf{W}$ to give a criterion:

$$\|\mathbf{W}\mathbf{X}_1\mathbf{T} - \mathbf{X}_2\|.$$

$\mathbf{W}$ is termed a weighting matrix. We shall allow $\mathbf{W}$ to be any given square matrix but in most applications it will be diagonal. Later (see Section 8.3.6), we shall discuss the possibility of estimating an optimal matrix $\mathbf{W}$. To allow for translations, we consider minimising:

$$\begin{aligned} S &= \|\mathbf{W}(\mathbf{X}_1 - \mathbf{1a}')\mathbf{T} - (\mathbf{X}_2 - \mathbf{1b}')\| \\ &= \|\mathbf{WC} - \mathbf{wc}' - \mathbf{X}_2 + \mathbf{1b}'\|, \end{aligned} \tag{8.1}$$

where

$$\mathbf{w} = \mathbf{W1}, \mathbf{C} = \mathbf{X}_1\mathbf{T}' \quad \text{and} \quad \mathbf{c}' = \mathbf{a}'\mathbf{T}.$$

Since we are allowing for translations of both matrices, we are free to select their initial origins as we think fit. We select $\mathbf{1}'\mathbf{X}_2 = \mathbf{0}$ and $\mathbf{1}'\mathbf{W}\mathbf{X}_1 = \mathbf{0}$. In Section 3.2.1, translation could be accounted for by the simple rank one matrix $\mathbf{1a}'$. In this section, it is evident from (8.1) that the translation terms define a matrix $\mathbf{wc}' - \mathbf{1b}'$ of rank two, leading to increased complexity.

Expanding (8.1) gives:

$$S = \|\mathbf{WC} - \mathbf{X}_2\| - 2\mathbf{w}'(\mathbf{WC} - \mathbf{X}_2)\mathbf{c} + N\mathbf{b}'\mathbf{b} + (\mathbf{w}'\mathbf{w})(\mathbf{c}'\mathbf{c}) - 2(\mathbf{b}'\mathbf{c})(\mathbf{1}'\mathbf{w})$$

which on differentiation with respect to $\mathbf{b}$ and $\mathbf{c}$ gives:

$$\begin{aligned} N\mathbf{b} &= (\mathbf{1}'\mathbf{w})\mathbf{c} \\ (\mathbf{WC} - \mathbf{X}_2)'\mathbf{w} &= (\mathbf{w}'\mathbf{w})\mathbf{c} - (\mathbf{1}'\mathbf{w})\mathbf{b}. \end{aligned}$$

Eliminating $\mathbf{b}$ from the second of the above equations gives:

$$\begin{aligned} (\mathbf{WC} - \mathbf{X}_2)'\mathbf{w} &= \left\{(\mathbf{w}'\mathbf{w}) - \frac{(\mathbf{1}'\mathbf{w})^2}{N}\right\}\mathbf{c} \\ &= (\mathbf{w}'\mathbf{N}\mathbf{w})\mathbf{c} \end{aligned}$$

giving estimates of the translation parameters:

$$\mathbf{b} = \frac{1}{N\lambda}(\mathbf{1}'\mathbf{w})(\mathbf{WC} - \mathbf{X}_2)'\mathbf{w}, \quad \mathbf{c} = \frac{1}{\lambda}(\mathbf{WC} - \mathbf{X}_2)'\mathbf{w}, \tag{8.2}$$

where $\lambda = \mathbf{w}'\mathbf{N}\mathbf{w}$.

The equations (8.2) give the translated matrices:

$$(\mathbf{X}_1 - \mathbf{1a}')\mathbf{T} = \mathbf{X}_1\mathbf{T} - \mathbf{1c}' \quad \text{and} \quad \mathbf{X}_2 - \mathbf{1b}' \tag{8.3}$$

resolving the problem of finding the optimal translation. These adjustments involve $\mathbf{T}$ (because $\mathbf{C} = \mathbf{X}_1\mathbf{T}$), so would seem to complicate the estimation of an optimal value of $\mathbf{T}$. The remainder of this section establishes some results that simplify the estimation of $\mathbf{T}$, independently of how it may be constrained.

Using (8.2) to evaluate $\mathbf{wc}' - \mathbf{1b}'$ gives:

$$\mathbf{wc}' - \mathbf{1b}' = \frac{1}{\lambda}\left[\mathbf{ww}' - \frac{(\mathbf{1}'\mathbf{w})}{N}\mathbf{1w}'\right](\mathbf{WC} - \mathbf{X}_2)$$
$$= \frac{1}{\lambda}\mathbf{Nww}'(\mathbf{WC} - \mathbf{X}_2)$$

which on insertion into (8.1) gives:

$$S = \|\mathbf{L}(\mathbf{WC} - \mathbf{X}_2)\|, \tag{8.4}$$

where $\mathbf{L} = \mathbf{I} - 1/\lambda\mathbf{Nww}'$.

Recalling that $\mathbf{C} = \mathbf{X}_1\mathbf{T}$, it follows from (8.4) that translations can be accounted for by adjusting $\mathbf{WX}_1$ and $\mathbf{X}_2$ to

$$\mathbf{WX}_1^* = \mathbf{LWX}_1, \quad \mathbf{X}_2^* = \mathbf{LX}_2,$$

so that (8.4) becomes:

$$S = \|\mathbf{WX}_1^*\mathbf{T} - \mathbf{X}_2^*\| \tag{8.5}$$

which is the basic Procrustean criterion (3.1) with $\mathbf{X}_1$ replaced by $\mathbf{WX}_1^*$ and $\mathbf{X}_2$ by $\mathbf{X}_2^*$ neither of which involve $\mathbf{T}$. Because $\mathbf{W}$ is assumed to be given, it follows that $\mathbf{T}$ may be estimated subject to any chosen constraints.

The initial choices of origin imply the following results for the new origins:

$$\mathbf{1}'\mathbf{X}_2^* = \mathbf{1}'\mathbf{X}_2 = 0 \quad \text{and} \quad \mathbf{1}'\mathbf{WX}_1^* = \mathbf{1}'\mathbf{WX}_1 = 0$$

and because $\mathbf{w}'\{\mathbf{I} - (1/\lambda)\mathbf{Nww}'\} = 0$, then $\mathbf{w}'\mathbf{X}_2^* = 0$ and $\mathbf{w}'\mathbf{WX}_1^* = 0$.

We now look at the residual sum-of-squares (8.4). First, we expand $\mathbf{L}'\mathbf{L}$ as:

$$\mathbf{L}'\mathbf{L} = \mathbf{I} - \frac{1}{\lambda}\mathbf{Nww}' - \frac{1}{\lambda}\mathbf{ww}'\mathbf{N} + \frac{1}{\lambda^2}\mathbf{ww}'\mathbf{Nww}'$$
$$= \mathbf{I} - \frac{1}{\lambda}\mathbf{ww}' + \frac{(\mathbf{1}'\mathbf{w})}{\lambda N}(\mathbf{1w}' + \mathbf{w1}'). \tag{8.6}$$

From the initial centering we have that $(\mathbf{WC} - \mathbf{X}_2)'(\mathbf{1w}' + \mathbf{w1}')(\mathbf{WC} - \mathbf{X}_2) = 0$ and hence from (8.4) and (8.6):

$$S = \text{trace}(\mathbf{WC} - \mathbf{X}_2)'\left(\mathbf{I} - \frac{1}{\lambda}\mathbf{ww}'\right)(\mathbf{WC} - \mathbf{X}_2)$$
$$= \|\mathbf{WC} - \mathbf{X}_2\| - \frac{1}{\lambda}\mathbf{w}'(\mathbf{WC} - \mathbf{X}_2)(\mathbf{WC} - \mathbf{X}_2)'\mathbf{w}. \tag{8.7}$$

This expression is valid for any setting of $\mathbf{T}$. In particular, when $\mathbf{T}$ is the transformation that minimises $\|\mathbf{WC} - \mathbf{X}_2\|$ when there is no allowance for translational effects; the second term is the reduction in the sum-of-squares given by fitting the best translations. The **T** that minimises (8.1), equivalently (8.5), will give a further reduction in the residual sum-of-squares. Although **T** is straightforward to estimate, we note that (8.7) shows that to estimate **W** for given **T** is more difficult—see Section 8.3.6.

In summary of this section, the optimal **T** is obtained by minimising (8.5), a simple one-sided two-sets Procrustes problem. This optimal setting of **T**, when inserted into (8.3), gives the required translated configurations; these will form the basis of any visualisations of the results. The result (8.7) gives a simple analysis of variance. The matrices $\mathbf{X}_1^*$ and $\mathbf{X}_2^*$ have interesting properties but, so far as we are aware, their only practical significance is that they simplify the calculation of the optimal **T**.

8.1.2 GENERAL FORMS OF WEIGHTING

This section diverges from our usual notation by replacing P_1 by P and P_2 by Q. **W** continues to denote a given weighting matrix that is not confined to being diagonal. The optimal estimation of translational effects is not discussed here, although to some extent they may be allowed for at the outset, by removing the weighted means from the columns of $\mathbf{X}_1$ and the unweighted mean from the columns of $\mathbf{X}_2$.

One of the most general forms of weighting is:

$$\sum_{i=1}^{N}\sum_{j=1}^{Q}\left(w_{ij}\sum_{k=1}^{P}(x_{1ik}t_{kj}) - x_{2ij}\right)^2, \tag{8.8}$$

where x_{1ik} is the i,kth element of $\mathbf{X}_1$. This degree of generality is sometimes useful, as when w_{ij} represents the replication of the elements in the cells of the target matrix $\mathbf{X}_2$. Equation (8.8) may be written:

$$\|\mathbf{W} * (\mathbf{X}_1\mathbf{T}) - \mathbf{X}_2\| \tag{8.9}$$

where $*$ denotes the Hadamard (i.e. element by element) product of two matrices. In (8.9) the weights are applied to the elements of the transformed $\mathbf{X}_1\mathbf{T}$. We could also incorporate weights in the following form:

$$\|(\mathbf{W} * \mathbf{X}_1)\mathbf{T} - \mathbf{X}_2\|. \tag{8.10}$$

where the elements of $\mathbf{X}_1$ are weighted before transformation.

The Hadamard product does not normally commute with the ordinary matrix product. That is:

$$\mathbf{W} * (\mathbf{X}_1\mathbf{T}) \neq (\mathbf{W} * \mathbf{X}_1)\mathbf{T}.$$

An important exception is when **W** is diagonal. Then $(\mathbf{W11}') * \mathbf{A} = \mathbf{WA}$ so that:

$$(\mathbf{W11}') * (\mathbf{X}_1\mathbf{T}) = \mathbf{WX}_1\mathbf{T} = ((\mathbf{W11}') * \mathbf{X}_1)\mathbf{T}$$

showing that (8.9) and (8.10) coincide for diagonal weighting matrices. It is because of this identity, that with the diagonal matrices representing scaling in

Section 8.3, we shall be spared from considering the greater generality that we must now consider in the context of a general weighting matrix. We first note that if $\mathbf{W}$ *is* diagonal then, provided we replace $\mathbf{W}$ by $\mathbf{W11}'$ in (8.9) and (8.10), they both become a special case of (8.30) in Section 8.3; recall that $\mathbf{T}$ may be replaced by $\mathbf{RTS}$ to give the general form of (8.30). Further, the problem of minimising (8.10) introduces nothing new, even when $\mathbf{W}$ is not diagonal, as we only have to replace $\mathbf{X}_1$ by $\mathbf{W} * \mathbf{X}_1$ in (8.18) or, more generally, in (8.19) and proceed as described in Section 8.3. The minimisation of (8.9) for general $\mathbf{W}$ is discussed below, but first we note that another common form of weighting replaces (8.8) by:

$$\sum_{i=1}^{N}\sum_{j=1}^{Q} w_{ij}\left(\sum_{k=1}^{P}(x_{1ik}t_{kj}) - x_{2ij}\right)^2 \tag{8.11}$$

or, in terms of Hadamard products $1'(\mathbf{W} * \{\mathbf{X}_1\mathbf{T} - \mathbf{X}_2\} * \{\mathbf{X}_1\mathbf{T} - \mathbf{X}_2\})1$ which again reduces to the cases considered in Section 8.3. When $w_{ij} = 0$, we must distinguish two very different situations. When $\mathbf{W}$ contains zero values and $\mathbf{X}_1$ and $\mathbf{X}_2$ are complete (8.11) merely gives zero weights to the corresponding terms. When values are missing from $\mathbf{X}_1$ and/or $\mathbf{X}_2$, then $\mathbf{W}$ may be used as an *indicator matrix* that designates the missing cells. An indicator matrix is *not* a weighting matrix; the details of this important special case are covered in Section 8.3. Special instances of (8.11) with diagonal weighting matrices are the minimising criteria:

$$\begin{aligned}&\text{trace}(\mathbf{X}_1\mathbf{T} - \mathbf{X}_2)\mathbf{W}_Q(\mathbf{X}_1\mathbf{T} - \mathbf{X}_2)'\\&\text{trace}(\mathbf{X}_1\mathbf{T} - \mathbf{X}_2)'\mathbf{W}_N(\mathbf{X}_1\mathbf{T} - \mathbf{X}_2)\\&\text{trace}(\mathbf{X}_1\mathbf{T} - \mathbf{X}_2)\mathbf{W}_Q(\mathbf{X}_1\mathbf{T} - \mathbf{X}_2)'\mathbf{W}_N\end{aligned}$$

where $\mathbf{W}_Q$ weights the columns of $\mathbf{X}_2$ and $\mathbf{X}_1\mathbf{T}$. Similarly, $\mathbf{W}_N$ weights the rows, while the third criterion applies 'product weights' $\mathbf{W}_{Ni}\mathbf{W}_{Qj}$ to the (i, j)th cells. The most important of these is $\mathbf{W}_N$ which is useful for expressing replications among the N cases. When the entries of $\mathbf{W}_N$ are integers w_i, the effect is the same as writing out the ith row of $\mathbf{X}_1$ and $\mathbf{X}_2$ w_i times. This is all very satisfactory and only leaves to be considered the minimisation of (8.9), or equivalently (8.8), for general $\mathbf{W}$.

We may minimise (8.8) for each value of j separately. In matrix form, the jth term may be written:

$$\|\mathbf{W}_j\mathbf{X}_1\mathbf{t}_j - (\mathbf{X}_2)_j\| \tag{8.12}$$

where $\mathbf{W}_j$ is the diagonal matrix formed from the jth column of $\mathbf{W}$ and $\mathbf{t}_j$ is the jth column of $\mathbf{T}$ and $(\mathbf{X}_2)_j$ is the jth column of $\mathbf{X}_2$. The minimum occurs when:

$$\mathbf{t}_j = (\mathbf{X}_1'\mathbf{W}_j^2\mathbf{X}_1)^{-1}(\mathbf{X}_1'\mathbf{W}_j(\mathbf{X}_2)_j) \tag{8.13}$$

and so to $\mathbf{T} = (\mathbf{t}_1, \mathbf{t}_2, \ldots, \mathbf{t}_Q)$. This solves the problem for general (unconstrained) $\mathbf{T}$ but each constrained form of $\mathbf{T}$ requires special consideration. With the constraint $\mathbf{t}_j'\mathbf{t}_j = 1$, $j = 1, 2, \ldots, Q$ we have the oblique Procrustes problem (Chapter 6) and if $\mathbf{T}$ is to be orthogonal we have to handle the additional constraints of the orthogonality of the columns.

8.2 Missing values

In this section we have a first look at the problems of handling missing values in two-sets Procrustes analysis; Chapter 9 discusses missing values in K sets. Thus, we consider the problem of minimising:

$$\|\mathbf{X}_1\mathbf{T} - \mathbf{X}_2\| \tag{8.14}$$

where $\mathbf{X}_1, \mathbf{X}_2$, or both have values missing in certain cells indicated by binary indicator matrices $\mathbf{W}_1$ and $\mathbf{W}_2$. Here $\mathbf{W}_i$ is zero except where the corresponding cell in $\mathbf{X}_i$ is missing, indicated by a unit.

First, suppose $\mathbf{X}_2$ has missing values, so we have to minimise:

$$\|\mathbf{W}_2 * (\mathbf{X}_1\mathbf{T} - \mathbf{X}_2)\|. \tag{8.15}$$

If the missing cells of $\mathbf{X}_2$ are filled in with putative values, we may minimise (8.14) and reduce the residual sum-of-squares by replacing the missing cells by $\mathbf{W}_2 * (\mathbf{X}_1\mathbf{T})$. When translation terms are to be eliminated, and $\mathbf{X}_1$ and $\mathbf{X}_2$ have been centred as described in Chapter 3, the updated version of $\mathbf{X}_2$ will need to be recentred. The process may be repeated until convergence. At convergence (8.15) will have been minimised and estimates of the missing values are available in $\mathbf{X}_2$.

Now, suppose $\mathbf{X}_1$ has missing values, so we have to minimise:

$$\|\mathbf{W}_1 * (\mathbf{X}_1\mathbf{T} - \mathbf{X}_2)\| \tag{8.16}$$

When $\mathbf{T}$ is an orthogonal matrix $\mathbf{Q}$, noting that $\|\mathbf{W}_1 * (\mathbf{X}_1\mathbf{Q} - \mathbf{X}_2)\| = \|\mathbf{W}_1 * (\mathbf{X}_2\mathbf{Q}' - \mathbf{X}_1)\|$, we may proceed as above by reversing the roles of $\mathbf{X}_1$ and $\mathbf{X}_2$. However, when $\mathbf{T}$ is not orthogonal, this procedure is not available. As before, we may fill in the missing cells of $\mathbf{X}_1$ with putative values and minimise (8.14). Then we must solve $\mathbf{W}_1 * (\mathbf{X}_1\mathbf{T}) = \mathbf{W}_1 * (\mathbf{X}_2)$ which gives a set of linear equations for the missing cells of $\mathbf{X}_1$. These equations may be solved row by row. Write $\mathbf{x}_{1i}$ for the m, say, values in the missing cells of the ith row of $\mathbf{X}_1$ and $\mathbf{x}_{2i}$ for the corresponding (non-missing) values in the cells of $\mathbf{X}_2$ and $\mathbf{T}_m$ for the m rows of $\mathbf{T}$ corresponding to the m columns in the ith row $\mathbf{X}_1$ that are missing. Then the missing values in the ith row of $\mathbf{X}_1$ satisfy the overdetermined linear equations:

$$\mathbf{x}_{1i}\mathbf{T}_m = \mathbf{x}_{2i}$$

with the least-squares estimate of $\mathbf{x}_{1i}$:

$$\mathbf{x}_{1i} = \mathbf{x}_{2i}\mathbf{T}_m'(\mathbf{T}_m\mathbf{T}_m')^{-1} \quad i = 1, 2, \ldots, P. \tag{8.17}$$

Being a least-squares estimate, substituting these estimates in $\mathbf{X}_1$ will reduce the residual sum-of-squares (8.14). Note that when $\mathbf{T} = \mathbf{Q}$, then the rows of $\mathbf{T}_m$ are orthonormal and then (8.17) reduces to $\mathbf{x}_{1i} = \mathbf{x}_{2i}\mathbf{Q}_m'$ agreeing with the procedure recommended above for handling orthogonal matrices. When translation terms are to be eliminated, and $\mathbf{X}_1$ and $\mathbf{X}_2$ have been centred as described in Chapter 3, the updated version of $\mathbf{X}_1$ will need to be recentred. The process may be repeated until convergence. At convergence (8.17) will have been minimised and estimates of the missing values are available in $\mathbf{X}_1$.

Finally, when $\mathbf{X}_1$ and $\mathbf{X}_2$ both have missing values, the procedures using $\mathbf{W}_1$ and $\mathbf{W}_2$ may be alternated. Even when $\mathbf{X}_1$ and $\mathbf{X}_2$ share one or more missing cells the process should converge satisfactorily.

8.3 Anisotropic scaling

Isotropic scaling was discussed at the beginning of Chapter 3 with the general solution (3.9). Anisotropic scaling comes in two forms, encapsulated in the minimisation of the following criteria:

$$\|\mathbf{X}_1\mathbf{RT} - \mathbf{X}_2\| \quad \text{and} \quad \|\mathbf{X}_1\mathbf{TS} - \mathbf{X}_2\|. \tag{8.18}$$

Isotropic scaling is given by a scalar s so it is immaterial on which side of $\mathbf{T}$ this is put. With anisotropic scaling there is a difference as to whether the scaling occurs before or after the transformation. Indeed, both scalings could be applied simultaneously by minimising:

$$\|\mathbf{X}_1\mathbf{RTS} - \mathbf{X}_2\|. \tag{8.19}$$

When the columns of $\mathbf{X}_1$ refer to variables, then (8.18) refers to their scaling. Usually the simple pre-scaling devices discussed in Chapter 2 will suffice. When the columns refer to the dimensions of a configuration, then the ambiguity of matching dimensions offers less justification for scaling. However, if one has confidence in the reality of reified dimensions, say deriving from factor analysis or PCA, then the possibilities of scaling might be deemed worthy of investigation.

Rather than treating $\mathbf{R}$ and $\mathbf{S}$ as scaling matrices we might regard $\mathbf{RTS}$ as a new form of constrained transformation matrix where $\mathbf{R}$, $\mathbf{T}$, and $\mathbf{S}$ have specified matrix types. For example, we might have $\mathbf{T} = \mathbf{Q}$, orthogonal, or $\mathbf{T} = \mathbf{P}$, orthonormal, and provided due note is taken of matrix dimensionalities, $\mathbf{RTS}$ then represents a more general type of transformation. Care has to be taken not to be too general with our specifications of $\mathbf{R}$ and $\mathbf{S}$. For example, the matrix $\mathbf{QS}$, where $\mathbf{Q}$ is orthogonal and $\mathbf{S}$ is symmetric, is no constraint at all, because any real square matrix can be written in this form; similarly, $\mathbf{R}$ and $\mathbf{S}$ cannot be general square matrices. Hence, in the following we confine our attention to when $\mathbf{R}$ and $\mathbf{S}$ are diagonal scaling matrices.

8.3.1 PRE-SCALING R

In minimising $\|\mathbf{X}_1\mathbf{RT} - \mathbf{X}_2\|$ for given $\mathbf{R}$, the estimation of $\mathbf{T}$ proceeds normally (replacing $\mathbf{X}_1$ by $\mathbf{X}_1\mathbf{R}$) for whatever constraints may be imposed on $\mathbf{T}$. For fixed $\mathbf{T}$, the estimation of $\mathbf{R}$ is given by the minimum of trace$(\mathbf{RX}_1'\mathbf{X}_1\mathbf{RTT}' - 2\mathbf{X}_2'\mathbf{X}_1\mathbf{RT})$, differentiation of which with respect to diagonal $\mathbf{R}$ gives:

$$\text{diag}(\mathbf{X}_1'\mathbf{X}_1\mathbf{RTT}') = \text{diag}(\mathbf{TX}_2'\mathbf{X}_1). \tag{8.20}$$

This is linear in the elements of **R** so may be solved directly. However (8.20) may be presented in the following form that is more convenient for computation:

$$[(\mathbf{X}_1'\mathbf{X}_1) * (\mathbf{T}\mathbf{T}')]\mathbf{r} = \mathrm{diag}(\mathbf{T}\mathbf{X}_2'\mathbf{X}_1)\mathbf{1}, \tag{8.21}$$

where $\mathbf{r} = \mathbf{R1}$ the vector form of the diagonal matrix **R**. The solution of the linear system (8.21) requires only the inverse of the symmetric Hadamard product appearing on the left-hand-side.

8.3.2 POST-SCALING S

In minimising $\|\mathbf{X}_1\mathbf{TS} - \mathbf{X}_2\|$ for given **T**, the estimation of **S** is given by the minimum of $\mathrm{trace}(\mathbf{ST}'\mathbf{X}_1'\mathbf{X}_1\mathbf{TS} - 2\mathbf{ST}'\mathbf{X}_1'\mathbf{X}_2)$, differentiation of which with respect to diagonal **S** gives:

$$\mathrm{diag}(\mathbf{T}'\mathbf{X}_1'\mathbf{X}_1\mathbf{T})\mathbf{S} = \mathrm{diag}(\mathbf{T}'\mathbf{X}_1'\mathbf{X}_2) \tag{8.22}$$

with its immediate solution. Writing (8.22) as $\mathbf{FS} = \mathbf{G}$, we have that $s_j = g_j/f_j$ for $j = 1, 2, \ldots, P_2$.

Comparing with the estimate (3.10) s of isotropic scaling, we see that:

$$s = \sum_{j=1}^{P_2} g_j \Big/ \sum_{j=1}^{P_2} f_j$$

which may be written as a weighted average of the scaling parameters of the individual dimensions:

$$s = \sum_{j=1}^{P_2} f_j s_j \Big/ \sum_{j=1}^{P_2} f_j \tag{8.23}$$

The denominator of (8.23) is merely $\mathrm{trace}(\mathbf{T}'\mathbf{X}_1'\mathbf{X}_1\mathbf{T})$ that appears in the denominator of (3.10); which often will be scaled to be P_1, further simplifying (8.23).

8.3.3 ESTIMATION OF T WITH POST-SCALING S

The estimation of **T** for fixed **S** is more of a problem than it was for fixed **R** (Section 8.3.1) and has to be treated specially for each type of transformation. When **T** is unconstrained, we have:

$$\|\mathbf{X}_1\mathbf{TS} - \mathbf{X}_2\| = \mathrm{trace}(\mathbf{ST}'\mathbf{X}_1'\mathbf{X}_1\mathbf{TS} - 2\mathbf{ST}'\mathbf{X}_1'\mathbf{X}_2 + \mathbf{X}_2'\mathbf{X}_2) \tag{8.24}$$

and differentiation with respect to **T** yields:

$$\mathbf{X}_1'\mathbf{X}_1\mathbf{TS}^2 = \mathbf{X}_1'\mathbf{X}_2\mathbf{S}$$

so that:

$$\mathbf{T} = (\mathbf{X}_1'\mathbf{X}_1)^{-1}\mathbf{X}_1'\mathbf{X}_2\mathbf{S}^{-1} \tag{8.25}$$

which is a minor variant of the estimate (3.10) where there is no scaling. However, when $\mathbf{T}$ is constrained, the occurrence of $\mathbf{S}$ introduces difficulties that do not occur when there is no scaling. Considering the case where $\mathbf{T} = \mathbf{Q}$, orthogonal, then differentiating (8.24) with respect to $\mathbf{Q}$ gives:

$$\mathbf{X}_1'\mathbf{X}_1\mathbf{Q}\mathbf{S}^2 - \mathbf{X}_1'\mathbf{X}_2\mathbf{S} = \mathbf{Q}\mathbf{\Lambda} \tag{8.26}$$

where $\mathbf{\Lambda}$ is a symmetric matrix of Lagrange multipliers imposing the orthogonality and normalisation constraints among the columns of $\mathbf{Q}$. Premultiplying by $\mathbf{Q}'$ gives:

$$\mathbf{Q}'\mathbf{X}_1'\mathbf{X}_1\mathbf{Q}\mathbf{S} - \mathbf{Q}'\mathbf{X}_1'\mathbf{X}_2\mathbf{S} = \mathbf{\Lambda}. \tag{8.27}$$

When $\mathbf{S} = \mathbf{I}$, the scaling vanishes and we should have the previously found solution to the orthogonal Procrustes problem again. This is indeed so, because then in (8.27) $\mathbf{Q}'\mathbf{X}_1'\mathbf{X}_1\mathbf{Q}$ and $\mathbf{\Lambda}$ are both symmetric and so therefore is $\mathbf{Q}'\mathbf{X}_1'\mathbf{X}_2$; the result then follows immediately from the necessary and sufficient conditions given in Section 4.2.

When $\mathbf{S}$ is any diagonal matrix, there seems to be no algebraic solution to (8.27). Lissitz, Schönemann, and Lingoes (1976) circumvent the problem by replacing the orthogonality constraint by a constraint requiring $\mathbf{Q}'\mathbf{S}\mathbf{Q} = \mathbf{I}$. One might try to use (8.27) as the basis for an algorithm which progressively updates a current value of $\mathbf{Q}$ until (8.27) becomes symmetric. We know of no algorithm of this kind. A different algorithmic approach has been suggested by Koschat and Swayne (1991) and was previously described in Section 5.4.1. Following the argument of Section 5.4.1 we write

$$\|\mathbf{X}_1\mathbf{T}\mathbf{S} - \mathbf{X}_2\| = \left\| \begin{pmatrix} \mathbf{X}_1\mathbf{T} \\ \mathbf{X}_0\mathbf{T} \end{pmatrix} \mathbf{S} - \begin{pmatrix} \mathbf{X}_2 \\ \mathbf{X}_1\mathbf{T}\mathbf{S} \end{pmatrix} \right\| \tag{8.28}$$

and choose $\mathbf{X}_0$ such that $\mathbf{X}_1'\mathbf{X}_1 + \mathbf{X}_0'\mathbf{X}_0 = \rho^2\mathbf{I}$ so that $\mathbf{S}'\mathbf{T}'(\mathbf{X}_1'\mathbf{X}_1 + \mathbf{X}_0'\mathbf{X}_0)\mathbf{T}\mathbf{S} = \rho^2\mathbf{S}'\mathbf{S}$ whenever $\mathbf{T}'\mathbf{T} = \mathbf{I}$. Thus when $\mathbf{T} = \mathbf{Q}$, orthogonal, or $\mathbf{T} = \mathbf{P}$, orthonormal, the quadratic term of (8.28) is independent of $\mathbf{T}$. Then, in the $(i+1)$th iterative step of the Koschat and Swayne algorithm for estimating $\mathbf{T}$, we have only to maximise trace $\mathbf{S}(\mathbf{X}_2'\mathbf{X}_1 + \mathbf{S}'\mathbf{T}_i'\mathbf{X}_0'\mathbf{X}_0)\mathbf{T}_{i+1}'$ where $\mathbf{T}_i$ is regarded as fixed. Thus, $\mathbf{T}_{i+1}' = \mathbf{V}\mathbf{U}'$ (Cliff 1966, see Chapter 4) derives from the SVD:

$$\mathbf{U}\mathbf{\Sigma}\mathbf{V}' = \mathbf{S}(\mathbf{X}_2'\mathbf{X}_1 + \mathbf{S}'\mathbf{T}_i'\mathbf{X}_0'\mathbf{X}_0) = \mathbf{S}(\mathbf{X}_2'\mathbf{X}_1 + \mathbf{S}'\mathbf{T}_i'(\rho^2\mathbf{I} - \mathbf{X}_1'\mathbf{X}_1)).$$

As in Section 5.4 we have ρ^2 at our disposal and typically may choose $\rho^2 = \|\mathbf{X}_1\|$ or $\rho^2 = \lambda$, the maximal eigenvalue of $\mathbf{X}_1'\mathbf{X}_1$. Thus we iterate on $\mathbf{T}_i$ until convergence; proof of convergence is given in Section 5.4.

In deriving their method, Koschat and Swayne (1991) considered this problem in the slightly different form of minimising:

$$\text{trace}(\mathbf{X}_1\mathbf{T} - \mathbf{X}_2)\mathbf{S}^2(\mathbf{X}_1\mathbf{T} - \mathbf{X}_2)' = \|(\mathbf{X}_1\mathbf{T}\mathbf{S} - \mathbf{X}_2\mathbf{S}\|, \tag{8.29}$$

which becomes (8.28) merely by replacing $\mathbf{X}_2\mathbf{S}$ by $\mathbf{X}_2$.

Algorithm 8.1

Step 1 Initialise $\mathbf{T}$ from the unweighted case or by random choice set $\rho^2 = \|\mathbf{X}_1\|$ or λ.

Step 2 Form $\mathbf{T}_1 = \mathbf{T}$ and $\mathbf{Z} = \mathbf{S}(\mathbf{X}_2'\mathbf{X}_1 + \mathbf{S}'\mathbf{T}_i'(\rho^2\mathbf{I} - \mathbf{X}_1'\mathbf{X}_1))$.

Step 3 Form SVD $\mathbf{Z} = \mathbf{U\Sigma V}'$ and set $\mathbf{T} = \mathbf{VU}'$.

Step 4 If $|\|\mathbf{T}_i\| - \|\mathbf{T}_i\||$ is greater than a threshold go to *Step 2*.

Step 5 Evaluate $\mathbf{X}_1\mathbf{TS}$ and $\|\mathbf{X}_1\mathbf{TS} - \mathbf{X}_2\|$.

The Koschat and Swayne algorithm allows us to estimate $\mathbf{T}$ in the presence of scaling $\mathbf{S}$ when $\mathbf{T}$ is orthogonal or orthonormal. When $\mathbf{T}'\mathbf{T} \neq \mathbf{I}$ the algorithm breaks down.

We note that (8.29) gives a link to the oblique axes problems discussed in Chapter 6. In (8.29) if we set (i) $\mathbf{T} = \mathbf{C}, \mathbf{S}^2 = (\mathbf{C}'\mathbf{C})^{-1}$ we recover (6.3) and if we set (ii) $\mathbf{T} = (\mathbf{C}')^{-1}, \mathbf{S}^2 = \mathbf{C}'\mathbf{C}$ we recover (6.4). Now, however, $\mathbf{S}$ is a function of $\mathbf{T}$ and special algorithms discussed in Chapter 6 are necessary.

8.3.4 SIMULTANEOUS ESTIMATION OF R, S, AND T

Minimising $\|\mathbf{X}_1\mathbf{RTS} - \mathbf{X}_2\|$ introduces no new problems. For fixed $\mathbf{T}$ and $\mathbf{S}$, $\mathbf{R}$ is estimated as in 8.3.1 for fixed $\mathbf{T}$ and $\mathbf{R}$, $\mathbf{S}$ is estimated as in 8.3.2. For fixed $\mathbf{R}$ and $\mathbf{S}$, the problems of estimating $\mathbf{T}$, under its associated constraints, are precisely as for estimating $\mathbf{T}$ when minimising $\|\mathbf{X}_1\mathbf{TS} - \mathbf{X}_2\|$, merely replacing $\mathbf{X}_1$ by $\mathbf{X}_1\mathbf{R}$ and using 8.3.3.

8.3.5 SCALING WITH TWO-SIDED PROBLEMS

Minimising the two-sided criterion $\|\mathbf{X}_1\mathbf{R}_1\mathbf{T}_1\mathbf{S}_1 - \mathbf{X}_2\mathbf{R}_2\mathbf{T}_2\mathbf{S}_2\|$ also introduces no new problems. For fixed $\mathbf{R}_2$, $\mathbf{T}_2$, and $\mathbf{S}_2$ one proceeds as in Section 8.3.4 and similarly for fixed $\mathbf{R}_1$, $\mathbf{T}_1$, and $\mathbf{S}_1$. Hence, the obvious alternating least-squares algorithm.

8.3.6 ROW SCALING

Clearly, scaling could also be attached to the rows of $\mathbf{X}$. Recall that in this case, translational effects cannot be handled by simple centring but need the methods of Section 8.1.1. For the moment we consider the case where the columns of $\mathbf{X}_1$ may or may not be centred and return to estimation of translation at the end of this section. We examine the problem of minimising

$$\|\mathbf{WX}_1\mathbf{RTS} - \mathbf{X}_2\| \tag{8.30}$$

for diagonal $\mathbf{W}$†. We have included the most general form on the right-hand side of $\mathbf{X}_1$ because this does not introduce any new difficulties. Given $\mathbf{W}$, the estimation

† Note that, here $\mathbf{W}$ represents a scaling factor to be estimated and not given weights.

of **R**, **T**, and **S** proceeds as in Section 8.3.4 with $\mathbf{X}_1$ replaced by $\mathbf{WX}_1$. Given **R**, **T**, and **S** and writing $\mathbf{Z} = \mathbf{X}_1\mathbf{RTS}$, the estimation of **W**, requires the minimisation of:

$$\|\mathbf{WZ} - \mathbf{X}_2\|$$

which gives the estimate, effectively the transpose of (8.22):

$$\mathbf{W} = (\text{diag}(\mathbf{ZZ}'))^{-1}\text{diag}(\mathbf{ZX}_2'), \tag{8.31}$$

showing that scaling the rows of $\mathbf{X}_1$ causes few problems. If, however we scale the rows of both $\mathbf{X}_1$ and $\mathbf{X}_2$, things need more careful consideration. The problem is to minimise

$$\|\mathbf{W}(\mathbf{X}_1\mathbf{RTS} - \mathbf{X}_2)\| = \text{trace}(\mathbf{X}_1\mathbf{RTS} - \mathbf{X}_2)'\mathbf{W}^2(\mathbf{X}_1\mathbf{RTS} - \mathbf{X}_2).$$

Given **W**, the estimation of **R**, **T**, and **S** proceeds as above but given **R**, **T**, and **S** the minimum occurs when $\mathbf{W} = \mathbf{0}$. The problem is therefore vacuous unless we put some constraint on permissible values of **W**. Assuming some constraint has been agreed, we note that the minimisation requires the minimum of $\sum_{i=1}^{N} r_i^2 w_i^2$, where r_i^2 denotes the sum of squares of the elements of the ith row of $\mathbf{X}_1\mathbf{RTS} - \mathbf{X}_2$, showing that for most normal constraints we can handle things one row at a time. Thus, if we set the total scaling to be unity, then $\mathbf{1}'\mathbf{W}\mathbf{1} = 1$ with which we may associate a Lagrange multiplier λ. Then, differentiation with respect to w_i gives $w_i r_i^2 = \lambda$ for $i = 1, 2, \ldots, N$. Applying the constraint gives $\lambda = \left(\sum_{i=1}^{N}(1/(r_i^2))^{-1}\right)$ so that:

$$w_i = \frac{1}{r_i^2} \Big/ \sum_{i=1}^{N} \frac{1}{r_i^2}. \tag{8.32}$$

If, however, we require that the sum-of-squares of the scalings is unity, then the constraint is $\mathbf{1}'\mathbf{W}^2\mathbf{1} = 1$. This constraint is linear in w_i^2 as is the criterion to be minimised. Minimisation of $\sum_{i=1}^{N} r_i^2 w_i^2$ subject to $\sum_{i=1}^{N} w_i^2 = 1$ is a linear programming problem with the rather trivial solution:

$$w_i = 1 \quad \text{for } \max(r_i^2) \quad \text{else } w_i = 0. \tag{8.33}$$

Row scaling with translation If we are concerned with eliminating translational effects in $\mathbf{X}_1$, with row scaling we have to use the method of Section 3.1.1. The problem is that the weights are unknown and have to be computed iteratively. For given **W** the translations and transformation parameters can be calculated using the results of Section 3.1.1. Then, in principle, new values of **W** could be computed by minimising (8.7) for diagonal **W** but a direct solution is not known. It seems more straightforward, though possibly less efficient, to use (8.31) or (8.32). Thus, the current estimate of $\mathbf{W}$ induces an update to the centring of $\mathbf{X}_1$. The whole process then iterates on $\mathbf{W}$ until convergence as shown in Algorithm 8.2.

Algorithm 8.2

Step 1 Initialise $\mathbf{W}$ (e.g. $\mathbf{W} = \mathbf{I}$).
Step 2 Centre $\mathbf{X}_1(\mathbf{1}'\mathbf{W}\mathbf{X}_1 = 0)$ and $\mathbf{X}_2(\mathbf{1}'\mathbf{X}_2 = 0)$. Calculate initial residual sum-of-squares S.
Step 3 Set $S_0 = S$. Solve (8.5), that is, minimise $S = \|\mathbf{W}\mathbf{X}_1^*\mathbf{T} - \mathbf{X}_2^*\|$ over $\mathbf{T}$.
Step 4 Estimate a new $\mathbf{W}$ (e.g. by using (8.31) or (8.32) or minimising (8.7)).
Step 5 If $S_0 - S > \varepsilon$ return to *Step 3*.
Step 6 Calculate translation parameters $\mathbf{b}$ and $\mathbf{c}$ from (8.2). Calculate $\mathbf{X}_1\mathbf{T}$ and Anova (8.7).

To sum up Section 8.3, we have seen that there are many possibilities for incorporating scaling. There are three types of scaling on the right and two on the left of $\mathbf{X}_1$, not to mention the myriad possibilities for choosing normalisations. This is rather a bewildering range of possibilities. For most practical purposes we have seen that scaling matrices are confined to diagonal form.

9
Generalised Procrustes problems

Chapters 3–8 have described the simplest forms of Procrustes problems with just two sets of configurations; now the generalisation to K sets is discussed. Similar, but non-Procrustean, methods are briefly reviewed in Chapter 13. Unless otherwise stated, it is assumed that all translations are taken care of by centring the configurations at their centroids and that any desired initial scaling (see Chapter 2) has been completed. Thus, it is assumed in the following that all matrices $\mathbf{X}_k$, for $k = 1, \ldots, K$ are centred at the origin. When the matrices are derived from any form of multidimensional scaling then the positions of their centroids are arbitrary, so then the assumption imposes no loss of generality. When, however, the matrices are data-matrices with the *same* variables then important information is likely to lie in differences between their means, which may be analysed by linear models and summarised by a MANOVA (see, for example, Hand and Taylor 1987). Note that even variables bearing the same name may differ, as is sometimes the case when two assessors give the same name to different sensory attributes. This may justify using Procrustes methods in what otherwise would be a straightforward MANOVA context. However, the classical context for Procrustes methods is when the different configurations arise either from innately *different* variables, for example, the free choice profiling of sensory experimentation (Williams and Langron 1984, Arnold and Williams 1986), or from the unmatched dimensions of different multidimensional scalings. Procrustes methods analyse for any information there may be in differences among configurations relative to their different centroids. This information is additional to, rather than a substitute for, any information there may be in differences between means.

It is assumed further that all configurations are embedded within the space of the configuration of highest dimension. Thus for two sets, it is assumed that if $P_2 < P_1$ then $\mathbf{X}_2$ is embedded in the same space as $\mathbf{X}_1$ by adding $P_1 - P_2$ columns of zeros to $\mathbf{X}_2$. This assumption imposes no real restriction, because it should be evident that better fits cannot be obtained by allowing all or part of the spaces occupied by the smaller configurations to be orthogonal to the largest space. A simple proof of this assertion follows from writing $\mathbf{X}_1 = (\mathbf{X}_{11}, \mathbf{X}_{12}, \mathbf{0})$ and $\mathbf{X}_2 = (\mathbf{X}_{21}, \mathbf{0}, \mathbf{X}_{22})$ where $\mathbf{X}_{11}$ and $\mathbf{X}_{21}$ have the same number of columns and $\mathbf{X}_{22}$ has fewer columns than $\mathbf{X}_{12}$. Then:

$$\|\mathbf{X}_1 - \mathbf{X}_2\| = \|\mathbf{X}_{11} - \mathbf{X}_{21}\| + \|\mathbf{X}_{12}\| + \|\mathbf{X}_{22}\|.$$

Embedding $\mathbf{X}_{22}$ (if necessary adding extra zero columns) into the same space as $\mathbf{X}_{12}$ gives $\mathbf{X}_1 = (\mathbf{X}_{11}, \mathbf{X}_{12})$ and $\mathbf{X}_2 = (\mathbf{X}_{21}, \mathbf{X}_{22})$. This operation preserves the configurations of $\mathbf{X}_1$ and $\mathbf{X}_2$ and gives:

$$\|\mathbf{X}_1 - \mathbf{X}_2\| = \|\mathbf{X}_{11} - \mathbf{X}_{21}\| + \|\mathbf{X}_{12}\| + \|\mathbf{X}_{22}\| - 2\text{trace}(\mathbf{X}_{12}'\mathbf{X}_{22}).$$

If the trace on the right-hand-side is positive, then the embedded configuration has smaller sum-of-squares; if the trace is negative, replace $\mathbf{X}_{22}$ by $-\mathbf{X}_{22}$, which again preserves the geometry of the configurations and reduces the sum-of-squares. In either case, the configurations embedded in the smaller space have smaller sum-of-squares than the unembedded configurations.

This chapter is organised into three main sections. The first is concerned with results that are applicable to all forms of transformation. The second discusses some of the more important special cases that use orthogonal and orthonormal matrices. The final main section discusses generalised pairwise Procrustes methods.

9.1 Results applicable to any transformation

9.1.1 GENERALISED PROCRUSTES CRITERIA AND SOME BASIC IDENTITIES

So far, we have considered Procrustes problems which match two matrices $\mathbf{X}_1$ and $\mathbf{X}_2$ using criteria mostly of the form $\|\mathbf{X}_1\mathbf{T}_1 - \mathbf{X}_2\mathbf{T}_2\|$. With K sets, the natural form of the norm to be minimised is

$$\sum_{i<j}^{K} \|\mathbf{X}_i\mathbf{T}_i - \mathbf{X}_j\mathbf{T}_j\| \tag{9.1}$$

where by implication all $\mathbf{T}_k$, $k = 1, 2, \ldots, K$ must have the same number of columns, R, say; often $R = \min(P_1, P_2, \ldots, P_K)$. It is essential that the matrices $\mathbf{T}_k$, are constrained in some way, else we have the trivial solution $\mathbf{T}_k = \mathbf{0}$, $k = 1, 2, \ldots, K$. We consider the precise form of constraints that might be adopted in Section 9.2 and in this section we merely take it as understood that appropriate constraints are in place.

The identity:

$$\sum_{i<j}^{K} \|\mathbf{X}_i\mathbf{T}_i - \mathbf{X}_j\mathbf{T}_j\| = K\sum_{k=1}^{K} \|\mathbf{X}_k\mathbf{T}_k - \mathbf{G}\| \tag{9.2}$$

where

$$\mathbf{G} = K^{-1}\sum_{k=1}^{K}(\mathbf{X}_k\mathbf{T}_k) \tag{9.3}$$

is useful. It provides a basis for algorithms and shows that the sum of the pairwise Procrustes solution may be expressed in terms of deviations from a single matrix $\mathbf{G}$. $\mathbf{G}$ is the average configuration of the initial matrices $\mathbf{X}_1, \mathbf{X}_2, \ldots, \mathbf{X}_K$ after transformation by $\mathbf{T}_1, \mathbf{T}_2, \ldots, \mathbf{T}_K$ and is known as the *group average* configuration. The group average as defined by (9.3) plays a central part in most forms of generalised Procrustes analysis. We may write (9.2) as the analysis of variance:

$$\sum_{k=1}^{K} \|\mathbf{X}_k\mathbf{T}_k\| = \sum_{k=1}^{K} \|\mathbf{X}_k\mathbf{T}_k - \mathbf{G}\| + K\|\mathbf{G}\| \tag{9.4}$$

or

$$T = S + G \tag{9.5}$$

where T is 'the total sum-of-squares', S is the 'residual sum-of-squares', and G represents the 'sum-of-squares attributable to the group average'.

Define
$$\mathbf{G}_k = \frac{1}{K-1}\sum_{i\neq k}^{K}(\mathbf{X}_i\mathbf{T}_i). \tag{9.6}$$

That is, $\mathbf{G}_k$ is the average configuration excluding the kth. We term $\mathbf{G}_k$ the k-excluded group average. Directly from this definition, it follows that:

$$K\mathbf{G} = \mathbf{X}_k\mathbf{T}_k + (K-1)\mathbf{G}_k \tag{9.7}$$

and $K(\mathbf{G} - \mathbf{G}_k) = \mathbf{X}_i\mathbf{T}_k - \mathbf{G}_k$. Using this result gives:

$$\begin{aligned}\sum_{k=1}^{K} \|\mathbf{X}_k\mathbf{T}_k - \mathbf{G}\| &= \sum_{k=1}^{K} \|(\mathbf{X}_k\mathbf{T}_k - \mathbf{G}_k) + (\mathbf{G}_k - \mathbf{G})\| \\ &= \sum_{k=1}^{K} \|(\mathbf{X}_k\mathbf{T}_k - \mathbf{G}_k) - \frac{1}{K}(\mathbf{X}_k\mathbf{T}_k - \mathbf{G}_k)\| \\ &= \left(\frac{K-1}{K}\right)^2 \sum_{k=1}^{K} \|\mathbf{X}_k\mathbf{T}_k - \mathbf{G}_k\|. \end{aligned} \tag{9.8}$$

A useful property of $\mathbf{G}_k$ is that it is independent of $\mathbf{X}_k$ and $\mathbf{T}_k$.

We have stated that the natural extension of two-sets Procrustes analysis is to minimise (9.1), which from (9.2) is the same as minimising S. Alternatively, we may wish to maximise G. When the matrices $\mathbf{T}_k$ are orthogonal, the total sum-of-squares $T = \sum_{k=1}^{K} \|\mathbf{X}_k\mathbf{Q}_k\| = \sum_{k=1}^{K} \|\mathbf{X}_k\|$ is not a function of the transformations. Then, minimising S and maximising G are equivalent; with other transformations this equivalence vanishes. This is seen clearly from the normal equations, which we now derive.

(i) Minimise S

Differentiating (9.1) with respect to $\mathbf{T}_k$ gives:

$$(\mathbf{X}_k'\mathbf{X}_k)\mathbf{T}_k - \mathbf{X}_k'\mathbf{G}_k = \mathbf{\Lambda}_T \tag{9.9}$$

where $\mathbf{\Lambda}_T$ is a Lagrangian term expressing the chosen constraint on $\mathbf{T}_k$.

(ii) Maximise G

Differentiating $\|\mathbf{G}\|$ with respect to $\mathbf{T}_k$, using (9.7) and noting that $\mathbf{G}_k$ is not a function of $\mathbf{T}_k$, gives:

$$(\mathbf{X}_k'\mathbf{X}_k)\mathbf{T}_k + (K-1)\mathbf{X}_k'\mathbf{G}_k = \mathbf{\Lambda}_T. \tag{9.10}$$

When $\mathbf{T}_k$ is orthogonal, $\mathbf{\Lambda}_T = \mathbf{T}_k\mathbf{\Lambda}$ where $\mathbf{\Lambda}$ is a symmetric matrix. Then, pre-multiplication of (9.9) and (9.10), by $\mathbf{T}_k'$ shows that $\mathbf{T}_k'(\mathbf{X}_k'\mathbf{G}_k)$ must be symmetric in both cases and by the result of Section 4.2, $\mathbf{T}_k$ derives from the singular value decomposition of $(\mathbf{X}_k'\mathbf{G}_k)$, giving the same result. However, in general, equations (9.9) and (9.10) are clearly different. We shall return to these results below when discussing algorithms.

The matrix $\mathbf{S}$ defined by $S_{ij} = \text{trace}(\mathbf{T}_i'\mathbf{X}_i'\mathbf{X}_j\mathbf{T}_j)$ appears frequently in the following. The different sums-of-squares appearing above may be expressed in terms of $\mathbf{S}$. For example, the equivalent expressions (9.4) and (9.5) may be written in yet a third way as:

$$\mathbf{1}'(\text{diag}\mathbf{S})\mathbf{1} = \left[\mathbf{1}'(\text{diag}\mathbf{S})\mathbf{1} - \frac{1}{K}\mathbf{1}'\mathbf{S}\mathbf{1}\right] + \frac{1}{K}\mathbf{1}'\mathbf{S}\mathbf{1}. \tag{9.11}$$

As will be seen, all three equivalent forms (9.4), (9.5), and (9.11), are useful in different contexts.

9.1.2 ALGORITHMS

We mainly discuss the minimisation of the residual sum-of-squares S and briefly return to the maximisation of G at the end of this section. The criterion (9.1), equivalently (9.2), is minimised when the one-sided Procrustes criterion $\|\mathbf{X}_k\mathbf{T}_k - \mathbf{G}\|$ is minimised, as described in Chapters 3–6, for $k = 1, \ldots, K$. The minimisation is done under whatever constraints may be put on the $\mathbf{T}_k$ and for each of the K sets. Of course, $\mathbf{G}$ is unknown and is to be estimated, but given putative values of the $\mathbf{T}_k$ (9.2) may be used as the basis of a simple alternating least-squares algorithm:

Algorithm 9.1

Step 1 Update the current setting of $\mathbf{G}$.
Step 2 Evaluate $\mathbf{T}_k(k = 1, \ldots, K)$ by solving the Procrustes fit to $\mathbf{G}$.
Step 3 Test for satisfactory convergence. If not converged, return to *Step 1*.

Initially, $\mathbf{G}$ may be formed by choosing the $\mathbf{T}_k(k = 1, \ldots, K)$ to be simple matrices of the correct type (e.g. unit matrices if the $\mathbf{T}_k$ are square). $\mathbf{G}$ may be updated either for each $k(k = 1, \ldots, K)$ or only after a complete cycle in which all K configurations have been matched to the current group average. Initial settings of $\mathbf{T}_k(k = 1, \ldots, K)$ may be avoided by setting $\mathbf{G} = \mathbf{X}_1$, or to the coordinates of one of the other configurations, and then starting with *Step 2* of the algorithm. Notice that operations (i) and (ii) must both reduce the residual sum-of-squares, which is bounded below by zero, and hence the algorithm must converge, at least to a local minimum. This is because for given $\mathbf{T}_k$, (9.2) is minimised when $\mathbf{G}$ is the mean given by (9.3) and when $\mathbf{G}$ is given, then the estimates of $\mathbf{T}_k$ minimise the two-sets Procrustes criteria.

Although the algorithm is expressed in a least-squares context, *Step 2* may be replaced by other optimality criteria as described in section 9.1.7.

In the context of generalised orthogonal Procrustes analysis, ten Berge (1977*a*) noted that if $\mathbf{G}$ happens to be zero, then *Step 2* of Algorithm 9.1 has no effect but that if the k-excluded group average $\mathbf{G}_k$ is used rather than $\mathbf{G}$ then this situation can be avoided; his improved algorithm is discussed in Section 9.2.1. Equation (9.8) shows that at the minimum, not only does each transformed configuration $\mathbf{X}_k\mathbf{T}_k$ have the best two-set Procrustes fit to the group average $\mathbf{G}$ but also the best two-set Procrustes fit to $\mathbf{G}_k$. The result suggests that ten Berge's improvements should also be effective for general $\mathbf{T}_k$, giving the following variant of Algorithm 9.1:

Algorithm 9.2

Step 1 Initialise $\mathbf{X}_k\mathbf{T}_k$, $k = 1, \ldots, K$.

Step 2 For $k = 1, \ldots, K$ (a) compute $\mathbf{G}_k$, (b) evaluate $\mathbf{T}_k$ by solving the Procrustes fit (9.9) to $\mathbf{G}_k$, and (c) update $\mathbf{X}_k\mathbf{T}_k$.

Step 3 Test for satisfactory convergence. If not converged, return to *Step 2*.

For any putative set of transformations $\mathbf{T}_k$ with k-excluded group averages $\mathbf{G}_k (k = 1, \ldots, K)$, minimising $\sum_{k=1}^{K} \|\mathbf{X}_k\mathbf{T}_k^* - \mathbf{G}_k\|$ over $\mathbf{T}_k^*$ in *Step 2* reduces the value of (9.2), that is, $\sum_{k=1}^{K} \|\mathbf{X}_k\mathbf{T}_k - \mathbf{G}\|$. An advantage of Algorithm 9.2 is that even if $\mathbf{G}$ is zero, not all of $\mathbf{G}_k$ can be zero (else every $\mathbf{X}_k\mathbf{T}_k$ will be zero, which we exclude) and so the algorithm proceeds. In Algorithm 9.1, $\mathbf{X}_k$ is fitted to $\mathbf{G}$ for $k = 1, \ldots, K$ and only then is $\mathbf{G}$ updated. We could have updated $\mathbf{G}$ for each value of k but this is not necessary; we do not know whether there would be any gain in efficiency. In Algorithm 9.2, a new $\mathbf{G}_k$ is used for each setting of k, thus implicitly defining a new $\mathbf{G}$. This guarantees a monotonic reduction in the criterion at every step. If $\mathbf{G}_k, k = 1, \ldots, K$, were based on the same $\mathbf{G}$, as in Algorithm 9.1, then examples can be constructed where the criterion value increases. It may be possible to show that this undesirable non-monotonicity is not a property of certain classes of the chosen transformations $\mathbf{T}_k$, but we know of no results along these lines. Thus, we prefer to offer Algorithm 9.2 in what may not always be its most efficient form. Note that, the original $\mathbf{X}_k$ is used for every Procrustes fit but also each $\mathbf{X}_k\mathbf{T}_k$ must be kept (and updated for each k) so that the current setting of $\mathbf{G}_k$ may be evaluated.

The above algorithms have been developed directly from the identities (9.1)–(9.8) and have not appealed directly to the normal equations (9.9) and (9.10).

It is evident that *Step 2* of Algorithm 9.2 corresponds to (9.9), confirming that Algorithm 9.2 does indeed solve the normal equations.

We turn now to the criterion of maximising G. To estimate $\mathbf{T}_k$ we have maximised $\|\mathbf{X}_k\mathbf{T}_k + \sum_{i\neq k}^{K} \mathbf{X}_k\mathbf{T}_k\|$, which may be written $\|\mathbf{X}_k\mathbf{T}_k + (K-1)\mathbf{G}_k\|$. Algorithm 9.2 may continue to be used but with *Step 2* replaced by solving this Procrustes problem, which is readily identified with the normal equation (9.10).

Minimisation of S and maximisation of G are not the only possible criteria. Often, it would make good sense to minimise S/T or maximise G/T. The normal equations are easily derived. For example, differentiation of S/T with respect to $\mathbf{T}_k$ gives:

$$\left(\frac{(K-1)}{T} - \frac{S}{T^2}\right)(\mathbf{X}_k'\mathbf{X}_k)\mathbf{T}_k - \frac{(K-1)}{T}\mathbf{X}_k'\mathbf{G}_k = \mathbf{\Lambda}_T. \qquad (9.12)$$

Inspection shows that (9.12) has the same general form as (9.9) and for fixed S and T we could solve the Procrustes problem (9.12) for $\mathbf{T}_k$ in *Step 2* of an algorithm similar to Algorithm 9.2. Unfortunately, the coefficients involving S and T are themselves functions of $\mathbf{T}_k$ and it is not clear that the setting of $\mathbf{T}_k$ so found would necessarily reduce the criterion S/T. The convergence properties of such algorithms need further investigation. Of course, when T is not a function of the transformations $\mathbf{T}_k$, $k = 1, 2, \ldots, K$ (which effectively restricts us to orthogonal matrices) the ratio and non-ratio criteria coincide and there is no problem.

9.1.3 MISSING VALUES WITHOUT SCALING

Missing values and initial scaling as discussed in Chapter 2 (not scaling factors which are discussed in Section 9.1.7) go together rather awkwardly. The first thing to be said is that missing values are essentially associated with measuring, or rather failure to measure, variables. With our emphasis on configurations, the pre-processing given by methods such as multidimensional scaling will already have handled any problems there may have been with missing information on variables or observed proximities. Then, the configurations passed on to Procrustes analysis will be complete. Nevertheless, missing observations on shape data and the possibility of using Procrustes methods on observed data, some of which may be missing, have to be handled.

With observed data, the likely lack of commensurability between variables, discussed in Chapter 2, may continue to be addressed by using simple transformations (e.g. logarithmic transformations) but normalisation is not readily available. We delay explaining the reasons for this difficulty until we have described a method for estimating missing values that is available in most circumstances.

Suppose $\mathbf{X}_k$ has M values missing in cells $(i_1, j_1), (i_2, j_2), \ldots, (i_M, j_M)$. As in Section 8.2, these may be given in an indicator matrix $\mathbf{W}_k$ zero, except for units in the missing-cell positions. The missing values may be estimated by a variant of the iterative EM algorithm (Dempster, Laird, and Rubin 1977) where M, rather than representing a step for maximum likelihood estimation, now stands for the least-squares Procrustes problem, while E is a step, described below, that gives expected values for the missing cells.

For fixed settings of $\mathbf{T}_h$ $(h = 1, \ldots, K)$ and $\mathbf{X}_h$ $(h \neq k)$ the terms of (9.1) that involve $\mathbf{X}_k$ requires the minimisation (see Section 9.1.2) of the criterion $\|\mathbf{X}_k\mathbf{T}_k - \mathbf{G}_k\|$ over the unknown values in given cells $(i_1, j_1), (i_2, j_2), \ldots, (i_M, j_M)$. Recall that $\mathbf{G}_k$ represents the k-excluded group average, which is independent of $\mathbf{X}_k$. We assume that the cells with unknown values contain putative values that we seek to update by minimising the criterion. Let us suppose that the updates are given in a matrix $\mathbf{X}$, which is zero except for the missing cells which contain values that we denote by $x_1, x_2, \ldots, x_M$. Thus we wish to find the values of $\mathbf{X}$ that minimise:

$$\|(\mathbf{X}_k - \mathbf{X})\mathbf{T}_k - \mathbf{G}_k\| = \|\mathbf{X}\mathbf{T}_k - (\mathbf{X}_k\mathbf{T}_k - \mathbf{G}_k)\|. \tag{9.13}$$

This is itself a Procrustes problem where now it is $\mathbf{X}$, rather than $\mathbf{T}_k$, that is to be estimated. Transposition would put (9.13) into our basic form (1.1). The constraint on $\mathbf{X}$ may be written:

$$\mathbf{X} = x_1\mathbf{e}_{i_1}\mathbf{e}'_{j_1} + x_2\mathbf{e}_{i_2}\mathbf{e}'_{j_2} + \cdots + x_M\mathbf{e}_{i_M}\mathbf{e}'_{j_M} = \sum_{m=1}^{M} x_m\mathbf{e}_{i_m}\mathbf{e}'_{j_m} \tag{9.14}$$

where, as usual, $\mathbf{e}_i$ represents a unit vector, zero except for its ith position. This function is linear in the parameters x_m so, in principle, the minimisation of (9.13) is a simple linear least-squares problem; unfortunately, the detailed formulae are somewhat inelegant. The terms in (9.13) that involve $\mathbf{X}$ are:

$$\text{trace}[(\mathbf{X}'\mathbf{X})(\mathbf{T}_k\mathbf{T}'_k) - 2\mathbf{X}'(\mathbf{X}_k\mathbf{T}_k - \mathbf{G}_k)\mathbf{T}'_k]$$

which we write:

$$\text{trace}[(\mathbf{X}'\mathbf{X})\mathbf{T} - 2\mathbf{X}'\mathbf{Y}] \tag{9.15}$$

where $\mathbf{T} = (\mathbf{T}_k\mathbf{T}'_k)$ is symmetric and $\mathbf{Y} = (\mathbf{X}_k\mathbf{T}_k - \mathbf{G}_k)$, is the current residual matrix.

The minimisation of (9.15) over x_r subject to the constraint (9.14) yields:

$$y_{i_r,j_r} = x_1(\mathbf{e}'_{i_r}\mathbf{e}_{i_1})(t_{j_r,j_1}) + x_2(\mathbf{e}'_{i_r}\mathbf{e}_{i_2})(t_{j_r,j_2}) + \cdots + x_M(\mathbf{e}'_{i_r}\mathbf{e}_{i_M})(t_{j_r,j_M})$$

which may be written:

$$y_r = \sum_{m=1}^{M} x_m(\mathbf{e}'_{i_r}\mathbf{e}_{i_m})(t_{j_r,j_m}), \quad r = 1, \ldots, M \tag{9.16}$$

or $\mathbf{y} = \mathbf{A}\mathbf{x}$, yielding $\mathbf{x} = \mathbf{A}^{-1}\mathbf{y}$. The matrix $\mathbf{A}$ is symmetric with elements $a_{rm} = (\mathbf{e}'_{i_r}\mathbf{e}_{i_m})(t_{j_r,j_m})$. This will be zero unless $i_r = i_m$ which would imply that x_r and x_m are in the same row of $\mathbf{X}$. When no row contains more than one missing cell, $\mathbf{A}$ is diagonal and then $x_m = y_{i_m,j_m}/t_{j_m,j_m}$.

In general, **A** contains diagonal blocks, each block being symmetric, with as many rows/columns as there are missing cells in the pertinent row of **X**. This is best illustrated by a small example, as follows. Suppose, for example, that:

$$\mathbf{W}_k = \begin{pmatrix} 0 & 1 & 0 & 0 & 0 \\ 1 & 1 & 1 & 0 & 0 \\ 0 & 0 & 0 & 1 & 0 \\ 0 & 0 & 1 & 0 & 1 \end{pmatrix}$$

so that the indices for the missing cells, numbered by rows, are as shown in Table 9.1.

From Table 9.1, using (9.16), the elements of **A** were found to be as in Table 9.2. The first unknown cell x_1 occurs by itself in the second column of the first row of $\mathbf{W}_k$ and so induces the single element t_{22} in Table 9.2. However, x_2, x_3, and x_4 all occur in the second row of $\mathbf{W}_k$, in columns 1, 2, and 3, respectively, and induce the three by three symmetric block. x_5 occurs as a single missing cell in the fourth column of third row and induce the diagonal value t_{44}. Finally, x_6 and x_7 both occur in columns three and five of the last row which induce the final two by two diagonal block. Note that the same values t_{ij} may be repeated in **A**.

Table 9.1. The indices of the missing cells.

m	i_r	j_r
1	1	2
2	2	1
3	2	2
4	2	3
5	3	4
6	4	3
7	4	5

Table 9.2. The matrix **A** derived from Table 9.1.

m	1	2	3	4	5	6	7
1	t_{22}						
2		t_{11}	t_{12}	t_{13}			
3		t_{12}	t_{22}	t_{23}			
4		t_{13}	t_{23}	t_{33}			
5					t_{44}		
6						t_{33}	t_{35}
7						t_{35}	t_{55}

Table 9.3. To allow for translations, the elements of Table 9.2 should be reduced by subtracting $1/n$ times the elements of this table.

m	1	2	3	4	5	6	7
1	t_{22}	t_{12}	t_{22}	t_{23}	t_{24}	t_{23}	t_{25}
2	t_{12}	t_{11}	t_{12}	t_{13}	t_{14}	t_{13}	t_{15}
3	t_{22}	t_{12}	t_{22}	t_{23}	t_{24}	t_{23}	t_{25}
4	t_{23}	t_{13}	t_{23}	t_{33}	t_{34}	t_{33}	t_{35}
5	t_{24}	t_{14}	t_{24}	t_{34}	t_{44}	t_{34}	t_{45}
6	t_{23}	t_{13}	t_{23}	t_{33}	t_{34}	t_{33}	t_{35}
7	t_{25}	t_{15}	t_{25}	t_{35}	t_{45}	t_{35}	t_{55}

So far we have not discussed the effects of translation, which will usually be required. One could proceed as above, and recentre after adjusting for the missing cells. This will certainly further reduce the residual sum-of-squares but, in general, can be done better. We assume that the current settings of $\mathbf{X}_k$ and hence $\mathbf{G}_k$ are centred. To preserve centring we replace $\mathbf{X}$ by its centred form $(\mathbf{I} - \mathbf{N})\mathbf{X}$. The effect of this is to replace the previous definition of a_{rm} by $a_{rm} = (\mathbf{e}_{i_r}'(\mathbf{I} - \mathbf{N})\mathbf{e}_{i_m})(t_{j_r, j_m})$. Thus, when $i_r = i_m$, $a_{rm} = (1 - 1/n)(t_{j_r, j_m})$ else $a_{rm} = -1/n(t_{j_r, j_m})$, so that $\mathbf{A}$ is now replaced by $\mathbf{A}^* = \mathbf{A} - 1/n\mathbf{T}^*$, where $\mathbf{T}^*$ is made up of those elements of $\mathbf{T}$ corresponding to the columns of the missing cells (e.g. the third column of Table 9.1). Thus, Table 9.2 would be reduced by $1/n$ times the elements of Table 9.3. This minor change gives $\mathbf{x} = \mathbf{A}^{*-1}\mathbf{y}$, so defining $\mathbf{X}$, and then $(\mathbf{I} - \mathbf{N})\mathbf{X}$ gives the required correction matrix. This is all that is necessary for handling translations. Recentring $\mathbf{X}$ derived from $\mathbf{x} = \mathbf{A}^{-1}\mathbf{y}$ and using $(\mathbf{I} - \mathbf{N})\mathbf{X}$ derived from $\mathbf{x} = \mathbf{A}^{*-1}\mathbf{y}$ will give different updates to the missing values, though both must converge to the same final estimates. A near exception to this rule occurs for orthogonal transformations when the two approaches hardly differ, see Section 9.2.1.

Algorithm 9.2 is easily modified to cope with missing values by adding the procedure described above to *Step 2* for updating the estimates of the values missing in $\mathbf{X}_k$. As described, slightly different updates will be required, depending on whether or not to allow for translational effects.

Earlier, we mentioned that there are difficulties in incorporating normalisation of configurations into the above missing value procedure. Throughout the iterations, the method described above maintains centring but not normalisation. Thus, even if we can normalise the initial configurations, this normalisation will be lost in subsequent iterations. In principle, normalisation could be imposed by satisfying the additional constraint $\mathrm{diag}(\mathbf{X}_k - \mathbf{X})'(\mathbf{X}_k - \mathbf{X}) = \mathbf{I}$ which, assuming that $\mathbf{X}_k$ is already in normalised form, implies that $\mathrm{diag}(\mathbf{X}'\mathbf{X} - 2\mathbf{X}'\mathbf{X}_k) = 0$. Unfortunately, this constraint leads to difficult optimisation problems, similar to those discussed in Appendix E, but harder.

A possibility is to use the method described above to estimate the missing values. We could then normalise the now complete data and repeat the process to obtain new estimates of the missing values. As a result, $\mathbf{X}_k$ would no longer be normalised but we could proceed iteratively until, hopefully, everything settles down. In Section 9.1.7 we discuss estimating missing values when scaling factors are included; normalisation is then less of a problem.

9.1.4 GENERALISED PROCRUSTES ANALYSIS WITH ISOTROPIC SCALING

Next we derive the modifications required to the criteria (9.1) and (9.2) when isotropic scaling factors are to be included. Now (9.1) and (9.2) are replaced by:

$$\sum_{i<j}^{K} \|\mathbf{s}_i\mathbf{X}_i\mathbf{T}_i - \mathbf{s}_j\mathbf{X}_j\mathbf{T}_j\| \tag{9.17}$$

$$= K\sum_{k=1}^{K} \|\mathbf{s}_k\mathbf{X}_k\mathbf{T}_k - \mathbf{G}\| \tag{9.18}$$

where
$$\mathbf{G} = K^{-1}\sum_{k=1}^{K}(s_k\mathbf{X}_k\mathbf{T}_k). \tag{9.19}$$

In general, all the identities given in equation (9.2)–(9.6) continue to hold with $\mathbf{T}_k$ replaced everywhere by $s_k\mathbf{T}_k$.

The difficulty that (9.17) has the pathological solution $s_k = 0$ for $k = 1, \ldots, K$, must be addressed by imposing some constraint and that usually adopted is the requirement that the total 'size' of the configurations is fixed after scaling and transformation, that is,

$$\sum_{k=1}^{K} \|s_k\mathbf{X}_k\mathbf{T}_k\| = \kappa, \text{ say.} \tag{9.20}$$

Often we shall choose $\kappa = \sum_{k=1}^{K} \|\mathbf{X}_k\|$ so that the total 'size' is the same before and after transformation; when the individual configurations are first normalised, we have $\kappa = K$. The important thing is that κ is fixed, not that total size is preserved; the actual choice of κ has only a trivial proportional effect on the estimate of the scaling factors. Note that we may write (9.20) as:

$$\sum_{k=1}^{K} \left\| s_k^* \frac{\mathbf{X}_k}{\|\mathbf{X}_k\|^{1/2}}\mathbf{T}_k \right\| = \kappa$$

where $s_k^* = s_k\sqrt{\|\mathbf{X}_k\|}$ so that even if we have not normalised in advance, we may do so by adjusting the estimated scaling factors. Throughout this section we use the constraint (9.20); other constraints on the scaling factors are discussed in Section 9.1.5.

Differentiating (9.17) with respect to the scaling factors, after associating a Lagrange multiplier with (9.20), leads immediately to:

$$s_k = \frac{\kappa}{K\|\mathbf{G}\|} \frac{\text{trace}(\mathbf{G}'\mathbf{X}_k\mathbf{T}_k)}{\|\mathbf{X}_k\mathbf{T}_k\|} \tag{9.21}$$

which is easily verified to satisfy (9.20). Comparison with (3.10) shows that s_k is proportional to the scaling factor for fitting $s_k\mathbf{X}_k\mathbf{T}_k$ to $\mathbf{G}$, the proportionality factor arising from the constraint (9.20). It follows from the corresponding result for the two-set case (Section 4.3) that s_k is positive. In (9.21), $\mathbf{G}$ is assumed to be in its final form, although in algorithms its iterates may be substituted. We may rearrange (9.21) to give:

$$\text{trace}(\mathbf{G}'\mathbf{X}_k\mathbf{T}_k) = \frac{\mu}{K} s_k \|\mathbf{X}_k\mathbf{T}_k\| \tag{9.22}$$

where $\mu = K^2\|\mathbf{G}\|/\kappa$.

The above may be expressed directly in matrix notation, as follows. First, we observe that (9.17) may be written:

$$KS = K\mathbf{s}'\text{diag}(\mathbf{S})\mathbf{s} - \mathbf{s}'\mathbf{S}\mathbf{s} \tag{9.23}$$

where $\mathbf{s} = vec(s_1, s_2, \ldots, s_K)$, and $\mathbf{S}$ is defined just before (9.11).

Also, the constraint (9.20) is:

$$\mathbf{s}'\text{diag}(\mathbf{S})\mathbf{s} = \kappa. \tag{9.24}$$

Differentiation of (9.23) with a Lagrangian term for the constraint (9.24) gives:

$$K\text{diag}(\mathbf{S})\mathbf{s} - \mathbf{S}\mathbf{s} = \lambda\text{diag}(\mathbf{S})\mathbf{s}. \tag{9.25}$$

or

$$\mathbf{S}\mathbf{s} = (K - \lambda)\text{diag}(\mathbf{S})\mathbf{s}. \tag{9.26}$$

Equation (9.26) is the matrix form of (9.22) and gives an eigenvector expression for $\mathbf{s}$. Premultiplying (9.25) by $\mathbf{s}$, we see from (9.22) and (9.24) that:

$$S = \lambda\kappa,$$

so S is minimised by the smallest value of λ satisfying (9.26) and so corresponds to the largest eigenvalue $(K - \lambda) = \mu$, say. The corresponding vector $\mathbf{s}$ should be normalised to satisfy (9.24).

In terms of λ and κ (9.23) may be interpreted as the analysis of variance:

$$\lambda\kappa(\text{Residual} = KS) = K\kappa(\text{Total} = K) - (K - \lambda)\kappa(\text{Group average} = KG) \tag{9.27}$$

where the group average interpretation follows from noting from (9.25) that $\mathbf{s}'\mathbf{S}\mathbf{s} = K^2\|\mathbf{G}\| = (K - \lambda)\kappa = \mu\kappa$.

Thus, provided the matrices $\mathbf{T}_k$, $k = 1, \ldots, K$ are known, the scaling factors may be estimated as the eigenvector satisfying (9.26), corresponding to the biggest eigenvalue μ (ten Berge 1977*a*), and scaled to satisfy (9.23). We give a more detailed discussion of the analysis of variance (9.27) in Chapter 10. The sum of the eigenvalues μ of (9.26) is K, so $\mu \geq 1$, the average eigenvalue, with equality only in the pathological situation when all the eigenvalues of (9.26) are equal.

We can modify the previous algorithm to include the estimation of isotropic scaling factors. Now, we supplement the initial simple settings of the $\mathbf{T}_k$, by setting $s_k = 1$.

Algorithm 9.3

Step 1 Update the current setting of **S**.
Step 2 Estimate the s_k, using (9.26) and compute a current setting of **G**.
Step 3 Minimise $\|s_k\mathbf{X}_k\mathbf{T}_k - \mathbf{G}\|$ to give new estimates of $\mathbf{T}_k$, $k = 1, \ldots, K$.
Step 4 Test for satisfactory convergence. If not converged, return to *Step 1*.

As in Algorithm 9.1 *Step 3* may be replaced by criteria other than least squares—see Section 9.1.6. In generalised orthogonal Procrustes analysis (GPA or GOPA), assuming the current value of **G**, Gower (1975) used (9.21) rather than (9.26) to update the scaling. The same could be done in the general context considered here. Because **G** incorporates the unknown scaling factors, this process does not give the optimal results for given $\mathbf{X}_k\mathbf{T}_k$, although it converges to the correct results as the $\mathbf{T}_k$ themselves converge. Using (9.26) ensures that the current optimal values of the scaling factors are found at every cycle of the iterations, so that fewer cycles should be needed to achieve convergence. To determine the relative efficiencies of the two methods, we have to balance the time taken to solve the eigenvalue problem (9.26), for which there are efficient algorithms, against the number of cycles needed for convergence when using (9.21). In practice, it seems that using (9.26) is usually the more efficient, though with modern computers the comparison is between two short times. Whichever method is adopted, the new values of s_k are used to calculate $s_k\mathbf{X}_k\mathbf{T}_k$ ($k = 1, \ldots, K$) and a new value of **G** to be used in *Step 3*; as in Algorithm 9.2, **G** may be replaced by the k-excluded group average, where now $\mathbf{G}_k = 1/(K-1)\sum_{i \neq k}^{K} s_i\mathbf{X}_i\mathbf{T}_i$. ten Berge, Kiers, and Commandeur (1993) report that this modification increases speed and reduces the occurrence of local minima. As in Algorithm 9.2, $\mathbf{G}_k$ would need to be updated for each value of k.

9.1.5 CONTRIBUTION OF THE kTH SET TO THE RESIDUAL SUM-OF-SQUARE

Writing $S_k = \|s_k\mathbf{X}_k\mathbf{T}_k - \mathbf{G}\|$, for the residual sum-of-squares of the kth set, and $\bar{S} = 1/K\sum_{k=1}^{K} S_k$, for the average residual sum-of-squares, we may obtain an expression for the difference $S_k - \bar{S}$, as follows.

Expanding S_k and using (9.19) and (9.22) gives:

$$S_k = \|\mathbf{G}\| + \left(1 - \frac{2\mu}{K}\right) s_k^2 \|\mathbf{X}_k \mathbf{T}_k\|$$

Averaging and using (9.20) gives:

$$\bar{S} = \|\mathbf{G}\| + \frac{\kappa}{K}\left(1 - \frac{2\mu}{K}\right).$$

Hence:

$$S_k - \bar{S} = \left(s_k^2 \|\mathbf{X}_k \mathbf{T}_k\| - \frac{\kappa}{K}\right)\left(1 - \frac{2\mu}{K}\right). \tag{9.28}$$

Now κ/K is the average size of the configurations, so (9.28) shows that when the square of the scaled and transformed kth configuration is close in size to the average, then its contribution to the residual sum-of-squares will be close to the average residual sum-of-squares. It follows that when S_k is less than or greater than $\bar{S}$ then the scaled and transformed configuration diverges from the average.

We may rewrite (9.28), by noting that the result given for $\|\mathbf{G}\|$ following (9.27) may be combined with the formula for $\bar{S}$ to give:

$$\bar{S} = \frac{\kappa}{K}\left(1 - \frac{\mu}{K}\right) \quad \text{and} \quad \text{hence } 1 - \frac{2\mu}{K} = \frac{2K}{\kappa}\bar{S} - 1$$

which on substitution into (9.28) gives

$$S_k - \bar{S} = \left(s_k^2 \|\mathbf{X}_k \mathbf{T}_k\| - \frac{\kappa}{K}\right)\left(2\frac{K}{\kappa}\bar{S} - 1\right). \tag{9.29}$$

With the initial scaling $\|\mathbf{X}_k\| = 1$ $(k = 1, \ldots, K)$, then $\kappa = K$ and when, further, $\mathbf{T}_k$ are constrained to be orthogonal (see Section 9.2.1) (9.29) simplifies to:

$$S_k - \bar{S} = (s_k^2 - 1)(2\bar{S} - 1) \tag{9.30}$$

showing that the closer is s_k to unity the better is the fit of the kth configuration to the average fit. Provided $\bar{S} > \frac{1}{2}$, the fit is better, the smaller is the scaling factor, when $\bar{S} < \frac{1}{2}$ the reverse is true. Thus s_k not only measures scaling but also the goodness-of-fit of the kth configuration (Collins 1991, Arnold 1992). Furthermore, with isotropic scaling, the iterative process of finding the optimal Procrustes rotation matrices is accelerated.

So, inclusion of isotropic scaling factors can be seen as a device that down-weights outliers. Alternatively, it may be considered as a substitute for initial scaling (Chapter 2) that corrects for differing spreads in the individual data matrices prior to the GPA procedure. The interaction between initial scaling and fitting optimal isotropic scaling factors (cf. Dijksterhuis and Gower 1991/2, p. 72, p. 80–2) is a complicated one that is further explored in Section 9.1.7.

9.1.6 THE CONSTRAINT (9.20) AND OTHER POSSIBILITIES

The constraint (9.20) is very mild. Having obtained scaling factors s_k as discussed above, it should be clear than any multiples of them would be equally acceptable. All that happens is that all configurations are scaled up or down by the same amount and this has no substantive effect. Hence, we may as well choose κ to have some convenient value and the obvious choice is $\kappa = K$, giving an average configuration size of unity. This choice of κ simplifies (9.29) by replacing κ/K by unity, as discussed above.

Although all quadratic constraints of the form (9.20) for different settings of κ are equivalent, there remains the question of why should we adopt a quadratic constraint in the first place; why not, for example, use a linear constraint such as:

$$\sum_{k=1}^{K} s_k \mathbf{a}'(\mathbf{X}_k\mathbf{T}_k)\mathbf{1} = \kappa?$$

Note that with centred configurations, we must exclude the possibility $\mathbf{a} = \mathbf{1}$. The minimisation of (9.17) and (9.18) subject to the above constraint leads not to an eigenvalue problem but to a matrix inversion problem. The estimated scaling factors will differ in a non-linear fashion from those given by the quadratic constraint, thus questioning the validity of both constraints. The position is closely analogous to that discussed by Healy and Goldstein (1966) in the context of the constraints appropriate for estimating Guttman optimal scores. Gower (1998) offered a resolution of the apparent conflict by noting that the *fundamental* expression of the appropriate criterion is in terms of a ratio, which implies eigenvector estimates rather than matrix inversion. The remark above on the lack of a substantive effect in replacing a given set s_k of scaling parameters by any proportional set, suggests that criteria expressed as ratios are also appropriate in the Procrustes context. Thus we might replace (9.18) by the ratio of the residual sum-of-squares to the total sum-of-squares:

$$\frac{S}{T} = \frac{\sum_{k=1}^{K} \|s_k\mathbf{X}_k\mathbf{T}_k - \mathbf{G}\|}{\sum_{k=1}^{K} \|s_k\mathbf{X}_k\mathbf{T}_k\|}. \tag{9.31}$$

This form of criterion underlines that scaling is invariant to the identification of the s_k up to a multiplicative constant. Like all ratios of quadratic forms, this criterion is minimised over $\mathbf{s}$ by solving an eigenvalue problem. Indeed, the eigenvalue problem is identical to (9.26) where the eigenvectors are arbitrarily scaled, but fixed for convenience by the mild quadratic constraint (9.20). Gower (1998) terms such constraints *weak constraints*, to distinguish them from *strong constraints* such as requiring $\mathbf{T}_k$ to be orthogonal or orthonormal or to be a matrix of direction cosines. Strong constraints have profound effects on the estimating equations. Rather confusingly, we could even minimise the ratio, scaling the eigenvectors of (9.26) to satisfy a linear relationship such as $\mathbf{s}'\mathbf{1} = 1$. This is quite distinct from minimising the numerator subject to the same, but now strong, linear constraint. Thus the ratio form of criterion is attractive. Unfortunately, we have already seen in Section 9.1.2 that, unless T is fixed, ratio criteria of this form lead to unresolved algorithmic problems when estimating the transformations $\mathbf{T}_k$.

To summarise so far, with isotropic scaling, the ratio form is a fundamental way of expressing the Procrustes criteria and it leads to the eigenvector estimators (9.26). Strong linear, or other, constraints are not needed to identify the scaling factors when using the ratio criterion. With this understanding, while recognising the fundamental nature of ratio criteria, it is acceptable to continue with writing Procrustes criteria as previously formulated and resolving identification problems by applying strong quadratic constraints, indeed, unless T is fixed, which essentially limits us to $\mathbf{T}_k = \mathbf{Q}_k$, orthogonal, we have no other option until algorithmic difficulties are overcome.

There is a possibility of combining the best of both approaches, applying the ratio criteria to the scaling factors and not to the matrix transformations. The remainder of this section develops this approach. When $\mathbf{T}_k = \mathbf{Q}_k$, orthogonal, the constraint (9.20) simplifies to $\sum_{k=1}^{K} \|s_k\mathbf{X}_k\| = \kappa = \sum_{k=1}^{K} \|\mathbf{X}_k\|$ which clearly preserves size after orthogonal transformation. For other cases when $\mathbf{T}_k$ is square, perhaps we should be working in an appropriate metric. For example, when $\mathbf{T}_k = \mathbf{C}_k$, direction cosines, (see Chapter 6 and Appendix D) the constraint would become trace $\sum_{k=1}^{K} s_k^2\mathbf{X}_k\mathbf{C}_k(\mathbf{C}_k'\mathbf{C}_k)^{-1}\mathbf{C}_k'\mathbf{X}_k' = \sum_{k=1}^{K} \|s_k\mathbf{X}_k\| = \kappa = \sum_{k=1}^{K} \|\mathbf{X}_k\|$, again preserving size. However, when $\mathbf{T}_k$ is rectangular with $R < P_k$ (9.20) expresses that the configurations transformed into the lower R-dimensional space should be the same size as the original configurations in $\max(P_1, P_2, \ldots, P_K)$ dimensional space. Is this a reasonable requirement? The situation is complicated because the transformations change the sizes of the configurations, so to confound scaling factors with these other size changes, makes interpretation difficult. Often, it will be better not to involve the transformations in the constraints on the scaling factors, suggesting that the choice

$$\mathbf{s}'\mathbf{s} = \|\mathbf{s}\| = \kappa \tag{9.32}$$

is worth considering. The choice $\kappa = K$ is attractive because with undifferential scaling we shall have $s_k = 1$, for $k = 1, \ldots, K$. With (9.32) replacing (9.24), the matrix form of (9.20), we have:

$$(K\ \mathrm{diag}(\mathbf{S}) - \mathbf{S})\mathbf{s} = \lambda\mathbf{s}. \tag{9.33}$$

Premultiplying by $\mathbf{s}$ shows that:

$$S = \lambda\kappa$$

as before, so now we require the smallest eigenvalue λ of (9.33). The corresponding eigenvector should be scaled to satisfy (9.32) so that $\mathbf{s}'(K\ \mathrm{diag}(\mathbf{S}) - \mathbf{S})\mathbf{s} = \lambda\kappa$. With this small change things proceed as with the constraint (9.20) or, equivalently (9.24).

Rather than as a constrained minimisation, we may develop this approach as minimising the ratio:

$$\frac{S}{\mathbf{s}'\mathbf{s}} = \frac{\mathbf{s}'(K\,\mathrm{diag}(\mathbf{S}) - \mathbf{S})\mathbf{s}}{\mathbf{s}'\mathbf{s}} \tag{9.34}$$

which, when minimised over $\mathbf{s}$, leads to the same eigenvalue problem (9.33) and when minimised over $\mathbf{T}_k$ leads to the usual minimisation of S. Thus, we may express the generalised Procrustes problem with the constraint (9.32) as a ratio minimisation problem with no explicit constraint.

In Section 9.1.5 results were obtained for assessing the contribution of the kth set to the residual sum-of-squares. These results were derived for minimising S under the constraint (9.20). Similar results apply to minimising S under the strong constraint (9.32) or minimising the ratio (9.34) with (9.32) treated as a weak constraint. Indeed, in the special case when $\|\mathbf{X}_k\| = 1$ and $\mathbf{T}_k = \mathbf{Q}_k$, orthogonal $(k = 1, \ldots, K)$, and we choose $\kappa = K$, then the constraints (9.20) and (9.32) coincide. It follows that under these conditions, the result (9.30) continues to apply under the constraint (9.32)

There is an even closer relationship between the two constraints. We may write:

$$S = \sum_{k=1}^{K} \|s_k\mathbf{X}_k\mathbf{Q}_k - \mathbf{G}_k\| = \sum_{k=1}^{K} \|s_k^*\mathbf{X}_k^*\mathbf{Q}_k - \mathbf{G}_k\|$$

where $s_k^* = s_k\sqrt{\|\mathbf{X}_k\|}$ and $\mathbf{X}_k^* = \mathbf{X}_k/\sqrt{\|\mathbf{X}_k\|}$ which is in normalised form. Thus:

$$\sum_{k=1}^{K} s_k^2\|\mathbf{X}_k\| = \sum_{k=1}^{K} (s_k^*)^2$$

showing that minimising S under constraint (9.20) is the same as minimising under the constraint (9.32) using the normalised data.

9.1.7 MISSING VALUES WITH SCALING

Section 9.1.3 discussed the estimation of missing values when there is no scaling. At first sight, the introduction of scaling factors introduces new difficulties. These arise from trying to accommodate the effects of the constraint (9.20), especially the rationale that the total 'size' of the configurations should be the same before and after transformation. With missing values, we do not know the size of the configurations before transformation. To make progress, ten Berge, Kiers, and Commandeur (1992) replaced this constraint by defining size over the non-missing cells, suitably centred. The problem is circumvented if we use the alternative rationale that the total size of the transformations after transformation should be fixed at κ, which as we have seen is arbitrary (Section 9.1.4) but often conveniently set as $\kappa = K$. It seems sufficient to set κ to its initial value based on whatever initial settings are chosen for the missing values. We may then proceed as in Section 9.1.3 with $\mathbf{X}_k\mathbf{T}_k$ replaced everywhere by $s_k\mathbf{X}_k\mathbf{T}_k$. The estimation of the isotropic scaling factors is as given in Section 9.1.4 and ensures that the chosen constraint is always maintained.

Alternatively, we could use a ratio form of criterion such as (9.31) or (9.34) discussed in Section 9.1.6. Minimising the ratio (9.31) is closer to adopting the constraint (9.20) but we have already seen that algorithms for minimising (9.31) over $\mathbf{T}_k$ await development. If a suitable algorithm were found, it is possible to develop a method for estimating one missing value, which may be used iteratively to cope with any number of missing values. If we adopt the ratio (9.34) or, equivalently, minimise S under the constraint (9.32), we may adopt the approach recommended in the previous paragraph for estimating missing values without any concerns over size. Of course, this criterion gives estimates of the scaling factors differing from those using the constraint (9.20) but, as argued in Section 9.1.7 (9.32) may be the better. Further, we recall the result of Section 9.1.6 which says that minimising S subject to (9.32) but with normalised data is the same as minimising S subject to (9.20) without reference to normalisation. To minimise S subject to (9.20) allowing for normalisation is also possible. This follows from noting that:

$$S = \sum_{k=1}^{K} \|s_k \mathbf{X}_k \mathbf{Q}_k - \mathbf{G}_k\| = \sum_{k=1}^{K} \|(s_k \rho_k)(\mathbf{X}_k \rho_k^{-1})\mathbf{Q}_k - \mathbf{G}_k\|,$$

with $\sum_{k=1}^{K} \|s_k \mathbf{X}_k\| = \sum_{k=1}^{K} \|(s_k \rho_k)(\rho_k^{-1}\mathbf{X}_k)\| = \kappa$. Hence, we may minimise S subject to (9.20), using our missing value procedure when necessary, without normalisation. The normalisation may be imposed at the end by setting $\rho_k = \sqrt{\|\mathbf{X}_k\|}$ so adjusting s_k to $s_k\sqrt{\|\mathbf{X}_k\|}$. This new parameterisation continues to satisfy (9.20). Thus, it turns out to be easier to handle normalisation with scaling factors than without. Perhaps this should be no surprise, because scaling and normalisation are inter-related as is evidenced by the arbitrary aliasing factor ρ_k, so the one can be allowed for in the other.

9.1.8 EXHIBITING THE GROUP AVERAGE

Having computed $\mathbf{G}$, how should it be exhibited on paper or computer screen? First we note that, irrespective of the nature of the $\mathbf{T}_k$, all configuration $s_k\mathbf{X}_k\mathbf{T}_k$, including that associated with $\mathbf{G}$, may be multiplied by an arbitrary orthogonal matrix $\mathbf{Q}$ without affecting (9.17) or (9.18). It is natural to choose $\mathbf{Q}$ so that $\mathbf{G}$ is referred to its principal axis. This involves a principal components analysis of $\mathbf{G}$ giving $\mathbf{Q}$ as the matrix of eigenvectors of $\mathbf{G}'\mathbf{G}$. Then we exhibit the first few dimensions of $s_k\mathbf{X}_k\mathbf{T}_k\mathbf{Q}$ and $\mathbf{GQ}$. However, this is by no means the only possibility. We could, for example, do a principal components analysis on the NK samples (cases) given by concatenating the K configurations $s_k\mathbf{X}_k\mathbf{T}_k$, thus replacing $\mathbf{G}'\mathbf{G}$ by the total dispersion matrix. Another possibility is to use a projection highlighting some projection-pursuit feature, either of the group average or of all the points.

Yet another possibility is to replace (9.18) by the minimisation of

$$K\sum_{k=1}^{K}\|s_k\mathbf{X}_k\mathbf{T}_k-\mathbf{H}\|,$$

where $\mathbf{H}$ is now specified to have rank (or dimension) h; typically one would set $h=2$ to give a two-dimensional representation. Rearranging the above gives:

$$\sum_{k=1}^{K}\|s_k\mathbf{X}_k\mathbf{T}_k-\mathbf{H}\|=\sum_{k=1}^{K}\|s_k\mathbf{X}_k\mathbf{T}_k-\mathbf{G}\|+K\|\mathbf{G}-\mathbf{H}\|. \quad (9.35)$$

With no constraint on $\mathbf{H}$, we have $\mathbf{H}=\mathbf{G}$, as before. Note that (9.35) gives the residual sum-of-squares about any $\mathbf{H}$, so that reducing $\|\mathbf{G}-\mathbf{H}\|$ reduces the residual sum-of-squares. If we regard $\mathbf{T}_k$ and s_k as fixed (and hence so is $\mathbf{G}$), then (9.35) is minimised by taking $\mathbf{H}$ as the rank h Eckart–Young approximation to $\mathbf{G}$. That is, if $\mathbf{G}=\mathbf{U\Sigma V}'$, then $\mathbf{H}=\mathbf{U}_h\mathbf{\Sigma}_h\mathbf{V}_h'$ where $\mathbf{U}_h$ and $\mathbf{V}_h$ represent the first h columns of $\mathbf{U}$ and $\mathbf{V}$, and $\mathbf{\Sigma}_h$ is a diagonal matrix of the h biggest singular values. The h-dimensional representation is given by a group average $\mathbf{GV}_h$, with individual configurations $s_k\mathbf{X}_k\mathbf{T}_k\mathbf{V}_h$ for $k=1,\ldots,K$. Indeed, when $\mathbf{T}_k$ are the results of a generalised Procrustes procedure with group average $\mathbf{G}$, then $\mathbf{H}$, so found, is equivalent to the principal component representation discussed above. If now we allow $\mathbf{T}_k$ and s_k to vary, then the reduced rank criterion (9.35) may be minimised by Algorithm 9.4. We assume that initial settings are available for the s_k and $\mathbf{T}_k$ giving an initial group average $\mathbf{G}$. For example, these could derive from a solution to the unconstrained problem.

Algorithm 9.4

Step 1 Compute the nearest $\mathbf{H}$ to the current setting of $\mathbf{G}$. This is the usual Eckart–Young solution given by the dominant rank-h component of the singular value decomposition of $\mathbf{G}$.

Step 2 Compute the K Procrustes solutions to $\|s_k\mathbf{X}_k\mathbf{T}_k-\mathbf{H}\|$ giving new settings for s_k and $\mathbf{T}_k$ and thus a new current $\mathbf{G}$.

Step 3 Test for convergence and if necessary return to *Step 1*.

Step 4 For each k, compute $s_k\mathbf{X}_k\mathbf{T}_k\mathbf{V}_h$ *and the group average* $\mathbf{GV}_h$.

Step 1 reduces the residual sum-of-squares by minimising the term $\|\mathbf{G}-\mathbf{H}\|$ of (9.35), the configurations remaining unchanged. *Step 2* further reduces the residual sum-of-squares by optimising the configurations about the current setting of $\mathbf{H}$. Thus the algorithm continually reduces the residual sum-of-squares until satisfactory convergence is achieved. *Step 4* gives the final representation relative to h orthogonal axes.

All the methods discussed in this section give possibly useful two-dimensional views of the final configurations but, nevertheless, the optimal solution often occurs in more dimensions. This disparity between the number of dimensions in the solution and the number of dimensions in the exhibited display has been at the root of some misunderstandings that are further discussed in Chapter 10.

9.1.9 OTHER CRITERIA

The above has been concerned solely with the least-squares approach. Just as with the two-sets problems, other criteria may be used. Nothing essentially new is required but for reference we give the modifications that may readily be substituted for the least-squares steps in the previously discussed algorithms.

The generalised form of the inner-product criterion (Section 3.1.2) is to maximise:

$$I = \text{trace} \sum_{i \neq j}^{K} (s_i s_j \mathbf{T}_i' \mathbf{X}_i' \mathbf{X}_j \mathbf{T}_j).$$

The contribution involving $\mathbf{T}_k$ is:

$$\text{trace}(s_k \mathbf{T}_k' \mathbf{X}_k' \mathbf{G}_k) \tag{9.36}$$

where $\mathbf{G}_k = 1/K - 1 \sum_{i \neq k}^{K} (s_i \mathbf{X}_i \mathbf{T}_i)$, the k-excluded group average.

The generalised form of the maximal group average criterion (Section 5.6) is to maximise $\|\mathbf{G}\|$, or equivalently to maximise:

$$KG = \left\| \sum_{i=1}^{K} (s_i \mathbf{X}_i \mathbf{T}_i) \right\|.$$

As already discussed in Section 9.1.1, the terms involving $\mathbf{T}_k$ may be written:

$$\|s_k \mathbf{X}_k \mathbf{T}_k + \mathbf{G}_k\|. \tag{9.37}$$

Both (9.36) and (9.37) are two-sets Procrustes problems with one matrix $\mathbf{G}_k$ and the other $s_k \mathbf{X}_k \mathbf{T}_k$. They may be used as updating steps in iterative algorithms such as 9.1 and 9.2 to optimise the generalised criteria. The details will depend on the choice of $\mathbf{T}_k$.

Note that $2I = KG - T$ which with $T = S + G$ gives $2I = (K - 1) - S$ so relating all the criteria. Essentially, I differs from G by excluding the 'squared' terms $\|s_k \mathbf{X}_k \mathbf{T}_k\|$. See Section 3.1.2 for comments on the role of these terms.

9.2 Special transformations

The above has given results for generalised Procrustes analysis, assuming that the $\mathbf{T}_k$ are general matrices that have been constrained in some, so far, unspecified way. Now we specify the constraints, so generalising the results of Chapters 4 and 5 for two-sets orthogonal and projection problems. It turns out that there is little to add to the general results of Section 9.1 and the results for the special cases of two-sets Procrustes problems discussed earlier, except for some possibilities for simplifications.

9.2.1 GENERALISED ORTHOGONAL PROCRUSTES ANALYSIS

In GPA all the transformation matrices $\mathbf{T}_k$ are orthogonal and hence will be written $\mathbf{Q}_k$. Furthermore, we assume that every $\mathbf{Q}_k$ and $\mathbf{X}_k$ has the same number of columns, if necessary by adding columns of zeros to any $\mathbf{X}_k$ that is deficient. The method seeks to rotate each configuration $\mathbf{X}_k$ to best match $\mathbf{G}$. With these modifications, Algorithms 9.1 and 9.2 are directly available. As with the two-sets case, we showed in Section 9.1.1 that the inner-product and maximal group average criteria amount to the same thing.

In Section 9.1.2, when discussing the least-squares algorithm for GPA with general $\mathbf{T}_k$, we noted that ten Berge (1977*a*) had pointed out that in some pathological situations, *Step 2* of Algorithm 9.1 has no effect and that Algorithm 9.2 is an improvement. In the orthogonal case, he suggests replacing step (ii) (c) by:

Step $2'$ (c) replace each $\mathbf{X}_k$ by $\mathbf{X}_k\mathbf{Q}_k$

Recall that, $\mathbf{G}_k$ is the k-excluded group average, which gives the mean of all the current $\mathbf{X}_i$, *excluding* $\mathbf{X}_k$. Thus, now the current setting of $\mathbf{X}_k\mathbf{Q}_k$, rather than $\mathbf{X}_k$, itself is used in each Procrustes fit. This works because the product of orthogonal matrices is another orthogonal matrix; the modification would not work with other types of matrix and would be totally meaningless for rectangular matrices. An almost identical modification may be made to *Step 3* of Algorithm 9.3 that incorporates isotropic scaling; this introduces no new problems and will not be discussed further. Step $2'$ ensures that $\mathbf{G}_k'\mathbf{X}_k\mathbf{Q}_k$ must be symmetric (see Section 4.2) and, therefore, so is:

$$[\mathbf{X}_k\mathbf{Q}_k + (K-1)\mathbf{G}_k]'\mathbf{X}_k\mathbf{Q}_k = K\mathbf{G}'\mathbf{X}_k\mathbf{Q}_k.$$

It follows from the necessary and sufficient conditions of Section 4.2 that $\mathbf{X}_k\mathbf{Q}_k$, the current setting of the kth configuration, best fits the current centroid, $\mathbf{G}$. *Step* $2'$ improves on the original algorithm where $\mathbf{X}_k\mathbf{Q}_k$ best fits the *previous* rather than the *current* setting of $\mathbf{G}$; ten Berge (1977*a*) reports that the algorithm modified in this way converges in fewer steps. It is, of course, possible that $\mathbf{G}_k = \mathbf{0}$, so that *step* $2'$ has no effect, but this cannot hold for all k, so the modified algorithm always gets started. It was shown in Section 9.1 that a similar modification is available for general matrices $\mathbf{T}_k$.

Section 9.1.3 gave a method for handling missing values in generalised Procrustes problems with K sets when excluding scaling parameters. With GPA there are considerable simplifications. In the notation of that section, when $\mathbf{T}_k$ is an orthogonal matrix, then $\mathbf{T} = \mathbf{I}$. Thus, $t_{ii} = 1$ and $t_{ij} = 0$, so then $x_m = y_{i_m, j_m}$, that is, $\mathbf{x} = \mathbf{y}$. As was observed by ten Berge, Kiers, and Commandeur (1993), this result follows directly from noting that the orthogonal matrix $\mathbf{Q}_k$ may be moved to the right-hand side of (9.13) to give $\|\mathbf{X} - (\mathbf{X}_k\mathbf{T}_k - \mathbf{G}_k)\mathbf{Q}_k'\| = \|\mathbf{X} - \mathbf{Y}\|$, which is clearly minimised by setting the non-zero cells of $\mathbf{X}$ to the corresponding values of $\mathbf{Y}$. Handling translation becomes particularly simple, because the matrix $\mathbf{A}$ is merely $\mathbf{I}$ and the adjustment $\mathbf{T}^*$ is $-1/n\,\mathbf{I}$. It follows that $(1 - 1/n)\mathbf{x}^* = \mathbf{y} = \mathbf{x}$. Thus, if we fit $\mathbf{x}$ without taking translation into account then we may equate this to $(1 - 1/n)\mathbf{x}^*$, to give the adjustments for the missing cells; the adjustments to the non-missing cells are then $-1/(n - 1)\mathbf{x}$. Note that this is not quite the same thing as centering $\mathbf{x}$, which would give $(1 - 1/n)\mathbf{x}$ for the missing cells and $(-1/n)\mathbf{x}$ for the non-missing cells, although as stated in Section 9.1.3 the two approaches can be expected to converge to the same final solutions.

Section 9.1.7 gave a general method for estimating missing values when isotropic scaling is included. With orthogonal transformations, similar modifications to those of the previous paragraph, simplify things. The main change is that now $\mathbf{T} = s_k^2\mathbf{I}$, so that $s_k^2\mathbf{x} = \mathbf{y}$. Also, $\mathbf{y}$ now incorporates a factor s_k. These minor changes are easily handled and make no material difference to the estimation of missing values.

Two sets as a special case of *K* sets In Section 4.3 we discuss an anomaly arising in two sets orthogonal Procrustes analysis in estimating scaling factors when fitting $\mathbf{X}_1$ to $\mathbf{X}_2$ compared with fitting $\mathbf{X}_2$ to $\mathbf{X}_1$. Rather than getting inverse estimates for the scaling factor s when $\|\mathbf{X}_1\| = \|\mathbf{X}_2\|$, both forms of the problem give the same estimate. In the two-sets case $\mathbf{X}_1$ and $\mathbf{X}_2$ appear asymmetrically but in the generalisations (9.1) and (9.17) all $\mathbf{X}_k$ appear symmetrically. Then, when $K = 2$ one is fitting neither to $\mathbf{X}_1$ nor to $\mathbf{X}_2$, but to their average, so the anomaly does not arise.

Kendall (1984) has proposed a form of orthogonal Procrustes analysis which on examination turns out to be equivalent to GPA for the case $K = 2$ and $P = 2$. The main difference is in expressing the criterion in ratio form rather than imposing the size constraint (9.20) but, as discussed in Section 9.1.5, for orthogonal matrices the two approaches do not affect the estimates. Kendall also requires that the orthogonal matrix be a rotation and not a reflection. This rotation is given in Section 4.6.2. The scaling factors are found to be:

$$s_i^2 = \frac{\kappa}{2\|\mathbf{X}_i\|} \qquad i = 1, 2 \tag{9.38}$$

but recall, it is only the ratio that has any substantive interest. The effect of (9.38) is the same as first normalising the two matrices. An explicit expression may be obtained for the special form taken by the residual sum-of-squares (9.17):

$$\|\mathbf{X}_1\mathbf{Q}_1 s_1 - \mathbf{X}_2\mathbf{Q}_2 s_2\| = (\|\mathbf{X}_1\| + \|\mathbf{X}_2\|)\left[1 - \frac{\text{trace}\,\boldsymbol{\Sigma}}{\sqrt{\|\mathbf{X}_1\|\,\|\mathbf{X}_2\|}}\right] \tag{9.39}$$

where trace $\boldsymbol{\Sigma}$ is given in Section 4.6.2. With normalised data (9.39) simplifies to $2[1 - \text{trace}\,\boldsymbol{\Sigma}]$. With Kendall's ratio form of the criterion, the value is given by the expression in square brackets in (9.39) which, normalised, becomes $(1 - \text{trace}\,\boldsymbol{\Sigma})$. The difference of a factor of 2 is merely the factor K that links (9.17) to (9.18). All these expressions are equivalent.

In summary, to avoid the asymmetric effects of scaling, first normalise $\mathbf{X}_1$ and $\mathbf{X}_2$ and either (i) proceed as in Chapter 4 without including any scaling factor, or (ii) use the generalised approach of this chapter, in which case it makes no difference whether or not scaling factors are introduced. Special case simple formulae (Section 4.6.2 and (9.39)) may save some computational time but are consistent with the generalisations for $K > 2$.

Weighting Formula (9.1) may be generalised for the case where $\mathbf{W}_i$ is an $n \times n$ diagonal matrix of weights to be attached to the rows of $\mathbf{X}_k (k = 1, \ldots, K)$. Then a weighted form of (9.2) is:

$$\text{trace} \sum_{k=1}^{K} (\mathbf{X}_k\mathbf{T}_k - \mathbf{G})'\mathbf{W}_k\mathbf{W}(\mathbf{X}_k\mathbf{T}_k - \mathbf{G}) \tag{9.40}$$

where $\mathbf{G} = \mathbf{W}^{-1} \sum_{k=1}^{K} \mathbf{W}_k\mathbf{X}_k$, and $\mathbf{W} = \sum_{k=1}^{K} \mathbf{W}_k$.

This relationship readily allows a weighted form of GPA. The only adjustment is that $\mathbf{Q}_k$ is now calculated from the singular value decomposition of $\mathbf{X}_k'\mathbf{W}_k\mathbf{W}\mathbf{G}$. A variant form in which $\mathbf{W}_k\mathbf{W}$ is replaced by $\mathbf{W}_k$ was discussed by Everitt and Gower (1981).

In Section 4.8.2 we discussed robust orthogonal Procrustes analysis for two sets. The main approach was that of Verboon and Heiser (1992), based on a re-weighted least-squares algorithm (IRLS) using majorisation of Huber, or Mosteller and Tukey, weights. Verboon and Gabriel (1995) give the extensions of this approach for K sets.

9.2.2 EXAMPLE OF A TYPICAL APPLICATION OF GPA IN FOOD SCIENCE

GPA is routinely used in food science, where a sensory panel of tasters assesses several properties of food items. The application is not restricted to food items, but we will focus on food data. A panel of K assessors typically expresses the intensity of a set of M properties of a set of N food items. The scores are expressed on a line-scale, and converted into numerals, ranging, say, from 0 to 100. The exact range is immaterial, and can be adjusted prior to analysis (see, for example, Chapter 2 and Dijksterhuis 1996, 1997).

An $\mathbf{X}_k$ represents the data matrix of the kth assessor, $k = 1, 2, \ldots, K$, with the N food items in its rows, and the M attributes in its columns. A data structure like this is a 3-mode structure and, in this case, it is presupposed that all the K assessors use the same M attributes, and that these are presented in the same order for each $\mathbf{X}_k$, $k = 1, 2, \ldots, K$. The sensory method that results in this type of data is known as 'conventional profiling'. A different data structure is obtained when the assumption of equal numbers of attributes is dropped. In this case there is no match between the M attributes in different $\mathbf{X}_k$, and the data is not referred to as 3-mode, but rather as K-sets (see Section 2.1 and Figure 2.2). This data structure typically stems from Free choice profiling (FCP), where the assessors are permitted to use an idiosyncratic set of M_k attributes; note the subscript k signalling idiosyncrasy here. Neither the label, nor the number of attributes is fixed in this case, leaving matrices $\mathbf{X}_k$ of different column order. GPA was the method which initially justified the collection and the analysis of FCP data (Arnold and Williams 1986).

A typical GPA of a set of sensory scores of a sensory panel is used to study the differences between the assessors in a panel by inspecting the individual configurations, after rotation of the $\mathbf{X}_k$ by $\mathbf{Q}_k$, or after a rotation and an optional isotropic scaling, $s_k\mathbf{X}_k\mathbf{Q}_k$. Note the use of the orthogonal transformation matrix $\mathbf{Q}$ which does not alter the distances between the points in the individual configurations $\mathbf{X}_k$; when an isotropic scaling s_k is included the *relative* distances are unaltered. The isotropic scaling factors themselves reflect the use of the scale by the assessors, with $s_k < 1$ signalling an assessor whose scores were deflated, to match the other assessors, and $s_k > 1$ signalling an assessor whose scores were inflated, $s_k = 1$ signals unscaled scores. The individual configurations $s_k\mathbf{X}_k\mathbf{Q}_k$ are plotted, together with the group average $\mathbf{G}$ (see Figure 9.1). In order to inspect the use of the attributes by the different panellists, the correlations of the attributes with the principal components of the group average $\mathbf{G}$ are plotted and inspected (see Figure 9.2). Note that the configurations depicted have all received the orientation of $\mathbf{G}$ *after* its reference to principal axes through the SVD, $\mathbf{G} = \mathbf{U\Sigma V}'$ (see Section 9.1.8). The exhibited dimensions are obtained by post multiplication of the above-mentioned configurations by the first few, here two, columns of $\mathbf{V}$.

The data in this example are from one sensory profiling session, the last of a range of seven, carried out by Byrne *et al.* (2001). The seven sessions took place in an experiment to develop a sensory vocabulary to describe warmed-over flavour in meat patties, for details of this study we refer to Byrne *et al.* (1999). The data were collected by eight assessors judging six products using 20 attributes. The GPA* of this data set explained 57.8, 18.4, 11.8, 7.11, and 4.76 percent of the variance in the five dimensions of the result. We will inspect the first two dimensions, representing 76.2% variance.

Figure 9.1 depicts the group average positions of the six food items (A, B, C, D, E, F), connected to the positions of the same food items for the individual assessors (called 'set' in the plot). We can see an overlap between the individual positions for the food items D and E in the bottom right-hand part of the plot. Other items appear more separated. It appears that the position of item C for assessor no. 8 ('set 8') is very close to food item A. This can mean that either assessor 8 judged these items to be generally very similar or a mistake was made in presenting the items to the assessors during testing.

The isotropic scaling factors for the 8 assessors are 0.91, 1.00, 0.96, 0.76, 0.80, 2.76, 1.10, and 1.52, respectively. This indicates that assessor six's scores have been inflated relative to the others by a factor of 2.76. This assessor obviously had used only a small portion of the line-scale.

* The software used was Senstools (OP&P Product Research, 1995–8).

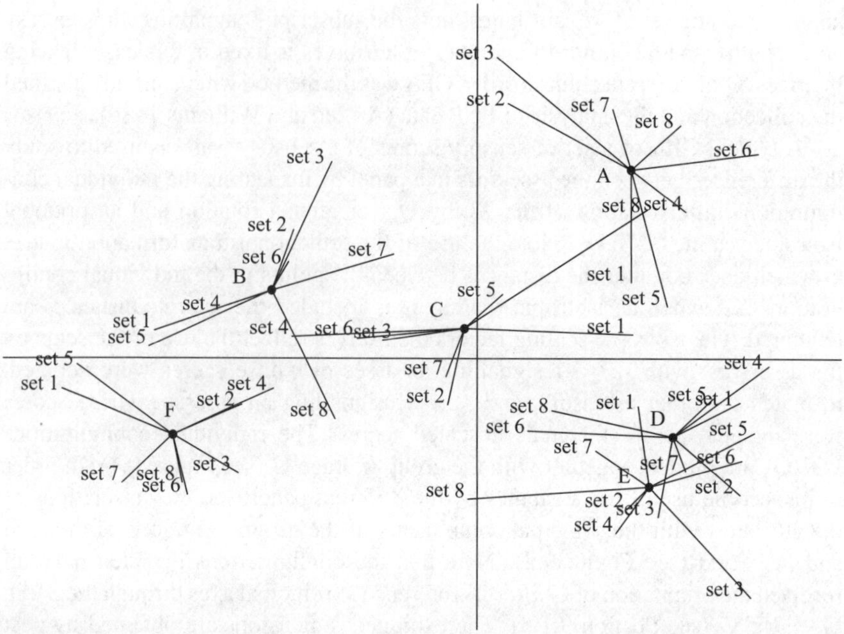

Fig. 9.1. GPA group average configuration, showing the positions of items in the individual configurations.

The inspection of the correlations of the individual attributes with the dimensions of the group average is done by making biplots like those in Figure 9.2. It can be inferred directly that assessor no. 2 showed a different use of the attribute 'roasted-odour'. The vector with label 'set 2 roasted-O' points in the direction of the food item B, while the vectors for most other assessors point in the direction of D and E, signalling these two items to possess the most 'roasted odour'. Assessor no. 1 ('set 1') has yet another deviating opinion concerning roasted odour, he/she scores A to have the most of this. For each separate attribute a biplot as in Figure 9.2 can be plotted.

9.2.3 GENERALISED PROJECTION PROCRUSTES ANALYSIS

Just as there are many forms of the two-sets projection Procrustes problem so are there in the generalisations to K sets. Some of these are discussed in the following.

Minimal residual Projection Procrustes Analysis (Green and Gower's form)
The generalisation to K sets is to minimise:

$$\sum_{i<j}^{K} \|s_i \mathbf{X}_i \mathbf{P}_i - s_j \mathbf{X}_j \mathbf{P}_j\| \qquad (9.41)$$

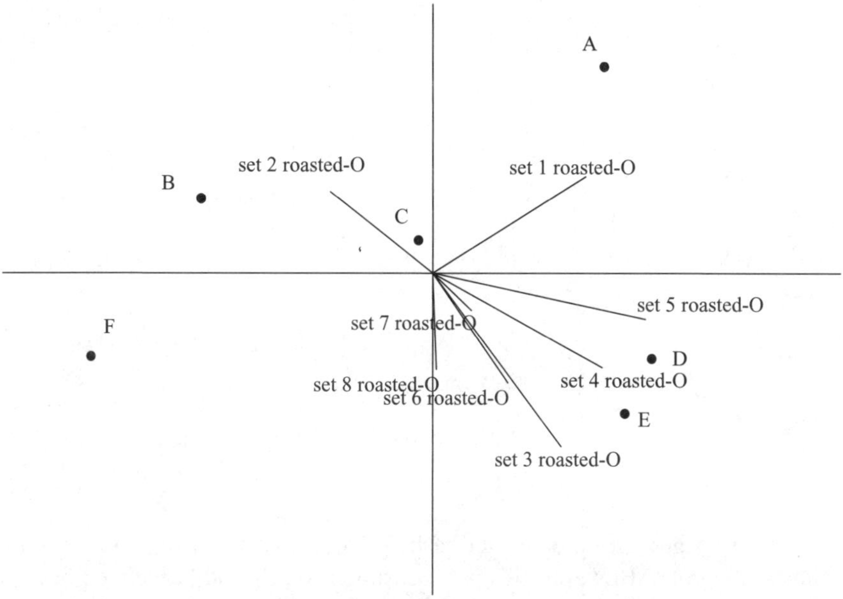

Fig. 9.2. Correlations of one of the attributes, roasted odour, for the 8 assessors, (bi-) plotted in the group average together with the group average configuration of the food items.

where each $\mathbf{P}_k$ has the same number R of columns, usually, but not necessarily, the same as the number of columns of the $\mathbf{X}_k$ of smallest dimensionality. This may be solved algorithmically in the same way as for generalised orthogonal Procrustes. As before one uses (9.2) to re-express (9.41) and successively fits each $\mathbf{X}_k$ to a working mean $\mathbf{G}$, now minimising the Projection Procrustes criterion as in Section 5.4 Algorithm 5.1.

There is an ambiguity over what might be acceptable constraints on the scaling factors. One could require that all dimensions of each $\mathbf{X}_k$, both the part $\mathbf{X}_k\mathbf{P}_k$ that is projected and the part $\mathbf{X}_k\bar{\mathbf{P}}_k$ that is orthogonal to the space of projection, are to be scaled. It turns out that there are at least four different constraints on the scaling of $\mathbf{s}$ that might be considered and these are shown in (9.42).

In (9.42) (i) is our standard scaling (9.20) (ii) scales so that the total size of the configurations in the full space remains unchanged, (iii) scales so that the projected configurations are scaled to have the same total size as the unscaled projected configurations while (iv) is the scaling that is unrelated to any property of the configurations. The second column expresses these constraints in standard algebraic form while the third column expresses the same constraints in matrix notation. The final column gives the eigenvalue equations to be solved under the constraints, assuming that all the elements of $\mathbf{S}$ are given.

$$
\begin{array}{llll}
\text{(i)} & \sum_{k=1}^{K} \|s_k \mathbf{X}_k \mathbf{P}_k\| = \kappa & \mathbf{s}'(\mathrm{diag}\mathbf{S})\mathbf{s} = \kappa & (K\,\mathrm{diag}\mathbf{S} - \mathbf{S})\mathbf{s} = \lambda(\mathrm{diag}\mathbf{S})\mathbf{s} \\
\text{(ii)} & \sum_{k=1}^{K} \|s_k \mathbf{X}_k\| = \kappa & \mathbf{s}'(\mathrm{diag}\|\mathbf{X}_k\|)\mathbf{s} = \kappa & (K\,\mathrm{diag}\mathbf{S} - \mathbf{S})\mathbf{s} = \lambda(\mathrm{diag}\|\mathbf{X}_k\|)\mathbf{s} \\
\text{(iii)} & \sum_{k=1}^{K} \|s_k \mathbf{X}_k \mathbf{P}_k\| & \mathbf{s}'(\mathrm{diag}\mathbf{S})\mathbf{s} & (K\,\mathrm{diag}\mathbf{S} - \mathbf{S})\mathbf{s} = \lambda(\mathrm{diag}\mathbf{S})\mathbf{s} \\
 & \quad = \sum_{k=1}^{K} \|\mathbf{X}_k \mathbf{P}_k\| & \quad = \mathbf{1}'(\mathrm{diag}\mathbf{S})\mathbf{1} & \\
\text{(iv)} & \sum_{k=1}^{K} s_k^2 = \kappa & \mathbf{s}'\mathbf{s} = \kappa & (K\,\mathrm{diag}\mathbf{S} - \mathbf{S})\mathbf{s} = \lambda \mathbf{s}
\end{array}
\tag{9.42}
$$

All these eigenvalue problem are easily solved; indeed, (i) and (iii) differ only in the scaling of $\mathbf{s}$. However, (iii) puts constraints on $\mathbf{P}_k$ and hence modifies the normal equations but, more seriously, although the total size of the configurations in the full space remains fixed, a part of $\mathbf{X}_k$ which is scaled in one iteration will move into the unscaled part at the next iteration. The existing scaling of that part should be undone, but this cannot be arranged compatibly with satisfying the constraints of (9.42) (iii). The difficulty is that the constraint (9.42)(iii) represents a special case of non-isotropic scaling (see Section 8.3). Further work is required to produce suitable algorithms for this case. With all these possibilities and complications, the simplicity of (iv) becomes overwhelmingly attractive (see Section 9.1.6).

Maximal group average projection Procrustes analysis (Peay's form) Peay (1988) introduced the problem of maximising the group average projection. For K groups this requires max $\| \sum_{k=1}^{K} \mathbf{X}_k \mathbf{P}_k \|$ or, with scaling, max $\| \sum_{k=1}^{K} s_k \mathbf{X}_k \mathbf{P}_k \|$. To achieve this we can adopt an iterative algorithm similar to those already discussed. Given some current set of projections we can try to find a projection $\mathbf{P}$, say, that increases the inner-product criterion by replacing $\mathbf{P}_k$ by $\mathbf{P}$ and leaving the remaining projections fixed. That is (9.37) requires that we find $\mathbf{P}$ such that $\|\mathbf{X}_k\mathbf{P} + (K - 1)\mathbf{G}_k\|$ is maximised—see Section 5.6. This replaces *Step 2* in Algorithm 9.3. With isotropic scaling the constraints of (9.42) are all potentially available and lead to eigenvalue equations for $\mathbf{s}$ similar to those given for minimal residual projection Procrustes analysis but with (diag $\mathbf{S} - \mathbf{S}$) replaced by $\mathbf{S}$. If the alternative constraint (iii) of scaling only in the projected space is adopted, then similar difficulties are encountered to those discussed above for Green and Gower's form.

Maximal inner product Procrustes analysis ten Berge and Knol (1984) have discussed this problem in the context of an inner-product criterion rather than the L_2-norm (see Section 9.1.9). Now (9.36) with $\mathbf{T}_k = \mathbf{P}_k$ gives the two-set inner-product criterion to be maximised at each step of the iterative algorithm. It was shown in Section 5.1 that this is equivalent to the orthogonal Procrustes problem. Thus, this step is simpler than for the least-squares solution.

9.3 Generalised pairwise methods

A different approach to handling K sets is to build up a matrix ${}_K\mathbf{M}_K$ whose i, jth element is the residual sum-of-squares $m_{ij}^2 = \|s_i\mathbf{X}_i\mathbf{T}_i - s_j\mathbf{X}_j\mathbf{T}_j\|$. Any triplet m_{ij}, m_{ik}, m_{jk} obeys the metric inequality $m_{ij} \le m_{ik} + m_{jk}$ but the m_{ij} are not necessarily Euclidean distances. Nevertheless, the matrix $\mathbf{M}$ can be analysed by multidimensional scaling methods to give a map with K points, each point representing one of the data sets. Neighbouring points represent configurations that are similar after being transformed by whatever restricted form of the $\mathbf{T}_i$ has been chosen. We call this approach generalised pairwise Procrustes analysis. No group average matrix is concerned in this approach but the centroid of the MDS configuration may be regarded as giving the position of an unknown average configuration. One may then discuss how close the different configurations are to this average. This approach is more flexible than the simultaneous K sets methods discussed above, for if the K points of a pairwise analysis cluster into two or more sets, then it would be natural to regard every set as having its own group average. Cluster analysis methods might be useful. This is not a possibility with the simultaneous methods discussed in Sections 9.1 and 9.2.

9.3.1 EXAMPLE OF GENERALISED PAIRWISE PROCRUSTES ANALYSIS

Gower (1970) gives an anthropometric example of GPA concerned with differences between six hominoid populations—Modern Homo Sapiens, Upper Palaeolithic Homo Sapiens, Middle East Neanderthal, Late (Pekin) Homo Erectus, and Australopithecus Africanus. Measurements were made on fossil skulls drawn from the six populations. The usual way of proceeding would be to evaluate Mahalanobis distances between the populations and to give graphical representations of the resulting distance-matrix, one point for each population. Unfortunately, fossil skulls are likely to be incomplete and we may have only an upper or lower jaw or part of the cranium. Eight different regions of the skull were identified, as follows, with the number of variables in each case shown in the final column.

1	Upper face	16
2	Upper jaw	15
3	Articular region	8
4	Balance	14
5	Basicranial region	12
6	Cranial vault	16
7	Lower jaw	16
8	Overall	16

Table 9.4. The residual sums-of-squares from orthogonal Procrustes analyses for each pair of configurations.

1							
2	1.0012						
3	1.0753	.5766					
4	1.0530	.6324	1.0997				
5	0.3485	.5736	0.6533	.5486			
6	0.8332	.5596	0.8034	.2582	.3466		
7	1.0275	.8155	0.4385	.5952	.5541	.3309	
8	0.8498	.2147	0.5483	.4580	.3504	.4155	.5075

We need a way of handling this situation. For each part of the skull, we may do a canonical variate analysis and get coordinates for each of the six populations. Thus, we have eight configurations and the question then arises as to whether the configurations agree. There are likely to be scale differences between the configurations, partly because they refer to different variables and partly because they are based on different numbers of variables. To eliminate this uninteresting effect, each of the configurations was first scaled to have unit sum-of squares. For every pair of scaled configurations we may do an orthogonal Procrustes analysis and tabulate the values found for the residual sums-of-squares, which the normalisation ensures form a symmetric matrix (see Section 4.3). The results are shown in Table 9.4.

A small residual sum-of-squares between two configurations indicates that the two configurations are similar, a big one that they are dissimilar. Therefore Table 9.4 has some of the characteristics of a distance matrix. Indeed it so happens that with these data the values shown may be treated as squared Euclidean distances in seven dimensions but this property is not true for all data. Both metric and non-metric multidimensional scalings of Table 9.4 are valid. Gower (1971) did both but the results were very similar, as verified by yet another Procrustes analysis. Here, we content ourselves with the principal coordinate analysis/classical scaling representation of Table 9.4. The first two (three) eigenvalues accounted for 62% (84%) of the variability, so it seems that a three dimensional representation is needed. This is shown in Figure 9.3.

It seems that there are considerable differences in the way that the populations are separated, depending on what part of the skull is used. Only the cranial vault and overall measurements tell a similar story, as is also evident from Table 9.4. It is fortunate that the fragmented nature of fossil skulls has forced us to examine the different parts independently and thus recognise the heterogeneous way in which different parts of the skull develop in different populations. Had we had complete skulls this may have passed unnoticed.

In the original paper, Gower also fitted the Mahalanobis distances using MDSCAL solutions in 3, 4 and 8 dimensions. These nonmetric solutions, together with the above metric Principal coordinate analysis were compared by pairwise

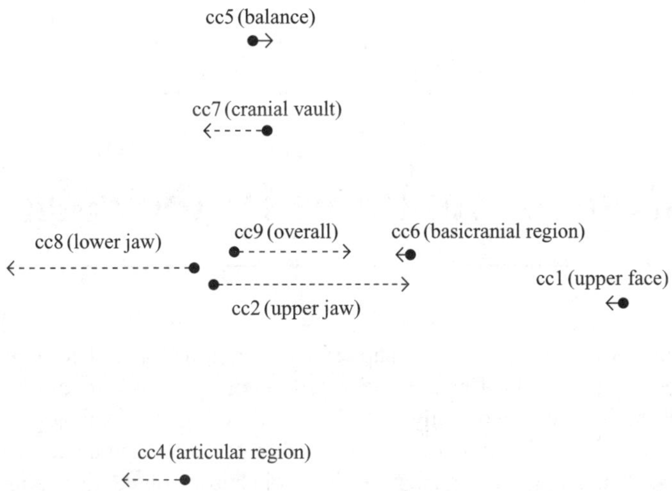

Fig. 9.3. Three dimensional Principal Coordinates Analysis of Table 9.4. The magnitude and sign of the third dimension is shown by the lengths and directions of the horizontal dotted lines (figure 4 from Gower 1971, p. 146).

Table 9.5. The residual sums-of-squares from_orthogonal Procrustes analyses for each of four methods of multidimensional scaling.

		1	2	3	4
1	PCO (3 dimensions)				
2	MDSCAL (3 dimensions)	0.066			
3	MDSCAL (4 dimensions)	0.027	0.091		
4	MDSCAL (8 dimensions)	0.043	0.120	0.048	

orthogonal Procrustes analysis to give the residual sums-of-squares shown in Table 9.5.

Similarly to the way Figure 9.3 was derived from Table 9.4, we could have represented the four methods of Table 9.5 as four points in a Euclidean space. We did not do this but it is clear from Table 9.5 that the best agreement was between PCO in three dimensions and MDCSAL in four dimensions; rather remarkably, the MDSCAL methods all gave poorer agreements between themselves.

Another example of a pairwise Procrustes analysis is given by Stoyan and Stoyan (1990). They are concerned with a two-dimensional shape analysis of nine landmark points on each of 20 hands. The emphasis is on Kendall's (1984) methodology but, as explained in Section 4.6.2, this is equivalent to a generalised orthogonal analysis with $K = 2$ and $P = 2$. They summarise their analyse of the 20×20 matrix of normalised Procrustes metrics not by a multidimensional scaling, as above, but by an average linkage hierarchical cluster analysis.

10
Analysis of variance framework

At several points in previous chapters, expressions have been given that decompose a total sum-of-squares into orthogonal components attributable to structures of interest—especially a residual sum-of-squares and a group average sum-of-squares. These may be presented in the familiar tabular form of an analysis of variance. In this chapter we shall concentrate on the least-squares versions of orthogonal and projection generalised Procrustes analysis. These fit into a Eulidean framework; as was pointed out by Gower (1995); other choices of $\mathbf{T}_k$ and other criteria do not share this advantage. The analysis of variance is useful not only for its usual purposes, which are discussed more fully in Chapter 12, but also as a framework for comparing and contrasting the different Procrustean criteria that arise or may arise.

10.1 Basic formulae

Three simple algebraic results are basic to the development of most forms of analysis of variance. Writing $\bar{x} = 1/N \sum_{i=1}^{N} x_i$ for the mean, these are:

$$\sum_{i=1}^{N}(x_i - y)^2 = \sum_{i=1}^{N}(x_i - \bar{x})^2 + N(y - \bar{x})^2 \tag{10.1}$$

with its important special case for $y = 0$:

$$\sum_{i=1}^{N}(x_i - \bar{x})^2 = \sum_{i=1}^{N} x_i^2 - N\bar{x}^2 \tag{10.2}$$

and

$$\sum_{i<j}^{N}(x_i - x_j)^2 = N\sum_{i=1}^{N}(x_i - \bar{x})^2. \tag{10.3}$$

These are expressed in terms of scalars but extend immediately to matrix forms by replacing x_i by a matrix $\mathbf{X}_i$ and replacing sums-of-squares by the Euclidean norm $\|\mathbf{X}\| = \text{trace}(\mathbf{X}'\mathbf{X})$. It is useful to show these results geometrically. Thus, in terms

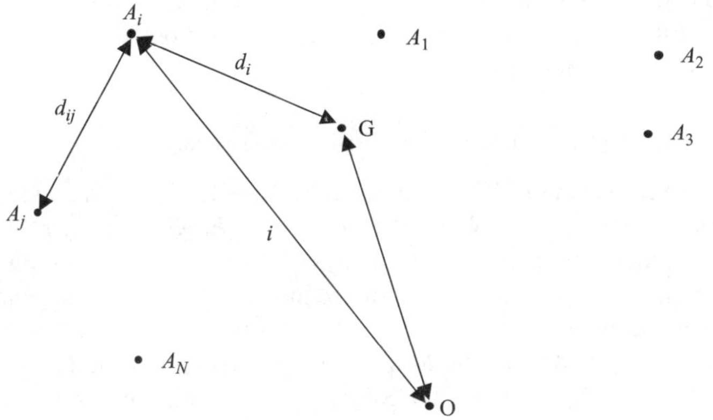

Fig. 10.1. N points A_i where $i = 1, 2, \ldots, N$ with centroid G referred to an arbitrary origin O. In the notation of this figure, equation (10.4) becomes $\sum_{i=1}^{N} \delta_i^2 = \sum_{i=1}^{N} d_i^2 + N\delta^2$ and equation (10.5) becomes $\sum_{i<j}^{N} d_{ij}^2 = N \sum_{i=1}^{N} d_i^2$.

of Figure 10.1, equation (10.1) generalises to:

$$\begin{aligned} &\sum_{i=1}^{N} \overline{\mathrm{A}_i\mathrm{O}}^2 = \sum_{i=1}^{N} \overline{\mathrm{A}_i G}^2 + N\overline{\mathrm{GO}}^2 \\ \text{or} \quad &\sum_{i=1}^{N} \|\mathbf{X}_i\| = \sum_{i=1}^{N} \|\mathbf{X}_i - \mathbf{G}\| + N\|\mathbf{G}\| \end{aligned} \tag{10.4}$$

and (10.3) generalises to:

$$\begin{aligned} &\sum_{i<j}^{N} \overline{A_i A_j}^2 = N \sum_{i=1}^{N} \overline{A_i G}^2 \\ \text{or} \quad &\sum_{i<j}^{N} \|\mathbf{X}_i - \mathbf{X}_j\| = N \sum_{i=j}^{N} \|\mathbf{X}_i - \mathbf{G}\|. \end{aligned} \tag{10.5}$$

These results generate all the forms of analysis of variance developed in this chapter.

10.2 Geometrical unification of orthogonal and projection Procrustes methods

We discuss how Procrustes methods based on orthogonal, **Q**, and projection, **P**, transformations may be embedded in a common Euclidean geometric framework. In 10.2.1 we discuss the general Euclidean representation of configurations and in Section 10.2.2 show how this space may be partitioned into subspaces associated with different kinds of orthogonal projection. Section 10.2.3 gives the basic algebraic representation of the geometry.

10.2.1 ANALYSIS OF VARIANCE OF CONFIGURATIONS

First we assume that the K $N \times P_k$ matrices $\mathbf{X}_k$ $(k = 1, \ldots, K)$ are represented by configurations of K sets of N points in $P = \max(P_1, P_2, \ldots, P_K)$ dimensions. The coordinates of the ith point, denoted by A_{ik}, in the kth configuration are given by $(x_{i1k}, x_{i2k}, \ldots, x_{iPk})$ where it is assumed that zero columns are appended when $P_k < P$ (see Sections 4.5, 5.7, 9.2.1, and Appendix C.1). The centroid of the kth configuration will be denoted by F_k and the centroid of all the points by G. When $P_k = P$ for $k = 1, 2, \ldots, K$ and the columns of each data matrix refer to the same P variables, then these centroid differences may contain valuable information on differences between means that is analysable by linear methods and summarised in a MANOVA; this aspect is well-covered in the literature (see, for example, Hand and Taylor 1984). The geometry is shown in Figure 10.2(a). The conventional analysis of variance may be written:

$$\sum_{k=1}^{K}\sum_{i=1}^{N}\overline{\mathrm{GA}}_{ik}^{2} = N\sum_{k=1}^{K}\overline{\mathrm{GF}}_{k}^{2} + \sum_{k=1}^{K}\sum_{i=1}^{N}\overline{\mathrm{F}_k\mathrm{A}}_{ik}^{2} \tag{10.6}$$

or $T = M + (T - M)$ and summarised in tabular form in Table 10.1, where the differences between means are referred to as the effect of 'translations'. A column for degrees of freedom has been included whose implications will be discussed later. Note that for each value of k (10.6) derives from (10.4) with $(\mathrm{G}, \mathrm{F}_k)$ substituted for (O, G).

In Table 10.1 the degrees of freedom are those standard for a multivariate Between and Within groups analysis. The TOTAL term may need some comment. There are a total of $N\sum_{k=1}^{K} P_k$ values in the K configurations. However, reference to Figure 10.2(a) confirms that Procrustes statistics are invariant

Table 10.1. Analysis of variance separating the translation term M from the total sum-of-squares T, the difference being attributable to the variance of the object scores about their mean values for each configuration.

Source of variation	Degrees of freedom	Sums-of-squares
Translations	$P(K-1)$	$M = N\sum_{k=1}^{K}\overline{\mathrm{GF}}_k^2$
Objects within configurations	$PK(N-1)$	$T - M = \sum_{k=1}^{K}\sum_{i=1}^{N}\overline{\mathrm{F}_k\mathrm{A}}_{ik}^2$
Total	$P(NK-1)$	$T = \sum_{k=1}^{K}\sum_{i=1}^{N}\overline{\mathrm{GA}}_{ik}^2$

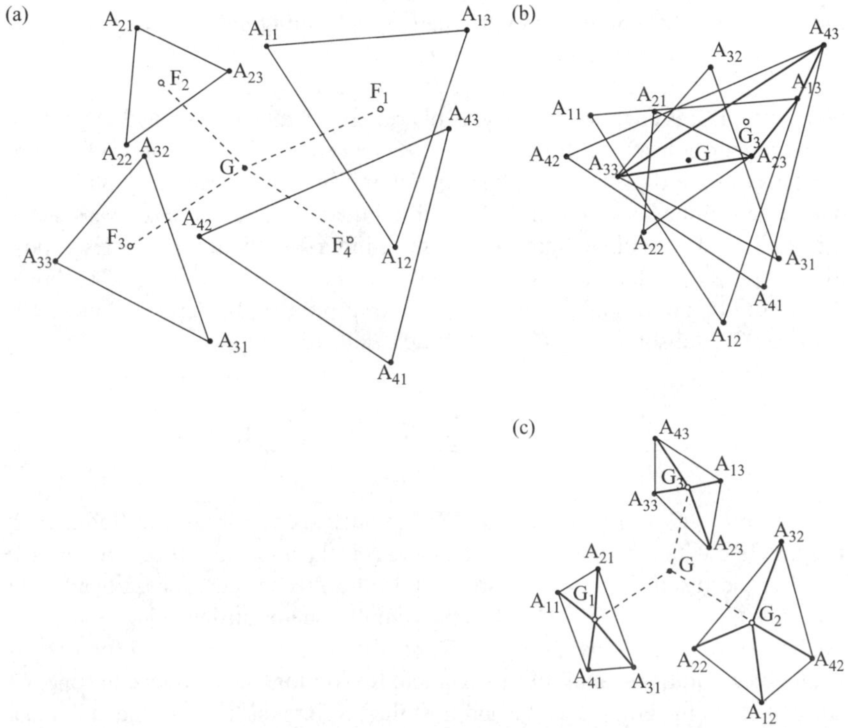

Fig. 10.2. $K = 4$ configurations, each of $N = 3$ points with overall centroid at G. In (a) the differences between the centroids F_1, F_2, F_3, and F_4 are the usual focus of interest in multivariate analysis. In (b) the configurations have been translated so that their centroids coincide at G. In Procrustes analysis it is the centroids G_1, G_2, and G_3 that are the focus of interest. Here, only G_3, the centroid of A_{13}, A_{23}, A_{33}, and A_{43}, is shown. Although no particular orientation is intended, note that to fit the other groups, group 4 would benefit from reflection as well as rotation. In (c) the four configurations in (b) are rotated and reflected so that sums-of-squares of the sets of points A_{1i}, A_{2i}, A_{3i}, and A_{4i} about their centroid G_i ($i = 1, 2, 3$) is minimised.

to a general translation—as opposed to independent translations for the individual configurations. This invariance can be simply handled by fixing the origin at the overall centroid G accounting for P degrees of freedom and leaving $N\sum_{k=1}^{K} P_k - P$ degrees of freedom for the total sum-of-squares. When P_k is constant, this becomes $P(NK - 1)$ as shown in Table 10.1.

Having isolated differences between the centroids F_k, there remain possible differences in orientation (including reflection) of the configurations. Differences in orientation are a property peculiar to multivariate data that has no univariate counterpart, apart from the trivial exception that univariate reflection may be viewed as a simple change of sign. Procrustean methods are concerned with the analysis of orientation. As explained in Section 3.2, the displacement between the centroids may be removed by translating each configuration so that all K sets have a common centroid, G, say, assumed at the origin. Thus, in Figure 10.2(b), F_1, F_2, F_3, and F_4 are translated to G. The total sum-of-squares of the configurations around G, is unaffected by any set of orthogonal transformations $\mathbf{X}_k\mathbf{Q}_k$ where $\mathbf{Q}_k$ has dimensions $P \times P$ for $k = 1, \ldots, K$. Indeed, the device of appending zero columns to a matrix with a deficiency of columns may be viewed as adopting just one of an infinite set of equally admissible initial orientations. For every set of orientations there will be a group average configuration with coordinates at positions G_i $(i = 1, 2, \ldots, N)$ the centroid of $A_{i1}, A_{i2}, A_{i3}, \ldots, A_{iK}$. The break down of the total sum of squares about the centroid may be derived from (10.4) with (G, G_i) substituted for (O, G) for each value of k, to give:

$$\sum_{k=1}^{K}\sum_{i=1}^{N}\overline{GA}_{ik}^2 = K\sum_{i=1}^{N}\overline{GG_i}^2 + \sum_{k=1}^{K}\sum_{i=1}^{N}\overline{G_iA}_{ik}^2$$

or $T - M = G + S$, and exhibited in the analysis of variance of Table 10.2. Here, in line with the notation of Chapter 9, T is the total sum-of-squares around G, S is the residual sum-of-squares within the N configurations around their centroids G_i and $G = T - M - S$ is the sum-of-squares attributable to the group average. In Table 10.2, the $(N-K)P$ degrees of freedom of Table 10.1 for 'objects within configurations' are broken down into two orthogonal components, one, G, attributable to the group average and the other, S, into variation of the individual configurations about the group average. At this stage, we make no assumption about how the orientation is chosen.

The degrees of freedom for 'orientations' depend on whether the transformations are orthogonal or refer to R-dimensional projections. If we examine the projection case, then the orthogonal case is included by setting $R = P$. The first column of a projection matrix $\mathbf{P}_k$ has $P - 1$ independent terms. The second column has to be normal and orthogonal to the first column and so has $P - 2$ degrees of freedom, and so on, until the Rth column which has $P - R$ degrees of freedom. The total number of degrees of freedom of is thus $(P - 1) + (P - 2) + \cdots + (P - R) = RP - \frac{1}{2}R(R + 1)$. For an orthogonal matrix, this becomes $\frac{1}{2}P(P - 1)$ degrees of freedom. For K sets, these quantities must be multiplied by K but we have to allow for the fact that the R-dimensional projections may be arbitrarily rotated in their R-dimensional space. Thus, we have to deduct $1/2R(R - 1)$ degrees of freedom, finally giving the degrees of freedom as shown in Table 10.2.

In principle, every term in the summations in the analysis of variance of Table 10.2 may be listed to examine for outliers. Done extensively this would be impracticable, although it might be useful in an interactive computer context. What is more practicable follows from noting that the double summations associated with S may be written in the two forms $S = \sum_{k=1}^{K}\sum_{i=1}^{N}\overline{G_iA}_{ik}^2 = \sum_{i=1}^{N}\sum_{k=1}^{K}\overline{G_iA}_{ik}^2$.

Table 10.2. Analysis of variance decomposing the total sum-of-squares T into a translation term M for configurations and terms representing the group average configuration, G, and deviations from the group average within configurations, S. (a) refers to orthogonal transformations and (b) to projections onto R dimensions.

Source of variation	Degrees of freedom	Sums-of-squares
Translations	$P(K-1)$	$M = N\sum_{k=1}^{K}\overline{\mathrm{GF}}_k^2$
Orientations:		
Group average	(a) $1/2P(P-1)(K-1)$ (b) $R(P-R)K+$ $1/2R(R-1)(K-1)$	$G = K\sum_{i=1}^{N}\overline{\mathrm{GG}}_i^2$
Individual deviations	(a) $PK(N-1)-$ $1/2P(P-1)(K-1)$ (b) $PK(N-1)-R(P-R)K-$ $1/2R(R-1)(K-1)$	By subtraction $S = \sum_{k=1}^{K}\sum_{i=1}^{N}\overline{\mathrm{G}_i\mathrm{A}}_{ik}^2$
Total	$P(NK-1)$	$T = \sum_{k=1}^{K}\sum_{i=1}^{N}\overline{\mathrm{GA}}_{ik}^2$

Thus, for each of the K sets, we may examine $\sum_{i=1}^{N}\overline{G_iA}_{ik}^2$ and for each of the N objects, we may examine $\sum_{k=1}^{K}\overline{G_iA}_{ik}^2$. This helps identify heterogeneity among the sets and objects, respectively (see Gower 1975, for an example).

10.2.2 PARTITIONING THE ANALYSIS OF VARIANCE INTO COMPONENTS REPRESENTING PROJECTIONS INTO INDEPENDENT SUB-SPACES

The configurations occupy P dimensions, each of which may be represented by a separate column in an extension of Table 10.2. The contributions in each column arise from an orthogonal projection onto one dimension of the P-dimensional space. These one-dimensional subspaces are mutually orthogonal. Rather than exhibit the sums-of-squares for individual dimensions, we may partition the P dimensions into two parts, one of R dimensions, termed the *exhibited space*, and the remaining $P-R$ dimensions, termed the *unexhibited space* to give Table 10.3; this terminology is explained below.

We distinguish between *signal* and *noise*. These concepts are intuitively appealing but difficult to pin down. Most Procrustes analyses display the group average optimally in $R = 2$ dimensions—because paper and computer screens are two-dimensional. Nobody would claim that all group averages are adequately represented in two dimensions, and indeed, efforts are often made to explore higher dimensions. Thus, our R-dimensional group average signal may be subdivided into R_1 dimensions that relate to the exhibited part of the display and $R_2 = R - R_1$ dimensions, that remain signal but are not displayed. Thus we have a signal exhibited in R_1 dimensions, a further part of the signal in R_2 dimensions; the remaining $P-R$ dimensions are attributable to noise. The role of these partitions in interpretation is discussed below.

Table 10.3. Analysis of variance as in Table 10.2 but partitioning every sum-of-squares into a part that is exhibited in R dimensions and an independent non-exhibited part in $P - R$ dimensions. The 'total' column is as in Table 10.2.

Source of variation	Sums-of-squares		
	Exhibited (R dimensions)	Unexhibited ($P - R$ dimensions)	Total (P dimensions)
Translations	M_R	$M_{(P-R)}$	M
Orientations			
Group average	G_R	$G_{(P-R)}$	G
Deviations	$T_R - M_R - G_R$	$T_{(P-R)} - M_{(P-R)} - G_{(P-R)}$	$T - M - G$
Total	T_R	$T_{(P-R)}$	T

The benchmark for noise is the residual sum-of-squares S in the analysis of variance. However, we have seen that not all of the sum-of-squares attributed to 'group average' need represent signal; much depends on the particular Procrustes method used and on the structure of the configurations $\mathbf{X}_k$. For example, in an orthogonal Procrustes analysis where each $\mathbf{X}_k$ is derived from a multidimensional scaling, then its dimensionality R will have already been determined to eliminate noise. The differences between the configurations contribute to S but the dimensionality of the group average is predetermined and in effect $R = P$. Figure 10.2(c) contains all the information supplied by a two-dimensional MDS analysis. When the initial configurations are given in terms of P, normalised, variables then a group average in R dimensions will have an element in the remaining $P - R$ dimensions that may be regarded as noise. In projection Procrustes methods, there is a suggestion that the group average of the R-dimensional projection is signal, again with $P - R$ dimensions of noise. Thus, we have the breakdown of Table 10.4.

The degrees of freedom to be associated with Tables 10.3 and 10.4 are omitted, but are the same as in Table 10.2. We may ask how these degrees of freedom might be distributed over the columns. When the projections are chosen randomly, it would be natural to distribute the degrees of freedom in the ratio $R_1 : R_2 : P - R_1 - R_2$. However, as we have seen, the projections are usually chosen to reflect the decreasing order of the eigenvalues of a components analysis of $\mathbf{G}$ or of some other matrix. Then, the larger components carry greater weight and should carry a greater proportion of the degrees of freedom than for the smaller components. There is no simple answer as to how to allow for this disparity. As a *very* crude approximation, one might adopt the asymptotic result from the multi-normal development of canonical variate analysis, which allows two degrees of freedom less for each successively smaller eigenvalue. Thus, for a total of ΔP degrees of freedom, the P sets of degrees of freedom associated

Table 10.4. Analysis of variance as in Table 10.3 but partitioning every sum-of-squares into a part that is exhibited in R_1 dimensions, a systematic (signal) but unexhibited part in R_2 dimensions, and a random (noise) part in the remaining $P - R$ dimensions, where $R = R_1 + R_2$.

Source of variation	Sums-of-squares		
	Exhibited (R_1 dimensions)	Unexhibited (R_2 dimensions)	Noise ($P - R$ dimensions)
Translations	M_{R_1}	M_{R_2}	$M_{(P-R)}$
Orientations			
Group average	G_{R_1}	G_{R_2}	$G_{(P-R)}$
Deviations	S_{R_1}	S_{R_2}	$S_{(P-R)}$
Total	T_{R_1}	T_{R_2}	$T_{(P-R)}$

with G would be distributed as $(\Delta + P - 1)$, $(\Delta + P - 3), \ldots, (\Delta - P + 3)$, $(\Delta - P + 1)$. This crude approximation assumes that R_1 and R_2 are based on a decomposition of G. There seems little justification for subdividing the degrees of freedom in the residual S in the same way and we suggest that the ratio $R_1 : R_2 : P - R_1 - R_2$ would be more appropriate; often it will suffice to pool. How degrees of freedom might be used is discussed briefly in Section 4.3 and again in Chapter 12.

We have not explicitly shown the effects of isotropic scaling in these analyses of variance but provided the constraint (9.20), or some variant of it, is adopted, there is little difficulty. This is because (9.20) ensures that the total sum-of-squares $T - M$ is invariant to scaling. Variants that might be adopted are influenced by the notion that scaling should not be associated with any noise that might be regarded as part of the system (Chapter 9, section on minimal Residual projection Procrustes Analysis). When the scaling (9.20) is applied only to the projections that are regarded as part of the signal, that is, in R or R_1 dimensions, in which case it is $T_R - M_R$ or $T_{R_1} - M_{R_1}$, respectively, that are unscaled, then there are the computational difficulties mentioned in Chapter 9, section 9.2.3 on minimal Residual Projection, but the total sum-of-squares remains constant and the analysis of variance of Table 10.4 remains valid.

Of course, isotropic scaling will reduce the value of S and increase G. The constraint (9.20) implies that there are $K - 1$ degrees of freedom associated with the K-scaling factors. Thus to allow for isotropic scaling, $K - 1$ should be deducted from the degrees of freedom for S shown in Tables 10.2, 10.3, and 10.4. The reduction in the residual sum-of-squares is not part of an orthogonal analysis of variance but may be used to test if there is any evidence that the introduction of scaling factors has any sizeable effect—see Chapter 12.

Clearly weighted Procrustes methods are called for when there is unequal replication. Weighting of this kind introduces no serious complications. The more elaborate forms of anisotropic scaling, discussed in Chapter 8, and written algebraically as $\mathbf{Q}_k\mathbf{S}_k$, $\mathbf{R}_k\mathbf{Q}_k$, $\mathbf{P}_k\mathbf{S}_k$, or $\mathbf{R}_k\mathbf{P}_k$ ($\mathbf{R}_k$ and $\mathbf{S}_k$, diagonal) may sometimes be expressed as terms in an analysis of variance but we have not pursued the details.

10.2.3 ALGEBRAIC FORM OF THE ANOVA

The 'translation' line in all the above ANOVA derives in the usual way from differences between the K means so, as in most of this book, it is assumed in this section that $\mathbf{X}_1, \mathbf{X}_2, \ldots, \mathbf{X}_K$ refer to deviations from their means and we concentrate on the *Orientation* part of the analysis. Thus, with the constraint (9.20)

$$T - M = \sum_{k=1}^{K} \|\mathbf{X}_k\| = \sum_{k=1}^{K} \|s_k \mathbf{X}_k \mathbf{Q}_k\| \tag{10.7}$$

for *any* orthogonal matrices $\mathbf{Q}_k$ ($k = 1, \ldots, K$). Equation (10.7) is appropriate, even when we are primarily concerned with projection Procrustes problems, because the P-dimensional geometry applies for both projection and rotation. Suppose $\mathbf{Q}_k$ is partitioned into H mutually independent projection matrices $\mathbf{P}_{kh}$ with R_h columns ($h = 1, \ldots, H$) where $\mathbf{P}_{kh}$ gives the projection of $\mathbf{X}_k$ onto the hth subspace of R_h dimensions. In Table 10.4, we have set $H = 3$ with $R_3 = P - R_1 - R_2$. In general $\mathbf{Q}_k = (\mathbf{P}_{k1}, \mathbf{P}_{k2}, \ldots, \mathbf{P}_{kH})$ and

$$T - M = \sum_{k=1}^{K} \|\mathbf{X}_k\| = \sum_{k=1}^{K} \|s_k \mathbf{X}_k \mathbf{Q}_k\| = \sum_{k=1}^{K}\sum_{h=1}^{H} \|s_k \mathbf{X}_k \mathbf{P}_{kh}\|. \tag{10.8}$$

For the R_h-dimensional part of the group average configuration we have that:

$$\overline{\mathbf{X}}_h = \frac{1}{K}\sum_{k=1}^{K} s_k \mathbf{X}_k \mathbf{P}_{kh}$$

so that

$$\sum_{k=1}^{K} \|s_k \mathbf{X}_k \mathbf{P}_{kh}\| = K\|\overline{\mathbf{X}}_h\| + \sum_{k=1}^{K} \|s_k \mathbf{X}_k \mathbf{P}_{kh} - \overline{\mathbf{X}}_h\| \tag{10.9}$$

dividing the orientation sum-of-squares in the R_h-dimensional projection into a part $K\|\overline{\mathbf{X}}_h\|$ for the group average and a part representing departures from the group average. This is the algebraic form of $T_{Rh} - M_{Rh} = G_{Rh} - S_{Rh}$.

10.2.4 AN EXAMPLE OF ANOVA

In this section we give a short example of a Procrustes ANOVA. The data from which this analysis was derived are discussed in Section 11.2 and in greater detail by

Table 10.5. ANOVA of coffee image data. Degrees of freedom are bracketed.

Source of variation	Sums-of-squares		
	Exhibited (two dimensions)	Unexhibited (six dimensions)	Total (eight dimensions)
Orientations			
Group average	40.29 (118)	19.60 (212)	68.89 (330)
Deviations	19.74	11.36	31.11 (286)
Total	69.04	30.96	100 (682)

Gower and Dijksterhuis (1994). Table 10.5 shows the important sums-of-squares. The translation term with 66 degrees of freedom has been excluded and we have not made any investigation of what part of the unexhibited space may be attributable to signal rather than noise (see below). The 330 degrees of freedom for the group average have been split between the exhibited and unexhibited parts as discussed above in Section 10.2.2.

The deviations (residual sum-of-squares) being noise should not differ in the exhibited and unexhibited parts but we cannot test for this formally as we do not know what degrees of freedom to assign to the two parts. Thus we take the residual mean square from the 'total' column as $31.11/286 = 0.109$. An F-test for the exhibited group average is $(40.29/118) \div (0.109) = 3.14$ on 118 and 286 degrees of freedom. For the unexhibited group average, the corresponding F-value is $(19.60/212) \div (0.109) = 0.85$. Thus the indication is that the exhibited part represents a definite signal while the unexhibited part cannot be differentiated from noise. Thus, the decision not to try to separate out any signal in the unexhibited space seems justified and there is little to support other than the two-dimensional displays that we show in Section 11.2.

10.3 Criteria as terms in the ANOVA

Tables 10.3 and 10.4 subsume several well-known Procrustean criteria and suggest others that have not been explored in the literature. In orthogonal Procrustes analysis, the term S in Table 10.3 is minimised or, equivalently, it is the sum-of-squares G that is maximised. Then, G_R is not the maximal value that could be attained within $R < P$ dimensions. By maximising G_R directly one obtains the largest projection of the group average configuration onto R dimensions (Peay 1988, see Chapter 9, section 9.2.3 on Maximal Group Average Projection Procrustes Analysis). Maximising G_R is not the same as minimising S_R, which is what Green and Gower (1979) do (Section 5.4). As a variant of Peay's method one could minimise $G_{(P-R)}$ rather than maximise G_R or, taking note of the MANOVA form of the table, one could maximise a variance-ratio such as G_R/S_R (see Section 9.1.6). In Section 9.1.8 we mention the possibility of minimising S

with a group average of specified rank; this is equivalent to minimising S under the constraint $G_{(P-R)} = 0$. Many other possibilities will suggest themselves to the reader.

The ANOVA also suggests a common algorithmic framework for this class of Procrustes problems. The table isolates all the terms that may be of interest for any set of rotations and projections. By identifying the entry in the table, or ratio of entries, that is to be optimised, it should be possible to cycle through a series of rotations and projections until the optimum is reached. Basic algorithms include those outlined in Chapters 3–5 but the details require further work.

Further work is also required to give guidance on determining a proper dimensionality R of any projection space. In the general case, the distributional underpinning for any significance testing of dimensionality is probably beyond the bounds of mathematical tractability. Chapter 12 discusses the few analytical results that are available and the possibilities of boot-strapping. Often it will have to suffice to select the various entries in Table 10.4 that might be attributable to noise and base acceptance of the partition on the observation that if R is correctly chosen then these entries should all be of comparable size, after due allowance is made for degrees of freedom and different dimensionalities. To be a little more precise, we may inspect the mean squares for heterogeneity; values markedly greater than the mean square associated with S suggest the presence of signal. In Chapter 12 we mention the possibility of using formal F-tests when comparing the ratios of mean squares. This is a crude approach that needs much improvement but it is better than choosing $R = 2$, merely because one can only conveniently inspect graphical displays in two dimensions.

Recalling that the ANOVAs of Tables 10.1–10.4 are valid for any set of orientations. It follows that an ANOVA exists for any set of configurations derived by optimising *any* criterion, such as the inner-product or L_1-norm criteria. The framework of the ANOVA highlights the projection spaces for noise and signal, the latter itself subdivided into exhibited and unexhibited parts, and concentrates attention on the various sub-spaces in which the performance of the chosen criterion might be evaluated. Of course, the nice additivity properties that go with sums-of-squares in orthogonal spaces no longer apply to partitions based on other criteria and it is then a little illogical to base interpretations on sums-of-squares (or ratios of sums-of-squares). Nevertheless, as a crude approximation, and used with caution, ANOVA may still help with interpretation.

10.4 Discussion

It is one thing to have a series of related Procrustes criteria and algorithms to fit them but it is another to know if and when each one is useful. Sometimes, articles in the literature seem to promote one Procrustes method rather than another on what seem to us to be rather dubious grounds. The ANOVA tables make it clear that the criteria for optimisation define different models that apply to different problems and should not be regarded as competitors. The new variants suggested by the unification discussed above identify further models of potential practical importance. Misunderstandings seem to arise partly from a tendency to focus on one particular item in the table without regard to the others and partly on comparisons across different criteria. Thus, one might find that one gets a bigger group average in R dimensions by maximising the trace of an inner product of residuals, than by minimising S. Such findings make little sense to us because if one wants to maximise a group average then one should do so; other criteria are irrelevant. A similar inconsistency arises when cross-comparisons are made with other methods that incorporate group averages into their models (see Chapter 13); one has to decide on the most appropriate model from the outset. Of course, one may argue about what is meant by a maximal group average. Is it a sum-of-squares or the volume of the configuration or some other measure of size? Further, can we distinguish signal from noise and so define size in a signal space, eliminating irrelevant irregularities? The thing is to decide just what characteristic of the configurations it is that one wants to optimise. The ANOVA table provides guidance in making such decisions and in the remainder of this section we shall try to show how.

We shall focus on three issues that seem to be implicit in the development of generalised Procrustes methods: (i) getting the biggest group average, (ii) separating signal from noise, and (iii) how to deal with unequal values of P_k. We shall confine our discussion to the least-squares formulation, because it underpins the analysis of variance and because defining size in terms of an L_2-norm $\|\mathbf{X}\|$ is almost universally accepted. A crucial property of our geometrical unification is that for *all* projection and rotation methods, the final configurations are orientations of the original configurations. This assertion manifests itself algebraically in Section 10.2.3 where $\mathbf{Q}_k$ is partitioned into independent orthonormal, projection, components. That is, all projections are incorporated within generalised rotations. It follows that when we compare our class of Procrustes problems, we are comparing different orientations of the K configurations.

The generalised orthogonal Procrustes criterion is simple and gives an obviously useful method for matching configurations that are independently oriented. It gives the biggest group average, G and, simultaneously, the smallest residual sum-of squares, S, whose contributions are summarised in the most simple form of ANOVA given in Table 10.2. When all P dimensions are regarded as signal, then no better can be done. We may, of course, seek the best R dimensions for exhibiting the group average. Principal components analysis provides the best R-dimensional projection, but the total size of G in the exhibited and unexhibited spaces does not change. Things are different when the $P - R$ unexhibited dimensions can be regarded as noise; we return to this case below.

Generalised orthogonal Procrustes analysis has been questioned when the dimensions P_k $(k = 1, \ldots, K)$ differ between configurations; in particular, there is some unease with the device of appending zero columns so that all configurations are embedded in the same maximal space. This unease seems to have been one of the reasons behind the interest in developing projection methods that work in terms of the minimal, or smaller, space shared by the K configurations. We have seen that this unease is misplaced because the zeros provide a valid initial orientation in the maximal space and because rotations of lower dimensional configurations in the maximal space are not a problem.

The ANOVA of Table 10.3 incorporates the modifications that accommodate R-dimensional projections. The many alternative projection methods described in Chapter 5 operate in the maximal space even though attention is focused on projections onto an R-dimensional sub-space. By concentrating on the R-space, attention is drawn away from what is happening in the $(P - R)$-dimensional space. Whether or not this matters, depends on the interpretation to be given to the R-dimensional sub-space.

Peay (1988) proposed his method of maximal group average projection (Sections 5.6 and Chapter 9, section on Maximal Group Average Projection) as one that gave an improved (bigger) estimate of the group average configuration in R dimensions; that is, it maximises G_R. When the R-dimensional space contains all the signal and coincides with the exhibited space (i.e. $R_2 = 0$), then the Peay-type approach described in Chapter 9, section on Maximal Group Average Projection is a useful one that may be regarded as a method for separating signal from noise. Compared with the orthogonal case, it will give a smaller value of G, a bigger value of G_R and a smaller value of $G_{(P-R)}$. However, when the exhibited space is only part of the R-dimensional signal space, then the R_1-dimensional group average configuration is indeed the biggest possible in R_1 dimensions but could give a misleading impression of the full R-dimensional group average.

With minimal error projection (Section 5.4) S_R is minimised. This method should be used with caution because it might be possible to find projections of the K configurations that give very good, even exact, R-dimensional fits but leave vast amounts of unexplained variation orthogonal to this projection space. The method may be thought of as finding what agreement there may be between the configurations but this agreement should always be contrasted with the degree of disagreement. Such information may be useful and was exactly what was required by Constantine and Gower (1982), although only for the case of oblique Procrustes analysis ($R = 1$); this application is discussed in Section 5.4.2. Originally Gower (1975) termed the group average of GPA the *consensus configuration* but has since avoided the term. The group average of minimal error projection is much more like a true consensus and, as with all consensuses, may be slight.

The key to choosing the most appropriate method is deciding on (i) what is signal and what is noise and (ii) what part of the signal is exhibited. The most desirable situation is when the exhibited space coincides with the signal. One has to beware of a natural tendency to regard variation in the R_2-dimensional space as noise when it might more realistically be part of the signal. Minimising S_R has shown us that the consensus signal may be small compared with non-consensus noise represented by $S_{(P-R)}$. Examination of mean-squares in the ANOVA, as discussed briefly in Section 10.3, gives some guidance as to what is noise and what is signal but help that might be supplied by inferential methodology (see Chapter 12) is at a basic level.

We have mentioned the place of principal components analysis in GPA for providing the exhibited part of the group average. This method is applicable to any R-dimensional group average. Indeed, any of the methods discussed in Section 9.1.8 are always available. A serious possibility that might be considered for showing a low-dimensional view of a high-dimensional group average configuration, is to use a non-metric multidimensional scaling method (see, for example, Kruskal and Wish 1978). By incorporating non-linear transformations, non-metric MDS provides an ideal way of exhibiting an R-dimensional configuration in R_1 dimensions. Methodology exists that allows the individual configurations to be exhibited together with the transformed group average, but it has to be remembered that all the centroid properties shown in Figure 10.2 are lost. In principle, the non-linear transformations of the $\mathbf{X}_k$ could be incorporated from the outset, to give a conventional Procrustes analysis of transformed values (see van Buuren and Dijksterhuis 1988). Thus, there are many ways of approximating the signal in R_1 dimensions but it should not be forgotten that the signal often occupies R dimensions.

11 Incorporating information on variables

The displays that go with Procrustes analyses show approximations to the configurations representing the K sets and the N cases whose exact forms occupy a P-dimensional Euclidean space. Throughout this book we have held that, in general, the P dimensions do not correspond to observed variables. In the strictest interpretation of this position, it is meaningless to ask how to incorporate information on the variables into the displays. In practice, we have seen in Chapter 2 that the columns of each $\mathbf{X}_k$ may relate directly to variables or they may derive from variables through some form of multidimensional scaling (see Figure 2.1). Then it is reasonable to ask if information on the variables can be included in the Procrustes displays, especially in what, in Chapter 10, we have termed the exhibited display.

Our approach to this is a simple one, based on generalised biplot theory (Gower and Hand 1996). The basic idea is that the initial configurations $\mathbf{X}_k$ may be referred to generalised coordinate axes representing the original variables. These axes may be simple orthogonal Cartesian coordinate axes, or they may be oblique or they may be curvilinear; for categorical variables, the generalised axes are sets of points, known as category level points. These generalised coordinate axes are termed *reference systems* by Gower (2002). Reference systems behave much like conventional coordinates but their construction is very different. Conventional coordinate axes are *given* constructs, that allow points to be positioned; reference systems start from a configuration of points and *then* construct a coordinate system, to which the given configuration, and additional points, may be referred. Details of how to construct reference systems are given by Gower and Hand (1996). Like conventional coordinate axes, the axes of a reference system are calibrated and labelled with a numerical scale; category level points are labelled by category names. The values of variables to be associated with any point P are given by the nearest scale-value on the axis of each quantitative variable, implying normal projection onto each axis, and for each categorical variable, by the label associated with the nearest of the set of category level points pertaining to that variable.

Reference systems composed of linear or non-linear coordinate axes (continuous variables) and category-level points (categorical variables) may be regarded as being rigidly embedded in the Euclidean space containing the configuration. In

Section 10.2.3, it was shown that all projection Procrustes methods may be treated as a generalised orthogonal rotation, part of which comprises the projection. It follows that after these types of Procrustes analyses, the embedded reference systems as well as the configurations themselves, coexist in the same P-dimensional space. Then, the reference system can be approximated in any sub-space from which biplot representations may be derived for all the orthogonal and projection Procrustes methods. In two-set problems, the relevant sub-space will be that containing $\mathbf{X}_2$. In K-sets problems the *exhibited space* is the centre of interest.

Within this basic framework, there are many possibilities, depending less on the particular Procrustes method used, than on the multivariate method of deriving $\mathbf{X}_k$ from $\mathbf{Y}_k$ and on the types of variable—quantitative, nominal categorical, ordered categorical. It would be an impossible task to describe all the variants in detail but, fortunately, the same basic principles are valid in all cases. In this chapter we give an overview of the more important special cases and illustrate methodology through an example. What is described here is only a beginning. Many more examples are needed together with the explicit formulae needed for the special cases of the general methodology.

11.1 Biplots

In this section we give a brief description of three of the routes through Figure 2.1. Section 11.1.1 discusses the situations where $\mathbf{X}_k$ derives directly from observed variables $\mathbf{Y}_k$. Section 11.1.2 and 11.1.3 discuss situations where $\mathbf{X}_k$ derives from metric or non-metric multidimensional scaling of a dissimilarity matrix $\mathbf{D}_k$ which itself derives from the observed variables. There are two main classes of multidimensional scaling methods—(a) where data matrices $\mathbf{Y}_k$ generate dissimilarity matrices $\mathbf{D}_k$ that subsequently may be optimally transformed (Section 11.1.2) and (b) where $\mathbf{Y}_k$ is optimally transformed before generating the dissimilarity matrices $\mathbf{D}_k$ (Section 11.1.3). In both cases, the MDS displays $\mathbf{X}_k$ derived from $\mathbf{D}_k$ are related to the variables of $\mathbf{Y}_k$.

11.1.1 LINEAR BIPLOTS

The most simple case is given by the classical multivariate set-up where each variable is represented by one of P orthogonal Cartesian axes; the cases then become a set of points in a P-dimensional Euclidean space. The axes may be calibrated in the usual way to give markers for the values of the variables. Suppose the cases are projected onto any subspace (often two-dimensional) of the full space. To predict the value of any variable for a point in the subspace is merely a matter of reading off the nearest marker on the appropriate Cartesian axes. Unfortunately, this means going out of the subspace, but it can be shown that the same effect may be obtained by reading off an appropriately located nearest marker on the projected Cartesian axes (Gower and Hand 1996) which are termed *biplot* axes. The positions of the markers on the biplot axes are such that when orthogonally projected back onto the original axes, the correct marker-values are recovered. Gower and Hand (1996) term this *back projection*.* Projection of the axes destroys their orthogonality but this is not a problem. The above procedure is used most commonly with the Principal Component biplot, where the relevant subspace is that spanned by the first few principal axes. The geometry, however, is valid for any subspace and, in particular, for the exhibited subspace containing an approximation to the group average configuration of the several variant forms of projection Procrustes analysis and of orthogonal Procrustes analysis.

When the columns of the $\mathbf{X}_k (k = 1, \ldots, K)$ refer to variables, we may follow the geometry described in the previous paragraph so that each of the K sets has its own Cartesian axes. The effect of orthogonal transformations (Chapter 4) is to rotate the Cartesian axes of each of the K sets. The same thing applies, as described in Chapter 9, to the Maximal Group Average Projection and Minimal Residual Projection Procrustes methods. In all these cases there is a subspace, usually two-dimensional, which shows the results of the Procrustes analysis and especially, the group average. Then the original Cartesian axes may be projected onto this subspace (and the markers back-projected) to give a representation of variables. A problem with this representation is that every variable appears K times, one projection for each set $\mathbf{X}_k$. When the variables match, the resulting confusion can be simplified by replacing the K axes by their mean, obtained by joining the origin to the centroid of the K unit-markers, one on each axis. When the K representations of the same basic variables align themselves in approximately the same direction the mean loses little information, otherwise a substantial loss of information can occur. Then, it is probably best to examine separately each final configuration $\mathbf{X}_k\mathbf{T}_k$ together with its own dedicated biplot axes. Alternatively, for each variable, we could represent its K representaions on separate plots of the same group average (see, for example, Figure 9.5). Indeed, these could be the only option with free choice profiling (Section 2.2.1), where the variables do not match, although the possibility should be kept in mind that aligned free-choice variables might indicate that they are measures of the same or similar things.

When the columns of $\mathbf{X}_k$ are derived by simple non-linear transformations from the observed basic variables, given as the columns of the data-matrix $\mathbf{Y}_k$, then linear calibration of the biplot axes will be in the units of the transformed variables. One may give a non-linear calibration in terms of the units of the original variables in precisely the same way as when a logarithmic scale is associated with a logarithmic transformation.

Another way of deriving the principal component biplot is to regress the values of each of the original variables on the coordinate positions in the subspace. The regression coefficients then give the coordinates of points which can be shown to lie in precisely the same direction from the origin as do the axes obtained by projection (see, for example, Gower and Hand 1996, Section 3.3.2). For this reason the regression approach is often recommended even when the points in the

* Ordinary orthogonal projection is useful for interpolating new cases into the displays of configurations.

subspace are not obtained by components analysis, but by some other method of multidimensional scaling. One such situation is when the points in the subspace are those of the group average derived by one of the forms of Procrustes analysis. When $\mathbf{X}_k$ has a non-linear relationship with $\mathbf{Y}_k$, it seems unwise to express this through linear regression.

11.1.2 BIPLOTS IN MULTIDIMENSIONAL SCALING OF TRANSFORMED DISSIMILARITIES

Section 2.1.3 outlines how a distance, or dissimilarity, matrix $\mathbf{D}_k$ may be derived from a data matrix $\mathbf{Y}_k$. Subsequently, $\mathbf{D}_k$ may be analysed by any form of multidimensional scaling, metric or non metric, to give a configuration $\mathbf{X}_k$. Gower and Hand (1996) show how to represent quantitative variables in the space of the configurations by a reference system of curvilinear coordinate axes. With classical scaling/principal coordinate analysis, the computations are straightforward and, as with linear biplots, after Procrustes analysis, the reference system may be back projected onto the exhibited space, to give calibrated curvilinear axes.

A categorical variable with L_k levels is represented by L_k labelled category level points and the level associated with the category pertaining to any case is given by the label of the nearest category level point. This implies that the whole space may be partitioned into convex neighbour regions, one for each category-level. These *neighbour regions* intersect the exhibited space to manifest themselves as convex *prediction regions* in the final Procrustes analysis. Cases lying within a prediction region are predicted as having the associated nominal category level. With ordered categorical variables, the category level points are collinear, giving prediction regions that are bounded by parallel straight lines; in this case the prediction regions may be defined by nominal markers on a linear axes. Thus, corresponding to the biplot axes of quantitative variables, we have convex prediction regions for categorical variables.

In Section 11.1.1, we saw how with linear biplots there were problems with handling the multiplicity of linear axes (K axes for each of the P variables) in the exhibited space. With prediction regions the position is much worse, because it is usually unhelpful, to put it mildly, to attempt to combine the P prediction regions for just one of the K configurations (see figures 4.6 and 4.7 in Gower and Hand 1996). In close analogy with quantitative variables (Section 11.1.1), for each categorical variable, we may find the centroid of the category level points for every level of that variable. This gives a set of average category level points for each variable and these will have their own neighbour regions that may be approximated as prediction regions in the exhibited space of any Procrustes analysis. It remains practicable to show only one set of prediction regions on any one diagram. A more manageable representation, that permits all categorical variables to be shown simultaneously, is to display the projected category level points together with their distances (or squared distance) from the true category level points. In principle, this allows the calculation of the distance of any point in the exhibited display from any category level point but, unfortunately, this is a cumbersome process.

All the above applies to classical scaling/principal coordinates analysis; with other methods of metric and with non-metric scaling, heavy computation is required and the reference system is directly represented in the space of $\mathbf{X}_k$ and may have suboptimal properties.

11.1.3 BIPLOTS IN MULTIDIMENSIONAL SCALING OF TRANSFORMED VARIABLES

Another form of multidimensional scaling is where optimal non-linear transformations of the variables of a data matrix $\mathbf{Y}_k$ are sought that generate a configuration matrix $\mathbf{Z}_k$, followed by finding a configuration $\mathbf{X}_k$ of specified low-dimensionality that either (a) directly approximates $\mathbf{Z}_k$ or (b) $\mathbf{X}_k$ generates distances that approximate distances generated by $\mathbf{Z}_k$. Non-linear principal components analysis, where the approximation is defined by minimising $\|\mathbf{Z}_k - \mathbf{X}_k\|$, is an example of (a) and the method described by Meulman (1992) is an example of (b). The variables of $\mathbf{Y}_k$ are transformed independently of each other. Permitted transformations include monotonic transformations for quantitative variables, the construction of optimal numerical scores for nominal categorical variables and properly sequenced scores for ordered categorical variables. Meulman (1992) used the least-squares scaling criterion to define optimal approximation among the two sets of distances but, in principle, any multidimensional scaling criterion could be used.

Unless the Multidimensional Scaling method is a projection method, like classical scaling/principal coordinates analysis, the configuration $\mathbf{X}_k$ will not be in a subspace of $\mathbf{Z}_k$. Then, the reference system associated with $\mathbf{Z}_k$ must be embedded in the same space as that holding $\mathbf{X}_k$. Two ways of doing this are (a) the regression method of fitting $\mathbf{Z}_k$ to $\mathbf{X}_k$ described in Section 11.1.1 and (b) the two-set orthogonal Procrustes rotation of $\mathbf{Z}_k$ to fit $\mathbf{X}_k$. Then, the linear axes of the rotated $\mathbf{Z}_k$ may be (back-)projected into $\mathbf{X}_k$ in the usual way to give a linear biplot.

Thus, we have the steps $\mathbf{Y}_k \to \mathbf{Z}_k \to \mathbf{X}_k$ followed by some form of Procrustes analysis of the K sets. Next, we wish to incorporate information on the variables in the Procrustes analysis. Each configuration $\mathbf{Z}_k$ carries with it, its own reference system. Nearly always, $\mathbf{Z}_k$ generates Pythagorean distances, so we may assume the reference system to be a set of Cartesian coordinate axes, labelled in terms of the values of the transformed scales of the original variables. Thus, with the transformed variables, the methods already discussed (Section 11.1.2) suffice and there is no new problem. In particular, the biplot axes are linear. As in Section 11.1.2, categorical variables define convex neighbour regions that intersect the exhibited space in convex prediction regions. The category level points are given by the coordinate values taken by the estimated optimal category scores found in $\mathbf{Z}_k$. Often, these will be one-dimensional but this is not a necessary constraint. With one-dimensional category scores, the prediction regions will be bordered by parallel lines.

It remains to be seen how to calibrate labels in terms of the original variables. In principle, if we have a formula giving the transformation $z_r = \tau_r(y_r)$ that takes the rth variable of the data matrix $\mathbf{Y}_k$ into the rth transformed variable in $\mathbf{Z}_k$, then the value of y_r that corresponds to a unit value of x_r may be found from the inverse transformation $y_r = \tau_r^{-1}(z_r)$. Then the calibration may be labelled in terms of unit steps of the original variables. Unless the transformations are linear, unit steps in the scales of the original variables do not correspond to unit steps of the scales of the transformed variables, so the calibration will have irregularly placed markers; this is a manifestation of the non-linear methodology. This approach is perfectly practicable but it does assume that the transformations $\tau_r(.)$ are available in some convenient analytical form. Gower, Meulman, and Arnold (1999) give an example with monotonic transformations used in conjunction with least-squares scaling. There the transformations were calculated as quadratic I-splines with a single internal knot, which would easily accommodate the inverse transformations. Unfortunately, the transformations were available only in graphical form, so the inverse transformations had to be done by a computerised form of visual interpolation.

In all the above, we have assumed that it is the configurations $\mathbf{X}_k$ that are the objects of the Procrustes analysis. For example, this would be appropriate if the MDS were regarded as a means of 'cleaning up' the information available in the observations of $\mathbf{Y}_k$. When this is not so, it might make as much sense to regard the $\mathbf{Z}_k$ to be the legitimate configurations, in which case we would have a much more simple problem.

11.2 An example: coffee images

The following example based on the work of Gower and Dijksterhuis (1994) illustrates some of the methodology, in the context of generalised orthogonal Procrustes analysis (GPA or GOPA). A panel of seven assessors ($K = 7$) assessed nine coffee packagings ($N = 9$) on eleven variables ($P = 11$). Of the eleven variables, five were categorical and six quantitative; quantitative scores were elicited by marking a point on a scale. Details of these variables are given in Table 11.1.

To obtain the configurations $\mathbf{X}_k$ we use principal coordinates analysis/classical scaling. This requires the definition of a distance between pairs of packagings. For quantitative variables we use the usual Pythagorean Euclidean distance d_{ij} defined by:

$$d_{ij}^2 = \sum_{h=6}^{11}(y_{ih} - y_{jh})^2$$

where, to ensure commensurability between variables, y_{ik} denotes the original observation centred and normalised to unit range. We must make a similar normalisation for the categorical variables and to this effect we adopt the Extended Matching Coefficient (EMC). This scores zero when there is a match (i.e. y_{ih} and y_{jh} denote the same categories, zero distance), else with a mismatch it scores

Table 11.1. The five questions with categorical responses are listed together with the six questions with quantitative responses. The codes given for the category levels are used in Figure 11.1 as labels. The names of the quantitative variables are italicised and used to label the higher-valued end of the linear biplot axes of Figure 11.1.

Questions on categorical variables	Category-levels
1. How often do you Drink this coffee?	Dn never
	Ds sometimes
	Dr regularly
	Do often
	Da always
2. What is the most suitable Moment for this coffee?	Mb with breakfast
	Mm the morning
	Ml with lunch
	Ma the afternoon
	Md after dinner
	Me the evening
3. What is the most suitable Occasion for this coffee?	Oh at home (each day)
	Ow at work
	Ov during vacations
	Or in a restaurant/cafe
	Op at week-ends/public holidays
	Od after dinner
4. Which Income-group buys this coffee?	Il low income
	Im middle income
	Ih high income
5. Would you Buy this coffee?	Bn never
	Bs sometimes
	Br rarely
	Bo often
	Ba always
6. What *price* are you willing to pay for this coffee?	
7. Amount of *odour*?	weak – strong
8. Amount of *taste* or *aroma*?	weak – strong
9. Full-flavouredness/*raciness*?	weak – strong
10. *Bitter*ness?	weak – strong
11. *Quality*?	weak – strong

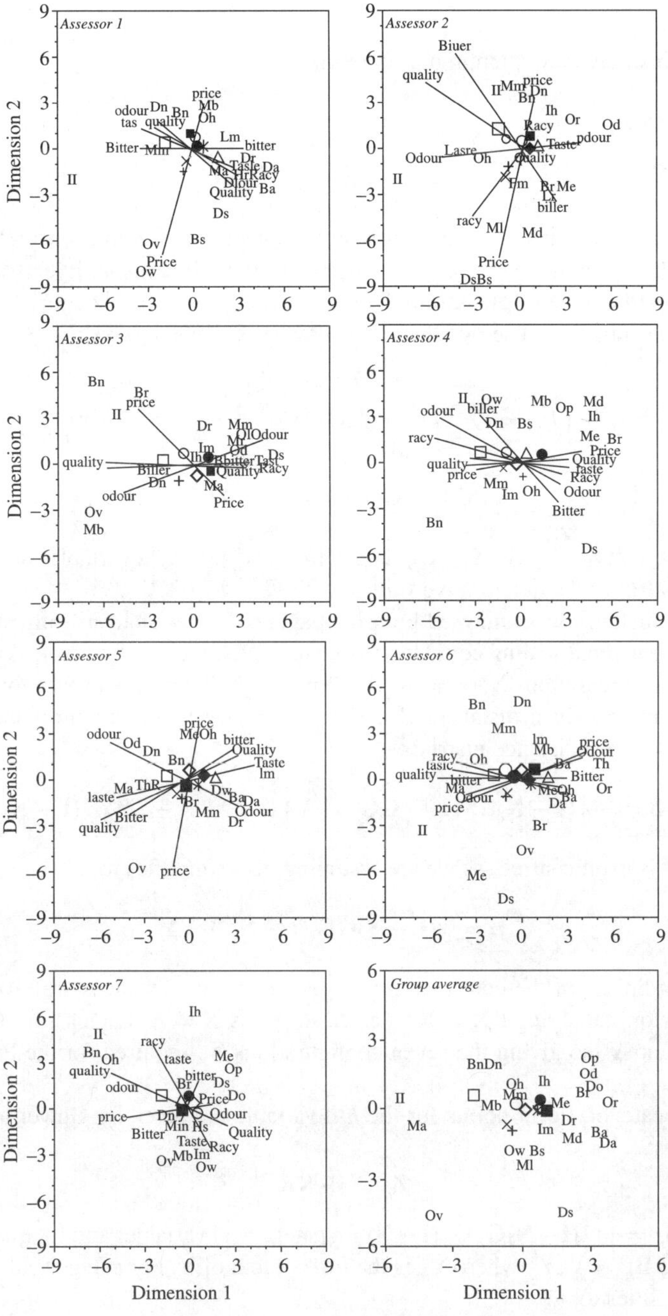

Fig. 11.1. Two-dimensional representations of the configurations for the seven assessors rotated to best fit their group average. The rotations were done in the maximal space of 26 dimensions. Everything is referred to the principal axes of the group average. The code for the category levels is given in Table 11.1. Quantitative variables are labelled at their higher values. The nine brands are denoted by the symbols Δ ∗ + × ○ □ ◇ ■ •.

unity. This may be written in matrix terms as:

$$\{d_{ij}^2\} = \sum_{h=1}^{5}(\mathbf{11}' - \mathbf{G}_h\mathbf{G}_h')$$

where $\mathbf{G}_h$ is the indicator matrix* for the hth categorical variable with N rows and L_h columns, the number of category levels for the hth variable. Every row of $\mathbf{G}_h$ is zero apart from a single unit in the column assigned to the category level taken by the relevant case. The two distances may be combined to give:

$$\begin{aligned}\{d_{ij}^2\} &= \sum_{h=1}^{5}(\mathbf{11}' - \mathbf{G}_h\mathbf{G}_h') + \sum_{h=6}^{11}(y_{ih} - y_{jh})^2 \\ &= (5\mathbf{11}' - \mathbf{GG}') + \mathbf{E11}' + \mathbf{11}'\mathbf{E} - 2\mathbf{YY}',\end{aligned}$$

where $\mathbf{G} = (\mathbf{G}_1, \mathbf{G}_2, \mathbf{G}_3, \mathbf{G}_4, \mathbf{G}_5)$ codes for the categorical variables and $\mathbf{Y}$ is the 9×6 matrix of the quantitative values, with $\mathbf{E} = \text{diag}(\mathbf{YY}')$.

This distance was analysed by principal coordinates analysis (any method of multidimensional scaling could have been used) for each of the seven assessors to give the configurations $\mathbf{X}_k (k = 1, \ldots, 7)$ in a high dimensional space that exactly reproduces the given distances d_{ij}. For each assessor, this requires the doubly centred squared-distance matrix:

$$\mathbf{B} = -\tfrac{1}{2}(\mathbf{I} - \mathbf{N})[(5\mathbf{11}' - \mathbf{GG}') + \mathbf{E11}' + \mathbf{11}'\mathbf{E} - 2\mathbf{YY}'](\mathbf{I} - \mathbf{N}).$$

When $\mathbf{Y}$ is itself centred, as we are assuming, this simplifies to:

$$\mathbf{B} = \tfrac{1}{2}(\mathbf{I} - \mathbf{N})\mathbf{GG}'(\mathbf{I} - \mathbf{N}) + \mathbf{YY}'$$

The coordinates of the configuration $\mathbf{X}$ generating the distances are given as the eigenvectors satisfying $\mathbf{BX} = \mathbf{X\Lambda}$ scaled so that $\mathbf{X}'\mathbf{X} = \mathbf{\Lambda}$. Each assessor, has his own $\mathbf{G}$ and $\mathbf{Y}$, so giving the seven configurations $\mathbf{X}_k$ required for the Procrustes analysis.

The category level points for the hth variable are given by Gower and Hand (1996) as:

$$\mathbf{Z}_h = \mathbf{B}_h\mathbf{X\Lambda}^{-1}$$

where $\mathbf{B}_h = 1/2(\mathbf{I} - \mathbf{N})\mathbf{G}_h\mathbf{G}_h'(\mathbf{I} - \mathbf{N})$ for categorical variables and for quantitative variables $\mathbf{B}_h = \mathbf{Y}_h\mathbf{Y}_h'$ where $\mathbf{Y}_h$ is the hth column of $\mathbf{Y}$. For categorical variables this simplifies to:

$$\mathbf{Z}_h = \tfrac{1}{2}(\mathbf{I} - \mathbf{N})\mathbf{G}_h\mathbf{G}_h'\mathbf{X\Lambda}^{-1}.$$

* Note that, in this section $\mathbf{G}$ and $\mathbf{G}_k$ refer to indicator matrices and not to group-average matrices as elsewhere in this book.

Because there are only L_h distinct categories, the distinct category level points are given by $\mathbf{Z}_h^{\bullet}$ with L_h rows, where:

$$\mathbf{Z}_h^{\bullet} = \frac{1}{2}\left(\mathbf{I} - \frac{i}{n}\mathbf{1}\mathbf{1}'\mathbf{L}_h\right)\mathbf{G}_h'\mathbf{X}\mathbf{\Lambda}^{-1}.$$

For quantitative variables:

$$\mathbf{Z}_h = \mathbf{Y}_h\mathbf{Y}_h'\mathbf{X}\mathbf{\Lambda}^{-1}$$

which has unit rank and so represents a set of N collinear points, as required of a linear biplot axis. One point is sufficient to draw such an axis and we note that the coordinates of a point representing a single unit in the scale of the normalised hth quantitative variable is given by $\mathbf{Y}_h'\mathbf{X}\mathbf{\Lambda}^{-1}$.

Figure 11.1 shows the results of the analysis of the coffee image data. The PCO configurations $\mathbf{X}_k (k = 1, \ldots, 7)$ each occupy 26 dimensions. In accordance with GPA, these were rotated orthogonally to fit best their group average. Then, everything was referred to the principal axes of the group average (see Section 9.1.8). An analysis of variance for this data was given in Section 10.2.4. Figure 11.1 shows the first two dimensions for each of the seven assessors, as well as the group average configuration in the lower right-hand corner. For each assessor, we have (i) the positions of each of the nine brands, (ii) the linear biplot axes for the five quantitative variables, and (iii) the projections of the category level points denoted by the code given in Table 11.1. This is an enormous amount of information that could not possibly be shown usefully on a single diagram. The linear axes are extended from their minimal to their maximal values at the labelled extremity. As can be seen, there is too much variation between the directions of the linear biplot axes to justify showing their average representation in the group average plot. However, we have averaged the category level points and these could be used, with dubious justification in this example, to interpolate new brands or proposed new brands. We would have liked to show the prediction regions for the categorical variables but there would have been five diagrams for each assessor, giving a total of 35, each with its own subsets of convex prediction regions. It would be hopelessly impracticable to show these on a single diagram for each assessor, let alone show a set of five averaged prediction regions for the group average. Prediction regions for a single configuration of cases (equivalent to the brands of this example) may be seen in Gower and Hand (1996) and Gower and Harding (1998).

12
Accuracy and stability

12.1 Introduction

Though Hurley and Cattell (1962) mention several attempts to develop a statistical test, no one test seems to have survived the 40 years between 1962 and the writing of this book. When they exist, statistics to test the significance of the results obtained using multivariate methods of data analysis are usually very complicated and rely on rather rigid, and often unrealistic, assumptions.

Generalised Orthogonal Procrustes Analysis (GPA) as a data analytical method was criticised by Huitson (1989) who warned against its 'uncritical' use. Based on random data, his arguments are not very convincing, as was lucidly pointed out by Arnold (1990) (see also Huitson 1990). Huitson passed over the possibility of the Procrustes Analysis of Variance (see Chapter 10) but he is right in asserting that there is no formal statistical way of testing the GPA group average. GPA has been under fire by others too; Stone and Sidel (1993) consider the transformations 'intractable', which may be true, seen from their own expertise (cf. Dijksterhuis and Heiser 1995). My laptop performs many actions indeed highly intractable to me, nevertheless I can write this text with its help.

Sibson (1979) and Langron and Collins (1985) studied formal statistical properties of Procrustes Analysis. Other authors used random data to find a measure of significance for the GPA result (Langeheine 1982, King and Arents 1991) or apply permutation tests Wakeling Raats, and MacFie (1992). There are many different approaches for addressing this problem, two of which have special interest: an analytical and a computational approach. In practice, however, the analytical approach may be so complex and based on unrealistic assumptions that in future, the relatively easily obtained computational results may supersede analytical investigations.

In the analytical approaches, assumptions are made about the distributions of the data matrices, from which asymptotic results are derived. In the computational approach, Monte Carlo and other computationally intensive methods are employed to infer significance. Dijksterhuis and Heiser (1995) give a brief overview of several publications on this topic.

12.2 Probability distributions of Procrustes loss values

Among the first papers using an analytical approach were Davis (1978) and Sibson (1979). Davis (1978) models the N samples as coming from separate multinormal distributions with known means and a common covariance matrix. He then derives the asymptotic distribution of the Procrustes residual sum-of-squares statistic,* showing it to be proportional to chi-squared $c\chi_\alpha^2$ on non-integer degrees of freedom where $c = \mu_2/2\mu_1'$, $\alpha = 2\mu_1'^2/\mu_2$ with μ_1' the asymptotic mean and μ_2 the asymptotic variance of NS. Davis (1978, appendix b) derives μ_1' and μ_2.

Sibson (1979) presented a perturbation theory for Procrustes statistics. When $\mathbf{X}_1$ is perturbed according to $\mathbf{X}_2 = \mathbf{X}_1 + \varepsilon\mathbf{Z}$, with $\mathbf{Z}$ an $(N \times p)$ matrix with elements i.i.d. $\mathcal{N}(0, 1)$, Sibson (1979) shows that the Procrustes statistic S under rotation is distributed according to $\varepsilon^2\chi^2 + O(\varepsilon^3)$ with $Np - 1/2p(p+1)$ degrees of freedom. The Procrustes statistic under rotation and isotropic scaling, S^* has the same chi-square distribution, with one fewer degrees of freedom. Interestingly, there is no linear term in ε, so provided ε is small, the approximation should be a good one. Sibson's degrees of freedom agree with those given by setting $K = 2$ in Table 10.2. The latter were obtained by accounting for all the free parameters in an orthonormal matrix, ignoring the non-linearity of the orthonormal constraints. It is satisfactory that the heuristic approach and Sibson's analytical results obtained for the orthogonal case with $K = 2$ agree, giving some confidence in the values of degrees of freedom given in Table 10.2 for the more general orthonormal case, as well as for general values of K.

Building on the results of Sibson (1979), Langron and Collins (1985) investigated the distribution of the Procrustes Statistic under perturbation of $\mathbf{X}_2$ when fitting it to $\mathbf{X}_1$. Their simulations showed that S can be approximated by a chi-square distribution even for large ε^2, while S^* behaves like a chi-square only for very small ε^2 (Langron 1981). In addition they show that when both $\mathbf{X}_1$ and $\mathbf{X}_2$ are perturbed, the values of S and S^* approximately double in size. Their simulations showed no significant departures from the respective chi-square distributions.

On the null hypothesis, all sums-of-squares in the analysis of variance (ANOVA) of Table 10.2 are distributed proportionally to χ^2 with degrees of freedom as specified, thus allowing the use of F-tests. Table 10.2 (a similar table is given by Langron and Collins' 1985 table 1, p. 281) may be used to test separately the translation, rotation, and scaling effects against the residual S using an F-distribution with the tabulated degrees of freedom. When scaling factors are included, the appropriate residual is S^*. To allow for the K scale factors constrained by (9.30), the degrees of freedom for S^* are $K - 1$ fewer than those for S in Table 10.2. A test for the effectiveness of the residuals is obtained by examining the reduction in the residual sum-of-squares obtained by introducing scale factors, as shown in Table 12.1. The degrees of freedom come directly from Table 10.2

* Davis (1978) uses the symbol m^2 which we denote by S for the residual sum-of-squares.

Table 12.1. ANOVA for the scaling factors in Least Squares Procrustes analysis (a) orthogonal case and (b) projection case.

Source	Degrees of freedom	SS	MS
Residual without scaling	(a) $PK(N-1)-1/2P(P-1)(K-1)$ (b) $PK(N-1)-R(P-R)K$ $-1/2R(R-1)(K-1)$	S	
Residual with scaling	(a) $PK(N-1)-1/2P(P-1)(K-1)-(K-1)$ (b) $PK(N-1)-R(P-R)K$ $-1/2R(R-1)(K-1)-(K-1)$	S^*	
Scaling	$K-1$	$S-S^*$	

with a reduction of $K-1$ when scaling is included. The test statistic is the ratio of the mean squares corresponding to $S-S^*$ and S^*. Similar tests may be associated with Tables 10.3 and 10.4, distributing the degrees of freedom across columns, as discussed in Section 10.2.2.

Of course, the above tests hold only under the distributional assumption made earlier. These results, although interesting, have rather limited value because (a) we know that underlying assumptions are usually unrealistic and (b) we have little idea how robust the results are, even for minor departures from the assumptions. Further, the results are based on Procrustes methods that minimise the residual sum-of-squares; they are unlikely to be valid for the other Procrustes criteria discussed in an analysis of variance context in Section 10.3.

The results of distributional studies are often very complicated. Though this does not need to be a problem, the more complex the mathematics the less likely it is that the results will have wide applicability. Nor is there much easily available and understandable software. In many applications of GPA the data sets can be relatively small and it is not known to what extent the asymptotic properties remain valid. The assumption of an underlying multivariate normal distribution for the data is nearly always violated, especially with the smaller data sets. Indeed, the concept that the N cases might be regarded as any kind of sample drawn from some probability distribution is often unrealistic. Thus, the validity of proposed tests is limited. The theory of Langron and Collins (1985) seems to have been little used in applications.

Apart from analytical results, perturbation has often been used in a Monte Carlo context. For example, Meyners, Kunert, and Qannari (2000) impose a simple error structure onto a known group average. Then they compare the known group-average with the simulated group averages given by GPA and by STATIS. The comparison is not straightforward as one has to devise a suitable yard-stick (see Chapter 14). They use two yardsticks (i) based on the two-set orthogonal Procrustes comparison and (ii) based on the RV coefficient. GPA/STATIS has to outperform STATIS/GPA on both yardsticks to be judged 'better'.

12.3 Resampling methods

This paragraph introduces two approaches to testing GPA results that do not involve complicated mathematical analysis, do not give asymptotic results, nor rely on distributional assumptions. The first approach will be called the 'random data' approach, and is used in a GPA context by Langeheine (1982) and King and Arents (1991). The other method is called the random permutation method (see, for example, Edgington 1987). Wakeling, Raats, and MacFie (1992) compared this method to the random data method. Dijksterhuis and Heiser (1995) present some background to using permutation tests for multivariate problems.

12.3.1 RANDOM DATA METHOD

Fitting matrices with random data was employed by Langeheine (1982) in the context of PINDIS individual difference scaling (Lingoes and Borg 1978 and see Section 13.1.2). For a range of data sets of different sizes estimated values (in this case Procrustes least square loss values, G_Q or G_S) can be computed. The constructed data sets contain random data, drawn from some distribution. Schöneman, James, and Carter (1979) observe that there is not much difference in the indices computed from such random data matrices when the data were sampled from a uniform, a standard normal, or an exponential distribution. It seems that random data remain random, whatever the underlying source. The idea is that the Procrustes loss shown by real data sets can be compared with the loss values from fitting random data sets of the same size. When the empirical value, that is, the value from the real data sets, is smaller than the loss value obtained by fitting random data sets, this gives credibility to the empirical result. It does not matter in principle if one uses loss values, as we do with Procrustes analysis, or r^2 fit values as Langeheine did. He reports the r^2 value between the $N \times P$ coordinates of $s_k\mathbf{X}_k\mathbf{Q}_k$ and the group average $\mathbf{G}$, averaged over the K sets. However, as Langeheine (1982) notes: 'statistical testing is reduced to [considering] the rather weak hypothesis whether an actual r^2 exceeds one expected from scaling random configurations'. Some interesting findings reported by Langeheine (1982) are that r^2 decreases as N increases, r^2 decreases as K increases, more so for small N than for larger N and that the higher the dimensionality of the configurations, the higher the r^2 (see Figure 12.1). Langeheine's (1982) results also show that the dimensional salience (dimension weighting) models show only a slightly better fit than the similarity (Procrustes) model. This illustrates that the gain in employing models more complex than Procrustes models may not be worth the extra trouble in fitting and interpreting their results.

King and Arents (1991) applied the random data method to infer the 'significance' of a GPA result. For a given empirical data set of order (K, N, P) they construct 50 data sets of the same size, filled with random numbers, drawn from a uniform distribution representing the same range of integers as the empirical

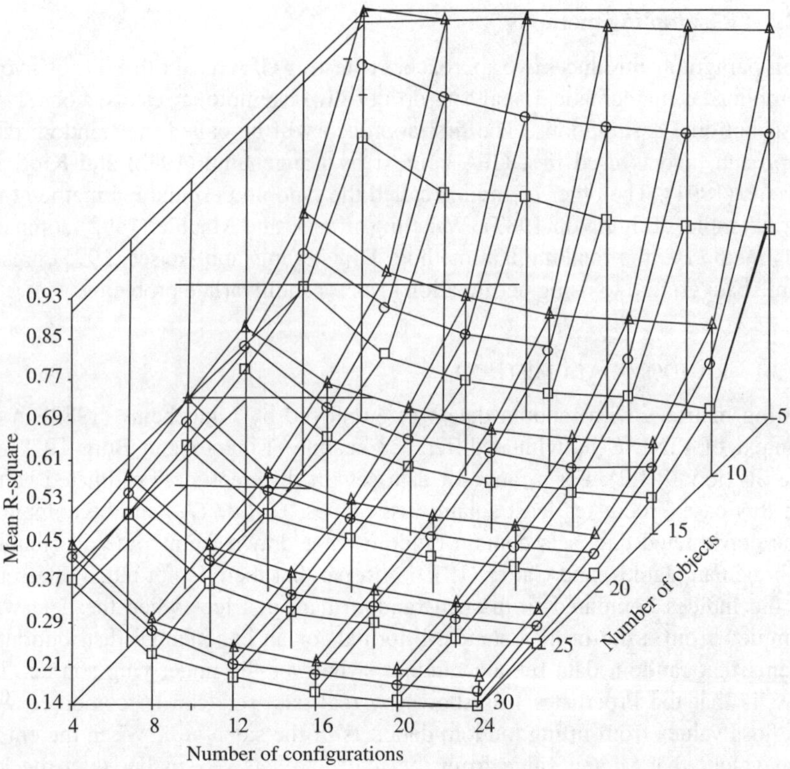

Fig. 12.1. The relationship between K, N, and the fit value r^2 (from Langeheine 1982, figure 1, p. 432). (□: in 2 dimensions; ○: in 3 dimensions; Δ: in 4 dimensions).

data set. In the case of K-sets data, where $P = P_k$, the same procedure is applied. They perform 50 separate GPAs and obtain a distribution of 50 values of explained variance using the group average criterion (Peay 1988). From this distribution the upper 95% point is used to assess the values from empirical data sets. This 95% point, U, is compared with the empirical fit value, the value actually observed. When *loss* values are used instead of *fit* values, the lower 5% point, is of interest. Of course other percentages can be used as well. Obviously to obtain a certain cut-off point, say α, at least α^{-1} analyses are needed.

Figure 12.2 shows some results from King and Arents (1991) presented graphically. For $K = 10$, data sets with $N \in [5, 10, 15, 20]$ and $P \in [5, 10, 15, 20]$ are constructed and the percentage variance explained is used to construct the surface in Figure 12.2. It can be seen that the fit decreases with increase of N, in accordance with the finding reported by Langeheine (1982). Furthermore the fit increases

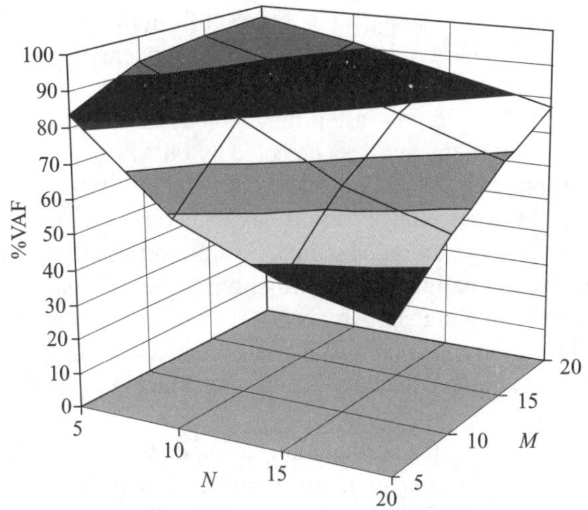

Fig. 12.2. Percentage variance explained by the group-average, for random data sets of different numbers of rows (N) and columns (M), $K = 10$ in this example (based on King and Arents 1991, table 2, p. 43).

with higher P, which is in accordance with Langeheine (1982) reporting increase of fit with higher dimensionality.

12.3.2 RANDOM PERMUTATION METHOD

Permutation tests are more often found in analysis of variance-type experiments (see, for example, Edgington, 1987) than in a MVA context. They are generalisations of Fisher's exact test for a (2×2) contingency table. In fact permutation tests are very general methods that can be applied in many situations. The idea of permutation tests is to destroy the correspondence between the rows of the data matrices. In the GPA framework this means that the rows of the matrices $\mathbf{X}_k$ are permuted. The number of variables P in a matrix $\mathbf{X}_k$ is of no importance for the permutation tests since only the rows are permuted; it is immaterial whether the number P_k of variables in each $\mathbf{X}_k$ varies or is constant. The set of all possible permutations of N rows of K matrices contains $(N!)^K$ permutations which will often be far too large a number to handle, so a much smaller random subset of permutations is used. Each permuted data set has a particular loss or fit-value, σ_s associated with it ($s = 1, \ldots, 100$, say). The value associated with the unpermuted data set is called the *empirical value*, σ_0. A certain significance value, can be attributed to the empirical value obtained. With a fit-value, like r^2 used in the previous section, the probability $p(\sigma_s < \sigma_0)$, gives this significance; with a loss-value the probability $p(\sigma_s > \sigma_0)$, gives it.

The null hypothesis is that σ_0 is not different from the other values σ_s. When this is true it means that GPA did not find any correspondence between the matrices $\mathbf{X}_k$. When p is small, say $p < 5\%$, the empirical value σ_0 lies in the tail of the distribution of σ_s values. In this case it may be judged that there is likely to be a correspondence between the matrices detected by GPA.

The method applied by Wakeling, Raats, and MacFie (1991) uses the permutation distribution of the percentage variance explained by the group average (Peay 1988 see Chapter 9, Section on Maximal Group Average Projection Procrustes Analysis). They found permutation distributions widely deviating from normality. This contrasts with King and Arents (1991) who report a normal distribution of this value. The difference may result from the fact that King and Arents analysed random data, while Wakeling, Raats, and MacFie (1991) used permuted real data.

The values obtained by the permutation test, and reported by Wakeling, Raats, and MacFie (1991) result in many more significant results than suggested by the random data approach of King and Arents (1991). Substitution of real data by random data will almost always increase the rank of the individual matrixes $\mathbf{X}_k$ (cf. Langeheine 1982), making the data sets unrepresentative of the empirical data set. Permutations of the empirical data set, however, retain all their distributional properties and hence result in better representative simulations, and in more realistic statistical significance values.

12.3.3 JACKKNIFING

Very probably Jackknifing has been applied in Procrustes analysis but we know of no examples. However, De Leeuw and Meulman (1986) discuss multidimensional scaling results where jackknifed MDS configurations are matched by orthogonal Procrustes analysis. At one level this may be regarded as a nice application of Procrustes analysis but at another level it is clear that jackknifed Procrustean group-averages might themselves be matched by using orthogonal Procrustes methods; this would be doubly nice.

12.4 Dimensionality of results

Though the dimensionality of the GPA result depends on the rank(s) of the original data matrices, this is obviously not the dimensionality in which one prefers to represent the results. The group average $\mathbf{G}$ of the GPA exists in the highest possible number of dimensions, depending on the maximal rank of the $\mathbf{X}_k$. For representational purposes $\mathbf{G}$ is referred to its principal axes. Figure 12.3 shows three scree graphs. The dotted line is typically found when random data are analysed. There are small differences between the eigenvalues (or the percentage explained variance %VAF) for successive principal components. The dashed line represents the case of equal eigenvalues, possible with structured or manipulated data but rarely approximated by real data. The line

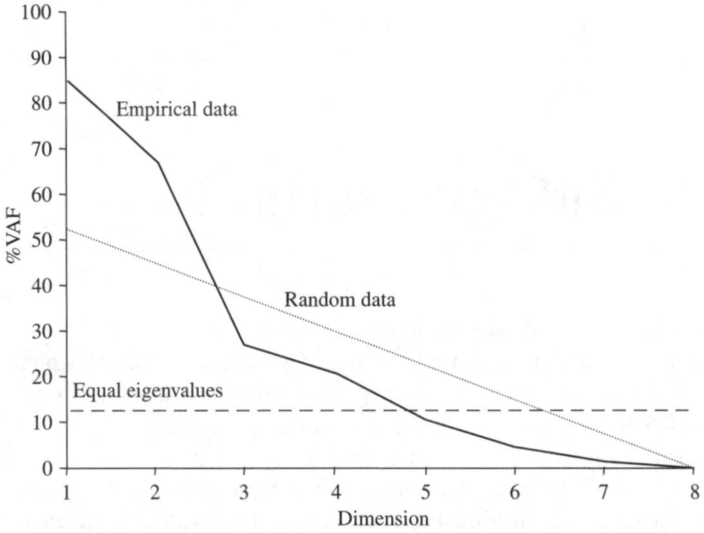

Fig. 12.3. Scree diagrams illustrating three different—fictitious—appearances (dotted lines: random data; dashed lines: the very rare case of equal eigenvalues; solid lines: empirical data).

labelled 'empirical data' shows the typical—albeit idealised—case of two principal components with a higher "percentage variance accounted for" than the higher—dimensional components, resulting in a choice for a two-dimensional representation of **G** that may be inspected.

13
Links with other methods

We begin this chapter by describing variants of the generalised Procrustes method and how to incorporate anisotropic scaling. Generalised Procrustes methods are examples of three-mode analyses (see general references given at the end of this chapter). Other well known methods for handling this type of data that are discussed in this chapter are Individual Differences Scaling (INDSCAL) (Section 13.2) and Structuration des Tableaux A Trois Indices de la Statisitique (STATIS), also known as Conjoint Analysis (Section 13.4), Procrustean Individual Difference Scaling (PINDIS) (Section 13.1.3), extensions of canonical correlation analysis to K sets of variables (Section 13.5) and generalisations of several methods of Multidimensional Scaling (Section 13.3). Sometimes these methods operate on K data matrices $\mathbf{X}_k$, $k = 1, \ldots, K$ and sometimes on derived distance matrices $\mathbf{D}_k$, $k = 1, \ldots, K$. In all cases some form of the group average concept is central to the development. We also include a brief discussion of the more tenuously related methods of various forms of Common Components Analysis, and three-mode components analysis (Sections 13.6 and 13.7).

13.1 Variant treatments of generalised Procrustes analysis

In Chapter 9 we presented generalised Procrustes analysis as a natural extension of two-sets Procrustes problems where the group average emerged as a simple consequence of the least-squares criterion (9.1). Several variant forms are possible based on transformations of the group average rather than of the individual configurations $\mathbf{X}_k$. It seems that such formulations are influenced by the role of the group average in Individual Difference models such as that of INDSCAL (Section 13.2). These models originate in psychometrics and postulate a group average, regarding the individual responses as being some simple transformation of the group response. Then, rather than being a simple algebraic consequence, as found in Chapter 9, the group average becomes a substantive parameter that requires estimation. The consequences of this change of emphasis are surprisingly profound.

In this section we do not specify the precise transformations concerned and so use our general notation $\mathbf{T}_k$. This is possible because, as will be seen, estimation of scaling and the calculation of group averages proceeds largely independently of the estimation of transformations.

13.1.1 ISOTROPIC SCALING OF THE GROUP AVERAGE

In an alternative formulation of isotropic scaling, one may scale a group average **C** rather than the individual $\mathbf{X}_k$ to give a criterion:

$$S = \sum_{k=1}^{K} \|s_k\mathbf{C} - \mathbf{X}_k\mathbf{T}_k\|. \tag{13.1}$$

Thus, the transformed configurations $\mathbf{X}_k$ are chosen to match *isotropically scaled* versions of a group average **C**. We cannot specify the form of the group average **C** as it has to be estimated; indeed it is premature to term **C** a group average. There is some indeterminacy in (13.1), because if s_k is multiplied by a constant, then **C** may be divided by the same constant without affecting the fit. Although there is this arbitrariness in the scaling, the solution $s_k = 0$ is not available as it was in Section 9.2. There, we had to adopt a strong constraint (9.7) to obtain an acceptable solution. Now, the estimate of $s_k\mathbf{C}$ is invariant to whatever constraint we put on the scalings s_k, so we are at liberty to choose a weak constraint to give the most convenient representation of all these equivalent solutions. Expanding (13.1) gives:

$$S = \mathbf{s}'\mathbf{s}\|\mathbf{C}\| + \sum_{k=1}^{K} \|\mathbf{X}_k\mathbf{T}_k\| - 2\,\text{trace}\,\mathbf{C}' \sum_{k=1}^{K}(\mathbf{s}_k\mathbf{X}_k\mathbf{T}_k).$$

Thus, if we proceed by normalising $\mathbf{s}'\mathbf{s} = K$, where **s** is the column-vector of scaling coefficients, differentiation with respect to **C** immediately gives the estimate:

$$\mathbf{C} = \frac{1}{K}\sum_{k=1}^{K}(s_k\mathbf{X}_k\mathbf{T}_k) \tag{13.2}$$

which is merely the classical group average **G** of (9.6). It is rather remarkable that even though the *criterion* scales **C** rather than the $\mathbf{X}_k$, nevertheless the *estimate* of **C** involves scaling the $\mathbf{X}_k$ in precisely the same way as in **G**. Differentiating S with respect to s_k, allowing for a Lagrangian $(\mathbf{s}'\mathbf{s} - K)$, gives;

$$\text{trace}(\mathbf{C}'\mathbf{X}_k\mathbf{T}_k') = \mu s_k \quad k = 1, \ldots, K$$

where because of (13.2), all K such equations may be gathered together into the single matrix eigenvalue equation:

$$\mathbf{S}\mathbf{s} = K\mu\mathbf{s} \tag{13.3}$$

where, as before, **S** is the inner-product matrix (9.11) with elements $\text{trace}(\mathbf{T}_k'\mathbf{X}_k'\mathbf{X}_k\mathbf{T}_k)$. For fixed $\mathbf{T}_k$ the eigenvalue μ that minimises S, maximises $\text{trace}\ \mathbf{C}'\sum_{k=1}^{K}(\mathbf{s}_k\mathbf{X}_k\mathbf{T}_k) = K\|\mathbf{C}\|$. Now, $K\|\mathbf{C}\| = (1/K)\mathbf{s}'\mathbf{S}\mathbf{s} = K\mu$, where the last step follows from (13.3). Thus, we require the maximum eigenvalue of

(13.3) to determine the associated eigenvector **s**. The eigenvector **s** is normalised to have sum-of-squares K. By putting the scaling factors with **G** the eigenproblem is slightly simpler than (9.10) found for the classical case, which has a diagonal matrix on the right-hand-side (i.e. $\mathbf{Ss} = \mu\, \mathrm{diag}(\mathbf{S})\mathbf{s}$). When the data are first normalised so that $\|\mathbf{X}_k\| = 1$ then $\mathrm{diag}(\mathbf{S}) = \mathbf{I}$ and then both forms of Procrustes analysis are identical, which is obvious because both lead to the same S with the same constraint $\sum_{k=1}^{K} \|\mathbf{s}_k \mathbf{X}_k\| = \mathbf{s}'\mathbf{s} = K$. We shall see that these results relate Generalised Orthogonal Procrustes Analysis (GPA) with a method known as STATIS (Section 13.4).

Possible algorithms are fairly obvious. One could start with all $s_k = 1$ and some **C**. Probably, the initial choice of **C** is unimportant but, for simplicity, one could start with the average of the $\mathbf{X}_k$. The main step is then to transform the current values of $\mathbf{X}_k\mathbf{T}_k$ to fit $s_k\mathbf{C}$ and then with the new $\mathbf{X}_k\mathbf{T}_k$ evaluate **S** and derive new s_k. Finally, recompute **C**.

Algorithm 13.1

Step 1 For given **s** and **C**, minimise $\|s_k\mathbf{C} - \mathbf{X}_k\mathbf{T}_k\|$ allowing for the constraints chosen for $\mathbf{T}_k$, $k = 1, \ldots, K$ under $\mathbf{T}_k$

Step 2 Evaluate **S** and thence new **s** and **G** and return to *Step 1*.

13.1.2 ANISOTROPIC SCALING OF THE GROUP AVERAGE

We may treat anisotropic scaling in a similar manner to that of the previous section by minimising the criterion of:

$$S = \sum_{k=1}^{K} \|\mathbf{CS}_k - \mathbf{X}_k\mathbf{T}_k\|, \tag{13.4}$$

where $\mathbf{S}_k$ is an $R \times R$ diagonal matrix of unknown scaling factors to be attached to **C** by the kth individual. Expanding (13.4) gives:

$$S = \mathrm{trace}\left\{\mathbf{C}'\sum_{k=1}^{K}\mathbf{S}_k^2\mathbf{C}\right\} + \sum_{k=1}^{K}\|\mathbf{X}_k\mathbf{T}_k\| - 2\mathrm{trace}\Big(\mathbf{C}'\sum_{k=1}^{K}(\mathbf{X}_k\mathbf{T}_k\mathbf{S}_k)\Big).$$

The lack of identifiability between each $\mathbf{S}_k$ and **C** may be resolved by imposing the weak constraint:

$$\sum_{k=1}^{K}\mathbf{S}_k^2 = K\mathbf{I} \tag{13.5}$$

so that S simplifies to :

$$S = K\|\mathbf{C}\| + \sum_{k=1}^{K}\|\mathbf{X}_k\mathbf{T}_k\| - 2\mathrm{trace}\left(\mathbf{C}'\left(\sum_{k=1}^{K}\mathbf{X}_k\mathbf{T}_k\mathbf{S}_k\right)\right). \tag{13.6}$$

Differentiation with respect to $\mathbf{C}$ gives:

$$\mathbf{C} = \frac{1}{K}\sum_{k=1}^{K}(\mathbf{X}_k\mathbf{T}_k\mathbf{S}_k), \tag{13.7}$$

the classical estimate of a group average as the mean of the transformed and scaled configurations.

Estimation of $\mathbf{S}_k$ requires R Lagrange multipliers associated with the R diagonal elements of the constraint (13.5). We imagine these to be collected in the diagonal of a matrix $\mathbf{\Lambda}$ to give the Lagrangian $(\sum_{k=1}^{K}\mathbf{S}_k^2 - K\mathbf{I})\mathbf{\Lambda}$. Differentiation with respect to $\mathbf{S}_k$ then gives:

$$\text{diag}(\mathbf{C}'\mathbf{X}_k\mathbf{T}_k) = \mathbf{S}_k\mathbf{\Lambda},$$

which on using (13.7) may be written:

$$\frac{1}{K}\text{diag}\left\{\mathbf{T}_k'\mathbf{X}_k'\left(\sum_{l=1}^{K}\mathbf{X}_l\mathbf{T}_l\mathbf{S}_l\right)\right\} = \mathbf{S}_k\mathbf{\Lambda}.$$

When $\mathbf{S}$ is diagonal, $\text{diag}\{\mathbf{AS}\} = \text{diag}\{\mathbf{A}\}\mathbf{S}$ and so the above becomes:

$$\frac{1}{K}\sum_{l=1}^{K}(\text{diag}\{\mathbf{T}_k'\mathbf{X}_k'\mathbf{X}_l\mathbf{T}_l\}\mathbf{S}_l) = \mathbf{S}_k\mathbf{\Lambda}. \tag{13.8}$$

Denoting $\text{diag}\{\mathbf{T}_k'\mathbf{X}_k'\mathbf{X}_l\mathbf{T}_l\}$ by $\mathbf{\Sigma}^{(kl)}$ (13.8) becomes:

$$\frac{1}{K}\begin{pmatrix} \mathbf{\Sigma}^{(11)} & \mathbf{\Sigma}^{(12)} & \cdots & \cdots & \mathbf{\Sigma}^{(1K)} \\ \mathbf{\Sigma}^{(21)} & \mathbf{\Sigma}^{(22)} & \cdots & \cdots & \mathbf{\Sigma}^{(2K)} \\ \vdots & \vdots & \ddots & & \vdots \\ \vdots & \vdots & & \ddots & \vdots \\ \mathbf{\Sigma}^{(K1)} & \mathbf{\Sigma}^{(K1)} & \cdots & \cdots & \mathbf{\Sigma}^{(KK)} \end{pmatrix}\begin{pmatrix} \mathbf{S}_1 \\ \mathbf{S}_2 \\ \vdots \\ \vdots \\ \mathbf{S}_K \end{pmatrix} = \begin{pmatrix} \mathbf{S}_1\mathbf{\Lambda} \\ \mathbf{S}_2\mathbf{\Lambda} \\ \vdots \\ \vdots \\ \mathbf{S}_K\mathbf{\Lambda} \end{pmatrix}. \tag{13.9}$$

Even though the two matrices on the right may be expressed as vectors (by post-multiplying by $\mathbf{1}$), (13.9) is not a simple eigenvalue problem. This is because of the R values $\boldsymbol{\lambda}_r$ in $\mathbf{\Lambda}$. However, by permuting the rows and columns, collecting together all terms associated with each λ_r we may arrange the left-hand side as a block diagonal matrix, the rth block of which gives:

$$\frac{1}{K}\begin{pmatrix} \sigma_r^{(11)} & \sigma_r^{(12)} & \cdots & \cdots & \sigma_r^{(1K)} \\ \sigma_r^{(21)} & \sigma_r^{(22)} & \cdots & \cdots & \sigma_r^{(2K)} \\ \vdots & \vdots & \ddots & & \vdots \\ \vdots & \vdots & & \ddots & \vdots \\ \sigma_r^{(K1)} & \sigma_r^{(K2)} & \cdots & \cdots & \sigma_r^{(KK)} \end{pmatrix}\begin{pmatrix} \sigma_r^{(1)} \\ \sigma_r^{(2)} \\ \vdots \\ \vdots \\ \sigma_r^{(K)} \end{pmatrix} = \lambda_r\begin{pmatrix} \sigma_r^{(1)} \\ \sigma_r^{(2)} \\ \vdots \\ \vdots \\ \sigma_r^{(K)} \end{pmatrix} \qquad r = 1, \ldots, R \tag{13.10}$$

where $\sigma_r^{(kl)}$ is the rth (diagonal) element of $\mathbf{\Sigma}_r^{(kl)}$ and $\sigma_r^{(k)}$ is the rth element of $\mathbf{S}_k\mathbf{1}$. Equation (13.10) is an ordinary eigenvalue problem. In accordance with (13.5), the eigenvectors must be normalised to have unit sum-of-squares. To decide which eigenvalue to select, we recall that we have to maximise trace$(\mathbf{C}'(\sum_{k=1}^{K}\mathbf{X}_k\mathbf{T}_k\mathbf{S}_k))$. We have the following steps:

$$\begin{aligned}\text{trace}\left(\mathbf{C}'\left(\sum_{k=1}^{K}\mathbf{X}_k\mathbf{T}_k\mathbf{S}_k\right)\right) &= \text{trace}\left(\sum_{k=1}^{K}\mathbf{C}'\mathbf{X}_k\mathbf{T}_k\mathbf{S}_k\right)\\ &= \sum_{k=1}^{K}\text{trace}\left(\text{diag}(\mathbf{C}'\mathbf{X}_k\mathbf{T}_k)\mathbf{S}_k\right) = \sum_{k=1}^{K}\text{trace}(\mathbf{S}_k\mathbf{\Lambda}\mathbf{S}_k)\\ &= K(\text{trace}\,\mathbf{\Lambda}) = K\sum_{k=1}^{K}\lambda_k.\end{aligned}$$

Thus we need the maximum eigenvalues and the corresponding eigenvectors of each of the R eigenvalue problems (13.10). Then $\mathbf{S}_k = \text{diag}(\sigma_1^{(k)}, \sigma_2^{(k)}, \ldots, \sigma_R^{(k)})$. One way of looking at this, is to write the eigenvectors of (13.10) in order, side-by-side to form a $K \times R$ matrix. The kth row of this matrix gives the diagonal elements of $\mathbf{S}_k$.

The steps outlined in the algorithm for isotropic scaling need little modification. One could start with all $\mathbf{S}_k = \mathbf{I}$ and some $\mathbf{C}$. Probably, the initial choice of $\mathbf{C}$ is unimportant but, for simplicity, one could start with the average of the $\mathbf{X}_k$. The main step is then to transform the current values of $\mathbf{X}_k\mathbf{T}_k$ to fit $\mathbf{C}\mathbf{S}_k$ and then with the new $\mathbf{X}_k\mathbf{T}_k$ evaluate $\mathbf{\Sigma}$ of (13.9) and then from (13.10) derive new $\mathbf{S}_k$. Finally, recompute $\mathbf{C}$.

Algorithm 13.2

Step 1 For given $\mathbf{C}$ and $\mathbf{S}_k$, minimise $\|\mathbf{C}\mathbf{S}_k - \mathbf{X}_k\mathbf{T}_k\|$ over $\mathbf{T}_k$, $k = 1, \ldots, K$ under the constraints chosen for $\mathbf{T}_k$.

Step 2 Evaluate $\mathbf{S}_k$ from (13.10) and then a new $\mathbf{C}$ and return to *Step 1*.

We know nothing about the performance of such an algorithm nor of the best strategy for choosing successive steps.

Before leaving this section, we note that the inclusion of scaling weights $\mathbf{S}_k$ in (13.4) shares features with the discussion of Chapter 8. There, there were no constraints on the elements $\mathbf{S}_k$, complicating the estimation of $\mathbf{Q}$; here the constraints (13.5) simplify the estimation process because the estimation of $\mathbf{S}$ depends solely on the term trace$\left(\mathbf{C}'\left(\sum_{k=1}^{K}\mathbf{X}_k\mathbf{T}_k\mathbf{S}_k\right)\right)$ in the expansion of (13.6) together with the constraint (13.5).

13.1.3 PROCRUSTEAN INDIVIDUAL DIFFERENCE SCALING

PINDIS (Lingoes and Borg 1978) refers to a set of Procrustes models of increasing complexity, as well as being the name of associated software. These models, as is evident in the name PINDIS, are very much in the Individual Differences tradition. They all regard the group average $\mathbf{C}$ as being given and seek to what extent various transformations of $\mathbf{C}$ agree with the given $\mathbf{X}_k$, $k = 1, 2, \ldots, K$. $\mathbf{C}$ may be derived theoretically or from previous work; often $\mathbf{C}$ will be the group average derived from GPA (possibly allowing for isotropic scaling). As will become evident, transformations allowed have already been discussed in the above and in Chapter 8. However, in Chapter 8 we were concerned only with two groups ($K = 2$) so things are now a little more complicated but the increased complexity is offset by not having to compute $\mathbf{C}$.

In our notation the most general form of the criteria minimised by PINDIS is:

$$\sum_{k=1}^{K} \|\mathbf{X}_k\mathbf{Q}_k - \mathbf{W}_k\mathbf{C}\mathbf{V}_k\mathbf{S}_k\|. \tag{13.11}$$

Here, $\mathbf{Q}_k$ and $\mathbf{V}_k$ are orthogonal matrices while $\mathbf{W}_k$ is a diagonal matrix of row weights and $\mathbf{S}_k$ is a diagonal matrix of column weights. We may write (13.11) as:

$$\sum_{k=1}^{K} \|\mathbf{X}_k - \mathbf{W}_k\mathbf{C}\mathbf{V}_k\mathbf{S}_k\mathbf{Q}_k'\| \tag{13.12}$$

and note that $\mathbf{V}_k\mathbf{S}_k\mathbf{Q}_k'$ is in the form of an SVD so may be replaced by any matrix $\mathbf{T}_k$. Thus (13.11) is of such generality that it is always possible to fit a model of the form $\sum_{k=1}^{K} \|\mathbf{X}_k - \mathbf{W}_k\mathbf{C}\mathbf{T}_k\|$ with general transformations $\mathbf{T}_k$ and that each term of the summation may be fitted independently of the others. This problem is covered by Section 8.1.6. Recall, that because of the row-scaling, a little extra care has to be taken if translational effects are to be included in the model. Thus we have only to use the methods of Section 8.1.6 to estimate $\mathbf{W}_k$ and $\mathbf{T}_k$ and then use the SVD $\mathbf{T}_k = \mathbf{V}_k\mathbf{S}_k\mathbf{Q}_k'$ to express the result in the form of (13.11) or (13.12).

The term *idiosyncratic* is used in the psychometric literature to distinguish parameters with a suffix k, referring to individuals, from parameters that are fixed over all groups—such as a group average. Thus, when $\mathbf{W}_k$ is to include translation, Lingoes and Borg refer to an *idiosyncratic origin*. We prefer to use the terms *local* and *global*. It is clear that (13.12) must be modified to include global parameters if the model is to be of any interest. It is the choice of global parameters that are the concern of PINDIS. In our notation, the main models considered are as follows:

(i) $\sum_{k=1}^{K} \|\mathbf{X}_k\mathbf{Q}_k - \mathbf{C}\|$

(ii) $\sum_{k=1}^{K} \|\mathbf{X}_k\mathbf{Q}_k - \mathbf{C}\mathbf{V}\mathbf{S}_k\|$

(iii) $\sum_{k=1}^{K} \|\mathbf{X}_k\mathbf{Q}_k - \mathbf{C}\mathbf{V}_k\mathbf{S}_k\|$

(iv) $\sum_{k=1}^{K} \|\mathbf{X}_k\mathbf{Q}_k - \mathbf{W}_k\mathbf{C}\|$

(v) $\sum_{k=1}^{K} \|\mathbf{X}_k\mathbf{Q}_k - \mathbf{W}_k\mathbf{C}\mathbf{S}_k\|$

These models give two nested sequences (i), (ii), (iii) and (i), (iv), (v) of increasing complexity.

In (i) there are no global parameters that require estimation, so each term of the summation may be minimised independently of the others as a simple two-sets orthogonal Procrustes problem (Section 4.1).

In (ii) $\mathbf{V}$ is a global parameter allowing the group average to be transformed orthogonally. For given $\mathbf{V}$, Section (8.1.2) allows the independent estimation of $\mathbf{Q}_k$ and $\mathbf{S}_k$, so the only new thing we have to consider is the estimation of the global parameter $\mathbf{V}$ given all the local parameters. After a little rearrangement, expanding (ii) gives:

$$\sum_{k=1}^{K} \|\mathbf{X}_k\| + \text{trace}\left(\mathbf{V}'\mathbf{C}'\mathbf{C}\mathbf{V}\mathbf{S}^2\right) - 2\text{trace}\left(\mathbf{V}'\mathbf{C}'\mathbf{A}\right),$$

where $\mathbf{S}^2 = \sum_{k=1}^{K}\left(\mathbf{S}_k^2\right)$ and $\mathbf{A} = \sum_{k=1}^{K} \mathbf{X}_k\mathbf{Q}_k\mathbf{S}_k$. Apart from terms not involving $\mathbf{V}$ this may be written as:

$$\|\mathbf{C}\mathbf{V}\mathbf{S} - \mathbf{A}\mathbf{S}^{-1}\|,$$

which may be minimised over $\mathbf{V}$ by the Koschat and Swayne Algorithm 8.1 described in Section 8.1.3.

Model (iii) is the one with the simple solution already discussed at the beginning of this section.

In (iv) there are no global parameters that require estimation, so the row-weights $\mathbf{W}_k$ may be estimated independently, using the methods of Section 8.1.6. (equation (8.14)) and the orthogonal matrices $\mathbf{Q}_k$ by orthogonal Procrustes analysis (Section 4.1).

Model (v) can be handled as for model (iv) but allowing for the greater complexity of the transformations $\mathbf{S}_k\mathbf{Q}_k'$, as described in Section 8.1.6.

Models (iv) and (v) both involve row-weights $\mathbf{W}_k$ which, when translation terms are to be fitted, require special treatment. No longer can we merely centre the matrices $\mathbf{X}_k$ and $\mathbf{C}$ but now must use the methodology of Section 3.2.2. Because $\mathbf{C}$ is global, any translation term associated with it must be estimated globally and this entails some adjustments to the results of Section 3.2.2.

The individual $\mathbf{X}_k$ are scaled and translated (to zero mean) versions of the observed configurations. Each is associated with an orthogonal matrix $\mathbf{Q}_k$ which has been regarded as a parameter requiring estimation. If these rotations are regarded as given, then $\mathbf{X}_k\mathbf{Q}_k$ defines a new given $\mathbf{X}_k^*$, say, and we have K fewer parameter-matrices to estimate. This would modify (iii) to the extent that Section 8.1.3 then gives the appropriate method for independently estimating the parameters $\mathbf{V}_k$ and $\mathbf{S}_k$.

Finally, we note that, in PINDIS it is conventional to compare the fits for different models $\mathbf{Z}$ by evaluating correlations $r^2(\mathbf{X}_k\mathbf{Q}_k, \mathbf{Z})$ where $\|\mathbf{X}_k\mathbf{Q}_k - \mathbf{Z}\|$ is minimised. In Chapter 3 we discussed relationships between correlation and least-squares but it seems to us more natural to compare the fits using the minimised residual sum-of-squares. If one is concerned with scale factors, these may be eliminated by evaluating the ratio with the residual sum-of-squares for the least restrictive model, say (iii).

13.2 Individual differences scaling

The aim of Individual differences scaling (INDSCAL) (Carroll and Chang 1970) is to represent the K matrices $\mathbf{D}_k$ of squared distances, one for each individual, in terms of a group average matrix $\mathbf{G}$, specified in some number R of dimensions and diagonal weight matrices $\mathbf{W}_k$. The term group average originated in individual differences scaling and its frequent usage in this book extends the terminology to include similar constructs in Procrustes analyses. The weights $\mathbf{W}_k$ refer to each of the R dimensions and are allowed to vary over the K individual sets. The distances are put into inner product form by the double centring operation $\mathbf{B}_k = -1/2(\mathbf{I}-\mathbf{N})\mathbf{D}_k(\mathbf{I}-\mathbf{N})$, where $\mathbf{N}$ equals $\mathbf{11}'/n$. Then the 'inner product' matrices $\mathbf{B}_k$ are matched to a weighted form of the unknown group average. Specifically, the problem is to determine $\mathbf{G}$ and $\mathbf{W}_k$ that minimise

$$\sum_{k=1}^{K} \left\| \mathbf{B}_k - \mathbf{G}\mathbf{W}_k^2\mathbf{G}' \right\|. \tag{13.13}$$

In (13.13) we have written the diagonal matrix in squared form to emphasise that its elements are positive and because, shortly, we need its square-root.

The computational process of minimising (13.13) is based on a special case of the CANDECOMP model $y_{ijk} = \sum_{p,q,r}^{P,Q,R} a_{ip}b_{jq}c_{kr}$ which may be fitted by ALS, at each stage keeping two sets of parameters, say b_{jq} and c_{kr}, fixed while fitting the other by ordinary linear least-squares. An interesting property of INDSCAL is that the estimate of $\mathbf{G}$ is fixed, in the sense that it cannot be rotated without giving a worse fit. However, inspection of (13.13) shows that there is some indeterminacy in the model for we may write $\mathbf{G}\mathbf{W}_k\mathbf{G}' = \mathbf{G}\mathbf{W}(\mathbf{W}^{-1}\mathbf{W}_k^2\mathbf{W}^{-1})\mathbf{W}\mathbf{G}'$ where $\mathbf{W}$ is an arbitrary diagonal matrix. Thus, we may assume the weight matrices are scaled so that:

$$\sum_{k=1}^{K} \mathbf{W}_k^2 = K\mathbf{I}. \tag{13.14}$$

The Procrustean analogue of (13.13) is based on configuration matrices $\mathbf{X}_k$, $k = 1, \ldots, K$, all with R columns. These may be given or they may be derived from distance matrices $\mathbf{D}_k$ and thence from their spectral decomposition $\mathbf{B}_k = \mathbf{X}_k\mathbf{X}_k'$. Again there is rotational indeterminacy that can be accommodated by considering $\mathbf{X}_k\mathbf{Q}_k$. Thus, we are lead to consider the Procrustean criterion:

$$S = \sum_{k=1}^{K} \|\mathbf{X}_k\mathbf{Q}_k - \mathbf{G}\mathbf{W}_k\|, \tag{13.15}$$

expressing that each (rotated) configuration approximates the weighted group average, subject to the constraint (13.14). The matrices $\mathbf{G}$, $\mathbf{Q}_k$, and $\mathbf{W}_k$ are to be estimated. Not only does (13.15) provide estimates of the group average $\mathbf{G}$ together with dimension weights $\mathbf{W}_k$, as with INDSCAL, but also individual configurations $\mathbf{X}_k\mathbf{Q}_k$ rotated into the position where they best match a scaled group average. As with INDSCAL itself, the estimated group average may not be rotated; indeed optimal rotations are included in the model and incorporated into the final estimates.

The minimisation of (13.15) is precisely the problem considered in Section 13.1.2 of the anisotropic scaling of a group average, except that now the transformations $\mathbf{T}_k$ are explicitly specified as orthogonal matrices $\mathbf{Q}_k$. Note that the constraint (13.14) is the same as (13.5). The algorithm is as described in that section but now we are in a position to supply the details of the step for calculating updates of the transformation: for fixed $\mathbf{G}$ and $\mathbf{W}_k$, each $\mathbf{Q}_k$, may be estimated from the SVD$(\mathbf{W}_k\mathbf{G}'\mathbf{X}_k) = \mathbf{U}_k\mathbf{\Sigma}_k\mathbf{V}_k'$ (Chapter 4) as $\mathbf{Q}_k = \mathbf{V}_k\mathbf{U}_k'$, $k = 1, \ldots, K$. (see also *Step 2* of the algorithm set out at the end of this section).

The scaling (13.14) implies that in exact fits of the INDSCAL model, we have that:

$$\frac{1}{K}\sum_{k=1}^{K}\mathbf{B}_k = \mathbf{G}\left(\frac{1}{K}\sum_{k=1}^{K}\mathbf{W}_k^2\right)\mathbf{G}' = \mathbf{G}\mathbf{G}',$$

so that an R-dimensional group average $\mathbf{G}$ is easily obtained from the first R terms of the spectral decomposition of the average of the inner-product matrices. This may provide an alternative way of initiating the algorithm.

As an alternative to the algorithm of Section 13.1.2, that avoids the eigenvalue calculations, one might try an ALS algorithm developed as follows. For fixed $\mathbf{G}$ and $\mathbf{W}_k$, each $\mathbf{Q}_k$, may be estimated from the SVD$(\mathbf{W}_k\mathbf{Q}'\mathbf{G}'\mathbf{X}_k) = \mathbf{U}_k\mathbf{\Sigma}_k\mathbf{V}_k'$ (Chapter 4) as $\mathbf{Q}_k = \mathbf{V}_k\mathbf{U}_k'$, $k = 1, \ldots, K$. Similarly, for fixed $\mathbf{W}_k$, and $\mathbf{Q}_k$, $\mathbf{G}$ may be estimated as $1/K\sum_{k=1}^{K}(\mathbf{X}_k\mathbf{Q}_k\mathbf{W}_k)$. The estimate of each $\mathbf{W}_k$, comes from the individual terms $\mathbf{W}_k\mathbf{G}'\mathbf{X}_k\mathbf{Q}_k = \mathbf{W}_k\mathbf{A}_k$ (say), coupled with the constraint (13.14). If we write $a_{ii}^{(k)}$ for the ith diagonal element of $\mathbf{A}_k$ and $w_i^{(k)}$ for the ith diagonal element of $\mathbf{W}_k$, then differentiation with a Lagrange multiplier λ gives;

$$a_{ii}^{(k)} = \lambda w_i^{(k)},$$

whence applying the constraint gives:

$$w_i^{(k)} = \frac{Ka_{ii}^{(k)}}{\left\{\sum_{k=1}^{K} {a_{ii}^{(k)}}^2\right\}^{1/2}}. \tag{13.16}$$

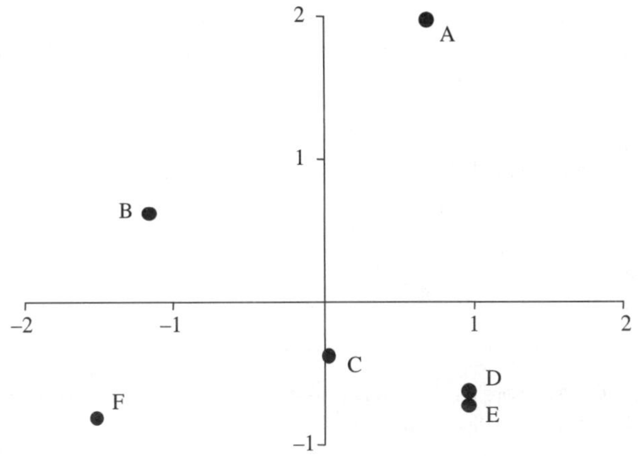

Fig. 13.1. INDSCAL group average of the six pork meat patties discussed in Section 9.2.2.

Consolidating these results, suggests an ALS algorithm incorporating the following three steps:

Algorithm 13.3

Step 1 For fixed $\mathbf{Q}_k$ and $\mathbf{W}_k$, set $\mathbf{G} = 1/K \sum_{k=1}^{K}(\mathbf{X}_k\mathbf{Q}_k\mathbf{W}_k)$.

Step 2 For fixed $\mathbf{W}_k$ and $\mathbf{G}$, evaluate $\text{SVD}(\mathbf{W}_k\mathbf{G}'\mathbf{X}_k) = \mathbf{U}_k\boldsymbol{\Sigma}_k\mathbf{V}'_k$ and set $\mathbf{Q}_k, = \mathbf{V}_k\mathbf{U}'_k$, for $k = 1, \ldots, K$.

Step 3 For fixed $\mathbf{Q}_k$ and $\mathbf{G}$, evaluate $\mathbf{A}_k = \mathbf{G}'\mathbf{X}_k\mathbf{Q}_k$ and (13.16) for $k = 1, \ldots, K, i = 1, \ldots, R$.

13.2.1 EXAMPLE OF INDSCAL

The same data set that was analysed in Chapter 9 using GPA is analysed here using INDSCAL. Eight assessors scored their perceived intensities of 20 properties of 6 different pork meat patties. These result in eight (6×20) data matrices that were normalised in the same way as in Section 9.2.2. Then the inner-product matrices $\mathbf{X}_k\mathbf{X}'_k$ were analysed by INDSCAL.* Figure 13.1 presents the INDSCAL group average configuration for the six food items (A, B, C, D, E, F). Figure 13.1 may be compared with Figure 9.4 which gives the group average found for GPA. Although the exact positions for the food items differ, the configurations are very similar.

In INDSCAL, rather than giving direct displays as in GPA, the relationships of the individual configurations to the group average are summarised by weights (termed *saliences*) to be attached to the axes of Figure 13.1; these are shown in Figure 13.2. The largest weights are those of assessors 1 and 5, associated with the

* The analysis was done by Renger Jellema using Henk Kiers' INDSCAL algorithm.

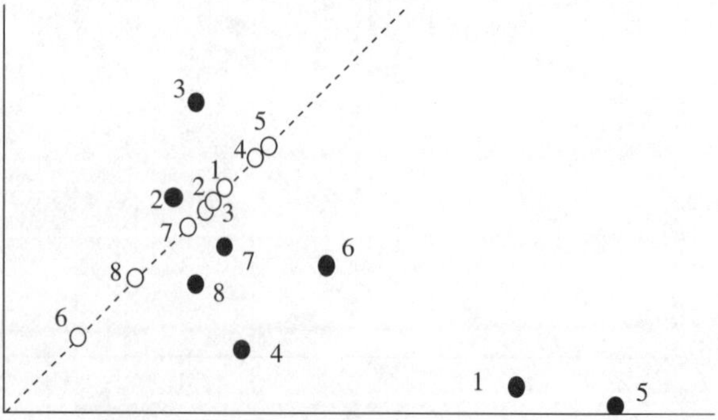

Fig. 13.2. INDSCAL weights (saliencies, •) and inverted GPA isotropic scaling factors (∘), the latter lie on the 45° dashed line.

horizontal axis, and assessor 3, associated with the vertical axis of the INDSCAL group average.

The dashed line of Figure 13.2 shows the inverses of the GPA scaling factors, which are seen to be roughly proportional to the INDSCAL weights (assessors 4 and 6 diverge most). This is consistent with the observation that the isotropic scaling factors in GPA are associated with the individual configurations, while in INDSCAL the saliences are attached to the group average. Thus we would expect an inverse relationship. Indeed, the term $s_k\mathbf{X}_k$ of GPA corresponds to $\mathbf{GW}_k^2\mathbf{G}'$ of INDSCAL and so we might expect an approximate inverse square root relationship.

Thus, both INDSCAL and GPA give similar information though there are differences in emphasis. GPA shows the full variation of the individuals (Figure 9.4) while INDSCAL presents its group average relative to a unique set of axes with associated two-dimensional saliences. These saliencies are similar to what would be produced by a GPA with anisotropic scaling (Section 13.1.2). Indeed, Section 13.1.2 offers a more direct comparison of INDSCAL and the Procrustes approach.

13.3 SMACOF etc.

Least-squares scaling is concerned with finding an approximation matrix $\mathbf{Z}$ in R dimensions which generates distances $\mathbf{\Delta}$ which minimise $\sum_{i<j}^{N}(d_{ij}-\delta_{ij})^2$. To generalise this to K sets of distance matrices we might minimise $\sum_{k=1}^{K}\left(\sum_{i<j}^{N}(d_{ijk}-\delta_{ijk})^2\right)$ but the minimum is achieved by minimising STRESS separately for each of the K sets. Thus we would merely do K separate least-squares scalings. To encapsulate the concept of a group average matrix we may replace δ_{ijk} by δ_{ij}, the elements of a matrix $\mathbf{\Delta}$ common to all sets. We may then write:

$$\sum_{k=1}^{K}\left(\sum_{i<j}^{N}(d_{ijk}-\delta_{ij})^2\right)=\sum_{k=1}^{K}\left(\sum_{i<j}^{N}(d_{ijk}-d_{ij})^2\right)+K\sum_{i<j}^{N}(d_{ij}-\delta_{ij})^2, \quad (13.17)$$

where $d_{ij} = K^{-1} \sum_{k=1}^{K} d_{ijk}$. It follows that to minimise the expression on the left hand side we only have to minimise $\sum_{i<j}^{N} (d_{ij} - \delta_{ij})^2$ which is a simple least-squares scaling problem. Thus the group average is obtained directly from the average of the distances over the K sets. These operations are available in the SMACOF program (Heiser and De Leeuw 1979) which also offers a form of the criterion with weights w_{ijk} and permits estimation of monotonic transformations of the original distances.

What this approach does not do is to give any comparison of the configurations for the individual sets with the group average. However this is easily remedied as follows. If we denote by $\mathbf{Z}$ a configuration that generates $\boldsymbol{\Delta}$ and by $\mathbf{Y}_k$ the configuration giving the least-squares scaling solution for the kth set then we may rotate each $\mathbf{Y}_k$ to fit the group average $\mathbf{Z}$ with the usual residual sum-of-squares. This residual sums-of-squares may be compared in an analysis of variance in a similar way to that discussed for generalised Procrustes analysis.

The least-squares scaling criterion is usually referred to as STRESS. The corresponding criterion where distances are replaced by squared distances is known as SSTRESS which is minimised by the ALSCAL program (Young, De Leeuw, and Takane 1976, 1978). Clearly (13.4) remains valid with distances replaced by squared distances giving a group average obtained from minimising SSTRESS for the averaged squared distances.

13.4 Structuration des Tableaux A Trois Indices de la Statisitique

STATIS is another method that focuses on finding a group average for a set of K matrices. The method is rather general as it may operate on (i) data matrices $\mathbf{X}_k$ with the same P variables observed by K different individuals (or one individual on K occasions) assessing the same N samples or (ii) it may operate on K different sets of variables of size $P_1, P_2, \ldots, P_K$ observed by the same individual in assessing N samples. The situations are rather different and engender somewhat different variations of the basic method. In all cases, STATIS operates on a symmetric $K \times K$ matrix $\mathbf{S}$, whose elements define the relationships between the K sets. The elements of $\mathbf{S}$ are given by the RV coefficient (Chapter 3). In case (ii) it will be seen that, because of the inequality of the numbers of variables, the term trace$(\mathbf{X}_k\mathbf{X}_l')$ required by the RV coefficient is undefined. Defining $\mathbf{C}_k = (\mathbf{X}_k\mathbf{X}_k')$ allows the RV coefficient based on trace$(\mathbf{C}_k'\mathbf{C}_l)$ to be formed. However $\mathbf{S}$ is defined, its principal eigenvector $\mathbf{s}$ may be used to form $\mathbf{C} = \sum_{k=1}^{K} s_k\mathbf{X}_k$ which may be interpreted as a kind of group average. If the number of columns varies among the matrices $\mathbf{X}_k$ then we form $\sum_{k=1}^{K} s_k\mathbf{C}_k$, which is an $N \times N$ matrix, and use its spectral decomposition $\mathbf{CC}'$ to give the group-average referred to principal axes.

Here we are primarily interested in case (i), operating on data matrices $\mathbf{X}_k$ which we may assume to be normalised as discussed in Chapter 2, giving $\mathbf{S} = \{\text{trace}(\mathbf{X}_k\mathbf{X}_l')\}$. We have seen in Section 13.1.1, equation (13.3) that this is precisely the matrix that arises when estimating the isotropic scaling factors that minimise S of (13.1). The difference is that the equation arises iteratively in Section 13.1.1 but in STATIS its use is non-iterative. In this respect, STATIS may be viewed as the first step in estimating isotropic scaling factors for the data matrices $\mathbf{X}_k$. Note the similar estimate of $\mathbf{s}$ given by the constraint (9.32) and the less directly related estimate of $\mathbf{s}$ given by (9.20). Meyners, Kunert, and Qannari (2000) recommend using the STATIS $\mathbf{s}$ as weights to compute the GPA group average (9.3) with $\mathbf{T}_k = \mathbf{Q}_k$. However, Meyners and co-workers use form (ii) of STATIS because they are concerned with possible unequal numbers of columns in the $\mathbf{X}_k$ and do not wish to pad out with zero columns. This suggests more of a concern with the projection case, $\mathbf{T}_k = \mathbf{P}_k$. What is clear from all this, is that the STATIS $\mathbf{s}$ has a close relationship with Procrustean isotropic scaling but the full details need clarification.

There are several major differences between the set-up for STATIS and that for Procrustes problems. First, in STATIS, the columns of $\mathbf{X}_k$ refer to variables and not to dimensions so $\mathbf{X}_k$ is a data matrix and not a configuration matrix. Second, no transformation matrix $\mathbf{T}$ is used by STATIS. Third, among the variants of STATIS, those that determine a 'compromise' variable rather than a group average of individuals predominate. In this respect STATIS has closer links with Common Components Analysis (Section 13.6.1) than with Procrustes Analysis.

The above gives a much simplified account of STATIS which, in general, allows weights for the cases and a general metric when evaluating the inner products. See Escoufier (1985) and Lavit (1988) for a full account of the method.

13.5 Links with canonical correlation

In this section, the columns of $\mathbf{X}_1$ refer to P_1 centred variables and the columns of $\mathbf{X}_2$ refer to P_2 centred variables. Canonical correlation (Hotelling 1936) requires the linear combination $\mathbf{X}_1\mathbf{v}_1$ of the first set of variables that have maximal correlation ρ with the linear combination $\mathbf{X}_2\mathbf{v}_2$ of the second set. Thus:

$$\rho^2 = \frac{(\mathbf{v}_1'\mathbf{X}_1'\mathbf{X}_2\mathbf{v}_2')^2}{(\mathbf{v}_1'\mathbf{X}_1'\mathbf{X}_1\mathbf{v}_1)(\mathbf{v}_2'\mathbf{X}_2'\mathbf{X}_2\mathbf{v}_2)} = \frac{(\mathbf{v}_1'\mathbf{Z}_{12}\mathbf{v}_2)^2}{(\mathbf{v}_1'\mathbf{Z}_{11}\mathbf{v}_1)(\mathbf{v}_2'\mathbf{Z}_{22}\mathbf{v}_2)}. \tag{13.18}$$

Differentiation with respect to $\mathbf{v}_1$ and $\mathbf{v}_2$ gives:

$$\begin{aligned} (\mathbf{v}_1'\mathbf{Z}_{11}\mathbf{v}_1)Z_{12}\mathbf{v}_2 &= (\mathbf{v}_1'\mathbf{Z}_{12}\mathbf{v}_2)\mathbf{Z}_{11}\mathbf{v}_1, \\ (\mathbf{v}_2'\mathbf{Z}_{22}\mathbf{v}_2)\mathbf{Z}_{21}\mathbf{v}_1 &= (\mathbf{v}_1'\mathbf{Z}_{12}\mathbf{v}_2)\mathbf{Z}_{22}\mathbf{v}_2. \end{aligned} \tag{13.19}$$

Eliminating $\mathbf{v}_2$ gives:

$$(\mathbf{Z}_{12}\mathbf{Z}_{22}^{-1}\mathbf{Z}_{21})\mathbf{v}_1 = \rho^2\mathbf{Z}_{11}\mathbf{v}_1.$$

Similarly, eliminatig $\mathbf{v}_1$:

$$(\mathbf{Z}_{21}\mathbf{Z}_{11}^{-1}\mathbf{Z}_{12})\mathbf{v}_2 = \rho^2\mathbf{Z}_{22}\mathbf{v}_2. \tag{13.20}$$

These are two-sided eigenvalue problems whose maximum eigenvalue, and corresponding eigenvectors give the maximal correlation and associated linear combinations. Because ρ^2 is defined in (13.18) as a ratio, the eigenvectors may be scaled in any way we please, usually, we adopt the scaling $\mathbf{v}_1'\mathbf{Z}_{11}\mathbf{v}_1 = \mathbf{v}_2'\mathbf{Z}_{22}\mathbf{v}_2 = 1$, in which case (13.18) gives:

$$\rho = \mathbf{v}_1'\mathbf{Z}_{12}\mathbf{v}_2,$$

and (13.19) becomes:

$$\begin{aligned} \mathbf{Z}_{12}\mathbf{v}_2 &= \rho\mathbf{Z}_{11}\mathbf{v}_1, \\ \mathbf{Z}_{21}\mathbf{v}_1 &= \rho\mathbf{Z}_{22}\mathbf{v}_2. \end{aligned} \tag{13.21}$$

which may be written

$$\begin{pmatrix} \mathbf{0} & \mathbf{Z}_{12} \\ \mathbf{Z}_{21} & \mathbf{0} \end{pmatrix}\begin{pmatrix} \mathbf{v}_1 \\ \mathbf{v}_2 \end{pmatrix} = \rho \begin{pmatrix} \mathbf{Z}_{11} & \mathbf{0} \\ \mathbf{0} & \mathbf{Z}_{22} \end{pmatrix}\begin{pmatrix} \mathbf{v}_1 \\ \mathbf{v}_2 \end{pmatrix},$$

or equivalently, in a form required later:

$$\begin{pmatrix} \mathbf{Z}_{11} & \mathbf{Z}_{12} \\ \mathbf{Z}_{21} & \mathbf{Z}_{22} \end{pmatrix}\begin{pmatrix} \mathbf{v}_1 \\ \mathbf{v}_2 \end{pmatrix} = (1+\rho) \begin{pmatrix} \mathbf{Z}_{11} & \mathbf{0} \\ \mathbf{0} & \mathbf{Z}_{22} \end{pmatrix}\begin{pmatrix} \mathbf{v}_1 \\ \mathbf{v}_2 \end{pmatrix}. \tag{13.22}$$

The scaling of the eigenvectors of (13.22) remains essentially arbitrary but the form of the equations requires $\mathbf{v}_1'\mathbf{Z}_{11}\mathbf{v}_1 = \mathbf{v}_2'\mathbf{Z}_{22}\mathbf{v}_2$ so if we set $\mathbf{v}_1'\mathbf{Z}_{11}\mathbf{v}_1 + \mathbf{v}_2'\mathbf{Z}_{22}\mathbf{v} = 2$ we retain our previous scaling. Rather remarkably, this single scaling of the eigenvectors of (13.22) subsumes the independent scalings of the vectors $\mathbf{v}_1$ and $\mathbf{v}_2$ of (13.21).

The above classical development of canonical correlation shows no obvious relationship with Procrustes Analysis. Let us now consider the minimisation of:

$$\|\mathbf{X}_1\mathbf{v}_1 - \mathbf{X}_2\mathbf{v}_2\|. \tag{13.23}$$

We remarked in Section 3.1 that this has a trivial zero minimum for $\mathbf{v}_1 = \mathbf{v}_2 = 0$, so some constraint on $\mathbf{v}_1$ and $\mathbf{v}_2$ is needed before the minimisation is of any interest. Consider a general quadratic size constraint:

$$\alpha\mathbf{v}_1'\mathbf{Z}_{11}\mathbf{v}_1 + \beta\mathbf{v}_2'\mathbf{Z}_{22}\mathbf{v}_2 + 2\gamma\mathbf{v}_1'\mathbf{Z}_{12}\mathbf{v}_2 = 1. \tag{13.24}$$

Minimising (13.23) subject to the constraint (13.24) associated with a Lagrange multiplier λ gives:

$$\begin{aligned} \mathbf{Z}_{11}\mathbf{v}_1 - \mathbf{Z}_{12}\mathbf{v}_2 &= \lambda(\alpha\mathbf{Z}_{11}\mathbf{v}_1 + \gamma\mathbf{Z}_{12}\mathbf{v}_2), \\ \mathbf{Z}_{22}\mathbf{v}_2 - \mathbf{Z}_{21}\mathbf{v}_1 &= \lambda(\beta\mathbf{Z}_{22}\mathbf{v}_2 + \gamma\mathbf{Z}_{21}\mathbf{v}_1), \end{aligned}$$

that is,

$$\begin{aligned} (1+\lambda\gamma)\mathbf{Z}_{12}\mathbf{v}_2 &= (1-\lambda\alpha)\mathbf{Z}_{11}\mathbf{v}_1, \\ (1+\lambda\gamma)\mathbf{Z}_{21}\mathbf{v}_1 &= (1-\lambda\beta)\mathbf{Z}_{22}\mathbf{v}_2. \end{aligned} \tag{13.25}$$

Provided $\alpha = \beta$, (13.25) coincides with (13.21) with $\rho = (1-\lambda\alpha)/(1+\lambda\gamma)$. It follows that the eigenvectors derived from (13.25) are those required for canonical correlation. The choice of $\alpha = \beta$ and γ must be consistent with (13.24) which gives:

$$\mathbf{v}_1'\mathbf{Z}_{11}\mathbf{v}_1 = \mathbf{v}_2'\mathbf{Z}_{22}\mathbf{v}_2 = \frac{1+\lambda\gamma}{2(\alpha+\gamma)}, \quad \mathbf{v}_1'\mathbf{Z}_{12}\mathbf{v}_2 = \frac{1-\lambda\alpha}{2(\alpha+\gamma)} = \frac{\rho(1+\lambda\gamma)}{2(\alpha+\gamma)},$$

from which it follows that $\min \|\mathbf{X}_1\mathbf{v}_1 - \mathbf{X}_2\mathbf{v}_2\| = \lambda$. The smallest value of λ gives the maximal value of ρ. Unless we choose $\alpha = \beta = 1/2$ and $\gamma = 0$, this scaling of the vectors differs from the one adopted above but has no effect on the value of the canonical correlation. Of special interest in the following, is the constraint:

$$\|\tfrac{1}{2}(\mathbf{X}_1\mathbf{v}_1 + \mathbf{X}_2\mathbf{v}_2)\| = 1, \qquad (13.26)$$

for which $\alpha = \beta = \gamma = 1/4$. The effect of (13.26) is to require the sum-of-squares of the group average to be of unit size. Thus, the least-squares formulation (13.23) with any quadratic constraint (13.24) gives us the linear combinations required for canonical correlation. Different choices of $\alpha = \beta, \gamma$ lead to different, but equally valid, scalings of the vectors.

In the above, we have considered only a single canonical correlation with its associated vectors. In the usual way, R dimensions may be fitted by taking $R - 1$ further eigenvalues and associated vectors assembled in the columns of $\mathbf{V}_1$ and $\mathbf{V}_2$. The criterion (13.23) is then replaced by:

$$\|\mathbf{X}_1\mathbf{V}_1 - \mathbf{X}_2\mathbf{V}_2\|, \qquad (13.27)$$

and the group average constraint by $\|\frac{1}{2}(\mathbf{X}_1\mathbf{V}_1 + \mathbf{X}_2\mathbf{V}_2)\| = 1$. The following orthogonality-like results follow as a consequence of the two-sided eigenvalue formulation:

$$\mathbf{V}_1'\mathbf{Z}_{11}\mathbf{V}_1 = \mathbf{V}_2'\mathbf{Z}_{22}\mathbf{V}_2 = \frac{1}{2(\alpha+\gamma)}(\mathbf{I}+\gamma\mathbf{\Lambda}), \quad \mathbf{V}_1'\mathbf{Z}_{12}\mathbf{V}_2 = \frac{1}{2(\alpha+\gamma)}(\mathbf{I}-\alpha\mathbf{\Lambda}),$$

where $\mathbf{\Lambda}$ is the diagonal matrix of the R smallest values of λ satisfying (13.25). The criterion (13.27) is manifestly in the form of a two-sided Procrustes problem where the transformations are constrained only by size considerations.

Next we consider generalisations of the above two-sets results to K sets of variables. Several generalisations of canonical correlation have been suggested (see Gower 1989 for a brief review, other chapters in Coppi and Bolasco (1989) and the references cited therein). Examination of the two-sets case suggests three lines of generalisation, all of which are consistent with the case $K = 2$: (i) we may require $\mathbf{v}_k'\mathbf{Z}_{kk}\mathbf{v}_k' = 1$ for $k = 1, \ldots, K$, (ii) we may require $\sum_{k=1}^{K} \mathbf{v}_k'\mathbf{Z}_{kk}\mathbf{v}_k = K$, and (iii) we may require $\mathbf{g}'\mathbf{g} = 1$ where $\mathbf{g} = 1/K \sum_{k=1}^{K} \mathbf{X}_k\mathbf{v}_k$. These constraints are understood to apply to each column of $\mathbf{V}$.

With the first type of generalisation (Kettenring 1971), we may construct the $K \times K$ matrix $\mathbf{R}$ with elements $\mathbf{v}_i'\mathbf{Z}_{ij}\mathbf{v}_j$ where $\mathbf{v}_i'\mathbf{Z}_{ii}\mathbf{v}_i' = \mathbf{v}_j'\mathbf{Z}_{jj}\mathbf{v}_j = 1$. Thus,

$\mathbf{R}$ is a correlation matrix and we may determine vectors $\mathbf{v}_k$ that maximise some property such as det $\mathbf{R}$, largest eigenvalue, $\mathbf{1}'\mathbf{R}\mathbf{1}$ etc. Such criteria do not generally have least-squares forms ($\mathbf{1}'\mathbf{R}\mathbf{1}$ is an exception) and their optimisation can be awkward because, effectively, we have K Lagrange multipliers, one for each set of variables. Also, there are no natural orthogonality properties as arise with two-sided eigenvalue formulations, so if we require an R-dimensional solution, orthogonality has to be imposed by constraining the Rth solution to be orthogonal to all preceding solutions.

Here we pursue only links with Procrustes analysis arising from generalisations of the two-sets criterion (13.27) subject to the constraints (ii) and (iii). We shall express the criterion in R dimensions. Thus, consider

$$S = \sum_{i<j}^{K} \|\mathbf{X}_i\mathbf{V}_i - \mathbf{X}_j\mathbf{V}_j\| = \text{trace}(K\mathbf{V}'\boldsymbol{\Delta}\mathbf{V} - \mathbf{V}'\mathbf{Z}\mathbf{V}), \tag{13.28}$$

where $\mathbf{Z} = \{\mathbf{Z}_{ij}\}$, $\boldsymbol{\Delta} = \text{diag}(\mathbf{Z}_{11}, \mathbf{Z}_{22}, \ldots, \mathbf{Z}_{KK})$ and $\mathbf{V}' = \mathbf{V}_1', \mathbf{V}_2', \ldots, \mathbf{V}_K'$. We shall write the constraint in the general form:

$$\text{diag}(\alpha\mathbf{V}'\boldsymbol{\Delta}\mathbf{V} + \beta\mathbf{V}'\mathbf{Z}\mathbf{V}) = \mathbf{I}, \tag{13.29}$$

which includes both (ii) and (iii). For the constraints (ii) set $\alpha = 1/K$ and $\beta = 0$ and for the constraints (iii) set $\alpha = 0$ and $\beta = 1/K^2$. Note that $\mathbf{V}'\mathbf{Z}\mathbf{V} = K^2\mathbf{G}'\mathbf{G}$. Differentiation of (13.28) with Lagrange multipliers $\gamma_1, \gamma_2, \ldots, \gamma_R$, and writing $\boldsymbol{\Gamma} = \text{diag}(\gamma_1, \gamma_2, \ldots, \gamma_R)$, gives:

$$K\boldsymbol{\Delta}\mathbf{V} - \mathbf{Z}\mathbf{V} = (\alpha\boldsymbol{\Delta}\mathbf{V} + \beta\mathbf{Z}\mathbf{V})\boldsymbol{\Gamma}. \tag{13.30}$$

Pre-multiplying the above by $\mathbf{V}'$ and using (13.28) and (13.29), immediately shows that $S = \text{trace}(\boldsymbol{\Gamma})$, so the criterion is minimised by selecting the R smallest values of γ_r that satisfy (13.30). We may rewrite (13.30):

$$\mathbf{Z}\mathbf{V} = \boldsymbol{\Delta}V\boldsymbol{\Phi}, \tag{13.31}$$

where $\boldsymbol{\Phi} = (K\mathbf{I} - \alpha\boldsymbol{\Gamma})(\mathbf{I} + \beta\boldsymbol{\Gamma})^{-1}$. The two-sided eigenvalue problem (13.31) is precisely in the form of (13.22), so apart from a scaling factor, the vectors associated with the constraints (ii) and (iii) are the same. As usual, the matrices $\mathbf{V}'\mathbf{Z}\mathbf{V}$ and $\mathbf{V}'\boldsymbol{\Delta}\mathbf{V}$ are diagonal. Thus we have orthogonality even though (13.29) only constrains the diagonal elements of these matrices. Thus (13.29) becomes:

$$\alpha\mathbf{V}'\boldsymbol{\Delta}\mathbf{V} + \beta\mathbf{V}'\mathbf{Z}\mathbf{V} = \mathbf{I},$$

which in combination with (13.31) gives the scaling:

$$(\mathbf{V}'\boldsymbol{\Delta}\mathbf{V})(\alpha\mathbf{I} + \beta\boldsymbol{\Phi}) = (\mathbf{V}'\mathbf{Z}\mathbf{V})(\alpha\boldsymbol{\Phi}^{-1} + \beta\mathbf{I}) = \mathbf{I}. \tag{13.32}$$

The scaling of the columns of $\mathbf{V}$ derives from substituting the successive eigenvalues ϕ_r of (13.31) into (13.32). Thus for scaling (ii), as required $\mathbf{V}'\mathbf{\Delta V} = K\mathbf{I}$ and for case (iii), $\mathbf{V}'\mathbf{ZV} = K^2\mathbf{G}'\mathbf{G} = K^2\mathbf{I}$.

In accordance with (9.2), we may rewrite (13.28) as:

$$K\sum_{k=1}^{K} \|\mathbf{X}_k\mathbf{V}_k - \mathbf{G}\|. \tag{13.33}$$

Consider now the correlation between one column $\mathbf{X}_k\mathbf{v}_k$ of $\mathbf{X}_k\mathbf{V}_k$ with a corresponding column $\mathbf{g}$ of $\mathbf{G}$. It makes no difference whether we use the constraints (ii) or (iii) as the vectors are proportional. For definiteness we take the constraints (iii), giving:

$$\rho_{k0}^2 = \frac{(\mathbf{g}'\mathbf{X}_k\mathbf{v}_k)^2}{(\mathbf{v}_k'\mathbf{Z}_{kk}\mathbf{v}_k)} = \frac{\varphi^2}{K^2}\mathbf{v}_k'\mathbf{Z}_{kk}\mathbf{v}_k,$$

whence from (13.31):

$$\sum_{k=1}^{K} \rho_{k0}^2 = \phi_r. \tag{13.34}$$

Thus, the vector $\mathbf{v}$ that minimises the Procrustean criterion (13.27) also maximises the sum-of-squares of the correlations of the linear combinations $\mathbf{X}_k\mathbf{v}_k$ with the group average. This seems very much to be in accord with the spirit of the original canonical correlation concept. As with the two-sets case, $\mathbf{v}_k$ may be replaced by $\mathbf{V}_k(k = 1, \ldots, K)$ to give R-dimensional solutions maximising the sum-of-squares of the RK correlations derived from the K sets of variables in each of the R dimensions.

Next we note that (13.33) suggests that we examine the minimisation of $\sum_{k=1}^{K} \|\mathbf{X}_k\mathbf{v}_k - \mathbf{f}\|$ for $\mathbf{f}'\mathbf{f} = 1$, where now there is no requirement that $\mathbf{f}$ is a group average. With a Lagrange multiplier δ we now have that the estimate of $\mathbf{f}$ is:

$$\mathbf{f} = (K/\delta)\mathbf{g},$$

which is proportional to the group average and gives the scaling of $\mathbf{v}$ as satisfying $\delta^2 = \mathbf{v}'\mathbf{Zv}$. Differentiating with respect to $\mathbf{v}$ and substituting for $\mathbf{f}$ gives:

$$\mathbf{\Delta v} = (1/\delta)\mathbf{Zv}, \tag{13.35}$$

again consistent with (13.22). The vectors again are proportionate so that equation (13.34) remains valid. Indeed, $\delta = \phi$ and while from (13.22) the scaling for the constraint $\mathbf{g}'\mathbf{g} = 1$ was $\mathbf{v}'\mathbf{Zv} = K^2$ here, for $\mathbf{f}'\mathbf{f} = 1$, it is $\mathbf{v}'\mathbf{Zv} = \delta^2$ which is consistent with the difference in scaling of the group mean. Despite the apparent generality allowed for $\mathbf{f}$ (apart from size), the two constraints lead to precisely the same results, apart from an overall scaling of K/δ.

The least-squares formulation establishes the link between canonical correlation and Procrustes analysis. It also allows alternating least-squares (ALS) algorithms to be developed. For example, for a given group average $\mathbf{G}$ we may compute by constrained least-squares the $\mathbf{V}_k$ that minimise $\|\mathbf{X}_k\mathbf{V}_k - \mathbf{G}\|$, $k = 1, \ldots, K$. Similarly, for given $\mathbf{V}_k$ we may update the group average. The process continues iteratively until adequate convergence. In itself, this might be regarded as a poor substitute for using direct eigenvalue algorithms to solve equations $\mathbf{Zv} = \lambda\boldsymbol{\Delta}\mathbf{v}$ with whatever scaling of the vectors is required. There are two reasons why the ALS procedure might be preferred. First, the order of $\mathbf{Z}$ is the total number of variables involved, which may be quite large. Second, and more important, is that when the variables are categorical, the estimation of optimal scores, possibly with ordinal constraints, may be incorporated into the least-squares process. One of the main computer programs for generalized canonical correlation analysis is OVERALS (Verdegaal 1986). For further details see, van der Burg and De Leeuw (1983), van der Burg (1988), and Gifi (1990), and the many references cited therein. Steenkamp, van Trijp, and ten Berge (1994) compare the performances, in a marketing context, of GPA analysis, INDSCAL and CANCOR (Carroll 1968).

13.6 Common components analysis

There is a sizeable literature on the common principal components of data matrices $\mathbf{X}_1, \mathbf{X}_2, \ldots, \mathbf{X}_k$ that share the same variables. The emphasis on variables, rather than configurations, coupled with no requirement for row-matching, suggests that this class of problems is rather far from Procrustes analysis. The principal components of $\mathbf{X}_1$ are the vectors $\mathbf{V}_1$ of the singular value decomposition $\mathbf{X}_1 = \mathbf{U}_1\boldsymbol{\Sigma}_1\mathbf{V}_1'$. Here, $\boldsymbol{\Sigma}_1$ is a diagonal matrix containing the singular values arranged in non-increasing order. The Eckart–Young theorem states that the best rank R approximation $\hat{\mathbf{X}}_1$ to $\mathbf{X}_1$ that minimises $\|\mathbf{X}_1 - \hat{\mathbf{X}}_1\|$ is given by setting to zero all but the first R elements of $\boldsymbol{\Sigma}_1$; equivalently we may take the first R columns of $\mathbf{U}_1$, and $\mathbf{V}_1$, with the first R diagonal elements of $\boldsymbol{\Sigma}_1$. Thus, with K sets, *and when the rows do match*, we may wish to minimise $\sum_{k=1}^{K} \|\mathbf{X}_k - \hat{\mathbf{X}}\|$ for $\hat{\mathbf{X}}$ of specified rank R. This is the same as minimising $\|\mathbf{G} - \hat{\mathbf{X}}\|$ where $\mathbf{G} = 1/K \sum_{k=1}^{K} \|\mathbf{X}_k\|$, so $\hat{\mathbf{X}}$ is given by the rank R singular value decomposition of the average data matrix $\mathbf{G} = \mathbf{U}\boldsymbol{\Sigma}\mathbf{V}'$. Thus, the first R columns of $\mathbf{V}$ may be viewed as average components of the K data-matrices.

The above is a true Procrustes problem, matching the K configurations to the best rank R configuration. Common components analysis is more concerned with matching spaces. To contrast the difference between Common components and Procrustean methods, suppose we have two configurations whose row-labels match (of course, if they do not match Procrustean matching is undefined). On the one hand, the two configurations may share common components by occupying the same two-dimensional subspace but be poor matches from the Procrustean point of view. On the other hand, the configurations may match exactly but occupy different subspaces of some higher dimensional space. The latter is of great Procrustean interest but of little interest from the point of view of common components. Common components analysis is concerned with the relationships between the subspaces containing the principal areas of variation. It is concerned with common subspaces, not with common configurations such as a group average.

Krzanowski (1979) supposed an R dimensional subspace common to all the data matrices and showed that if the columns of $\mathbf{V}_k$ give the first R principal components associated with $\mathbf{X}_k$ then the eigenvectors of $\sum_{k=1}^{K} \mathbf{V}_k\mathbf{V}_k'$ give the best vectors that span the common space. The criterion maximised is the sum-of-squares of the cosines of the maximal angles between the common subspace and the spaces spanned by each $\mathbf{V}_k$.

Flury (1988) considered the situation in which some subset of eigenvectors, not necessarily those associated with the biggest eigenvalues, are common to the K data-matrices. In this approach the actual eigenvectors have to match, not just span a common subspace. Thus, one is looking for clusters of K vectors, one from each set of variables. For Schott (1991) some set of eigenvectors span a common subspace but need not match within the subspace.

We have already mentioned the link between Common components analysis and STATIS (Section 13.4). A related method, Analyse factorielle multiple (Escofier and Pages 1988), considers the same set-up as STATIS but evaluates a weighted PCA for the concatenated set $\mathbf{X} = (\mathbf{X}_1, \mathbf{X}_2, \ldots, \mathbf{X}_K)$ and then relates the components of the individual sets to the global components. In this method, as in STATIS, and unlike the Common components methods discussed previously, the groups may be defined on different variables but as with Procrustean methods, the same individuals are associated with the rows of each $\mathbf{X}_k$ so the row-labels match.

13.7 Other methods

The psychometric literature is full of a host of three-way and three mode methods of data analysis. Some of these may be thought of as analysing K matrices and some may be thought of as fitting triple product models. Most have little relationship with Procrustean methods but perhaps we should mention three mode PCA and three mode factor analysis. These methods are concerned with three-way arrays and fit models of the following type:

$$x_{ijk} = \sum_{p,q,r=1}^{P,Q,R} a_{ip} b_{jq} c_{kr} g_{pqr} \quad \text{for } i = 1 \ldots I,\ j = 1 \ldots J,\ k = 1 \ldots K$$

where P, Q, R are much smaller than I, J, K. The three-way array $\{g_{pqr}\}$ is termed the core matrix and acts as a kind of group average. When $\{x_{ijk}\}$ refers to an i, jth element of a data matrix $\mathbf{X}_k$, then the core matrix provides a summary. Note, that the core matrix is still a three-way array, albeit a, hopefully, much smaller one than $\{x_{ijk}\}$. For further information, see Tucker (1966), Kroonenberg (1983), and Kroonenberg and De Leeuw (1980).

For further unformation on multiway and multimode models, see the general references of Arabie, Carroll, and De Sarbo (1987), Coppi and Bolasco (1989), Gifi (1990), Law *et al.* (1984), and Lebart, Morineau, and Piron (1995).

14
Some application areas, future, and conclusion

The origins of Procrustes analysis were discussed in Chapter 1, together with a very brief overview of later developments. Previous chapters have covered many of the developments of the past 50 years, together with examples of applications; much new material has also been included. In this chapter, we briefly list some of the fields where we have found applications of Procrustes analysis. This list is bound to be incomplete, as applications have to be sought in the vast non-statistical literature and are easily missed. In moving into new application areas, new considerations call for new developments and new methodology. We highlight areas where new methodology is currently being developed or where we see a clear need for further work.

14.1 Application areas

Originally Procrustes analysis stems from psychometrics, and in particular from factor analysis. It was used to match a newly derived factor structure either to one that was previously determined, or to an hypothesised structure. Structures could be referred either to orthogonal or oblique axes (see Chapter 6). Applications of Procrustes analysis now occur in such diverse fields as molecular biology, sensory analysis, market research, shape theory, image analysis, and a host of other fields. One thing that stands out when considering these applications is that they fall into two main classes. In the first class are the multidimensional matching problems that have been the main thrust of this book where dimensions often have no substantive meaning; indeed, this has been seen as a primary justification for using Procrustes methods. In the second class are problems, where the objects to be matched exist physically in two, or sometimes three, dimensions. In other words the dimensions on which the objects are measured have direct physical meanings. Then, although orientation may be of little significance, special considerations apply concerning the spatial integrity of the configurations.

14.1.1 PERSONALITY PSYCHOLOGY/FACTOR ANALYSIS

In the factor analytic tradition, Procrustes matching is applied in personality psychology, where a hypothesised structure serves as a target matrix to which a data matrix, for example, with results on psychological tests, is matched. In these applications, the rows of the matrices refer to individuals, and the columns to scores on test-items. This gives a two-sets Procrustes problem. K-sets problems could arise when comparing different psychological tests on the same set of individuals; the variables in the sets are the results obtained by the individuals (rows) on the items of the different tests.

Another method for studying personality is the repertory grid (see Slater 1976, 1977). Here *constructs* elicited from a person are scored or ranked on responses, termed *elements*; the constructs may be treated as variables. This gives a two-way table classified by constructs and elements. A collection of these tables, perhaps made on different occasions, so giving a measure of the progress of some treatment regime, may be analysed by Procrustes methods. In Frewer, Salter, and Lambert (2001) repertory grid methodology is used to elicit individual patients' concerns about medical treatments, and to construct idiosyncratic questionnaires. The answers to the concerns in these questionnaires provide a variant of Free Choice Profiling data (Section 2.2.1) that are typically analysed by Generalised Orthogonal Procrustes Analysis (GOPA or GPA).

14.1.2 SENSORY ANALYSIS

GOPA has become popular in sensory analysis of the perceived intensities of certain properties of samples, often of food. It is applied as an individual difference method to correct for the differences in scoring between the assessors in a sensory panel (see, for example, Langron and Collins 1984, Arnold and Williams 1985). The paper by Banfield and Harries (1975) is probably the first in which Procrustes analysis is applied in a sensory context. These authors use two-set Procrustes analysis to match the scores of pairs of judges. Hames and MacFie (1976) used the generalised pairwise variant of this method (Section 9.3) to study the judging of meat by several assessors. The previous approach differs from the GOPA method used in most later studies of sensory data. GOPA is described by Gower (1975) who applied it to the raw data sets from three assessors who visually scored the quality of beef. GOPA has become one of the standard methods for analysing data from a sensory panel (cf. Dijksterhuis 1996), providing information on all three modes in the data, the individuals, the products (food), and the variables (descriptive terms whose perceived intensities form the data). An example of this use of GOPA is given in Section 9.2.2. One could visualise the use of a projection variant of orthogonal Procrustes analysis for this use; indeed, the method suggested by Peay (1988) has been used in this field (Dijksterhuis and Punter 1990). The discussion of Section 10.4 gives some reasons for caution in the uncritical use of projection methods.

Free choice profiling (FCP) (Section 2.2.1), in which the assessors in a sensory panel are allowed to use their own terms to describe the products, was suggested by Williams and Langron (1984). This results in K individual data matrices, one for each assessor, with columns representing different variables. The estimation of absolute differences between assessors is beyond reach but their views on relative differences may be analysed by GOPA.

14.1.3 MARKET RESEARCH

Another application area where GOPA finds applications is in market research. The method seems to have spread from sensory science to marketing applications. A method akin to FCP to score and analyse data from several assessors using idiosyncratic sets of attributes was suggested by Steenkamp *et al.* (1994) in a paper that also contains a review comparing GOPA, generalised canonical analysis (CANCOR, Carroll 1968) and INDSCAL (Carroll and Chang 1970).

The images that consumers have of certain products can be matched to each other using GOPA. Leemans *et al.* (1992) explored the possibilities of GOPA for studying the positioning of novels based on consumers' perceptions. Gower and Dijksterhuis (1994) used GPA as part of a method to match consumers' images of different types of coffee (see Section 11.2).

14.1.4 SHAPE ANALYSIS

Now we come to matching-problems in two or three dimensions. A major development arises from the study of shape. Whole books are written on shape theory (e.g. Bookstein, 1991, Dryden and Mardia 1998, Kendall *et al.* 1999, Lele and Richtsmeier 2001). In this section we give a simplified overview of Kendall's shape theory; other aspects of shape analysis are discussed in Section 14.1.5.

In a survey of his statistical theory of shape, Kendall (1989) says that his original interest in shape was stimulated by a problem in archaeology, concerning the alignment of standing stones. Every set of three stones defines a triangle and the question is whether or not there is an excess of flat triangles (i.e. with an angle approximating 180°). However, most applications of shape analysis are found in biology as with the example in Section 4.4, where the shapes of juvenile and adult macaque monkey's skulls are compared. Naturally, biological shapes occur in two or three dimensions.

It seems that questions of inference are a primary concern in statistical shape space studies. Thus, answers are sought to questions such as: what is the evidence that this shape differs from that shape?, are shape estimators consistent?, what is the effect of anisotropic distribution of errors?, are there an excess of flat triangles among a set of standing stones? In most of the applications with which we have been concerned, it would be surprising if the configurations were very similar. Estimation of an average configuration is a primary interest, together with an indication of in what respects different configurations differ from the average. Inferential questions are secondary although not to be disregarded. For example, we might ask if there is any evidence that the configurations fall into two, or more, distinct groups.

A raw data matrix $\mathbf{X}$ that includes size as well as shape information defines a configuration termed *form*. After translation of the centroid of $\mathbf{X}$ to the origin and normalisation, to eliminate size, such that $\|\mathbf{X}\| = 1$, the NP values of $\mathbf{X}$ define the coordinates of a point in $P(N-1)$-dimensional Euclidean space. The normalisation $\|\mathbf{X}\| = 1$ implies that the points lie on a hypersphere of unit radius, termed the *sphere of preshapes*.* Rotations $\mathbf{XQ}$ of the configuration $\mathbf{X}$ define a manifold on the hypersphere of all points that are considered to have the same shape. This manifold is termed a *fibre* (or two *fibres*, if rotation and reflection are distinguished, as they usually are). The hypersphere is composed of non-intersecting fibres representing configurations of different shapes. A typical point on a fibre, in its $N \times P$ form, is an *icon* of the shape. Suppose we have configurations $\mathbf{X}_1$ and $\mathbf{X}_2$ on two fibres, then the distance between them may be measured either by their chord-distance or by their monotonically related great circle distance. When $\mathbf{X}_2$ is in a fixed position on its fibre, we can ask where is the nearest point on the fibre representing the shape of $\mathbf{X}_1$? When measured along a chord, this shortest distance is the square-root of $\|\mathbf{X}_1\mathbf{Q} - \mathbf{X}_2\|$, thus giving a direct link to two-set orthogonal Procrustes analysis, usually in its two-dimensional form.

Because biological specimens have a 'right way up', reflection is not usually admitted, so $\mathbf{Q}$ is restricted to being a proper rotation (i.e. $\det(\mathbf{Q}) = 1$) leading to special adaptations of the ordinary unconstrained orthogonal Procrustes analysis (see Sections 4.6.1 and 4.6.2). An interesting exception to this disregard of reflection occurs in problems of bilateral symmetry. For example, it is well known that the left- and right-hand sides of the human face may not be entirely symmetrical. To examine this possibility, it is natural to match a face to its own reflection. Mardia, Bookstein, and Moreton (2000) develop the required Procrustean methodology.

To analyse shape, we need to replace each fibre by a single point, thus giving a *shape space*. One way of doing this is to use the lower triangular coordinate parameterisation described in the footnote. This gives a unique point on each fibre but has two disadvantages. First, it depends on identifying the first, second, third ... points and hence on the ordering of the rows of $\mathbf{X}$ and, second, the nearest point on another fibre is unlikely to have the desired lower triangular representation. None of this matters in Kendall's problem of stone alignments, because the vertices of his triangles are unlabelled and he does not wish to compare one triangle with another, except in terms of flatness. In Procrustean matching problems, labelling of points is paramount and their matching to each other, as well as to an average or target, is the central issue. Hence we assume the centred pre-shape representation.

The usual way to pass from a set of fibres to a shape space of unique shape points depends on fixing one point $\mathbf{G}$, say, on its fibre. In two-set Procrustes problems $\mathbf{G}$ will normally correspond to $\mathbf{X}_2$ and in K-sets problems to some kind of group

* After normalisation, the origin may be shifted to the first point, the second point placed in a two-dimensional plane, the third point in three dimensions, and so on up to P dimensions. Then, the first P rows of $\mathbf{X}$ have a triangularly placed set of zero coordinates, so reducing the number of non-zero parameters to $P(N-1) - \frac{1}{2}P(P-1)$. However, we may still arrange that $\|\mathbf{X}\| = 1$ so the points remain on a unit hypersphere. There are several other possible parameterisations.

average (see below). All other points are set to the optimal positions $\mathbf{X}_k\mathbf{Q}_k$ on their fibres, which minimise the Procrustean fit to $\mathbf{G}$. These K points are then orthogonally projected onto the tangent plane to the hypersphere at $\mathbf{G}$ giving the *Procrustean tangent coordinates* of each shape. Suppose we write $\mathbf{x} = \text{vec}(\mathbf{X}_k\mathbf{Q}_k)$ and $\mathbf{g}$=vec($\mathbf{G}$) so that $\mathbf{x}$ and $\mathbf{g}$ give the coordinates on the hypersphere in the form of column-vectors. Then, projection $\mathbf{T}$ onto the tangent plane is merely projection orthogonal to the vector $\mathbf{g}$, giving Procrustean tangent coordinates $\mathbf{t} = \text{vec}(\mathbf{T})$ as:

$$\mathbf{t} = (\mathbf{I} - \mathbf{g}\mathbf{g}')\mathbf{x} = \mathbf{x} - \alpha\mathbf{g}$$

where $\alpha = \mathbf{g}'\mathbf{x} = \text{trace}(\mathbf{G}'\mathbf{X}_k\mathbf{Q}_k)$ which is the sum of the singular values of $\mathbf{G}'\mathbf{X}_k$. The tangent coordinates that correspond to the $N \times P$ configuration matrix $\mathbf{X} = \mathbf{X}_k\mathbf{Q}_k$ are given by:

$$\mathbf{T}_k = \mathbf{X}_k\mathbf{Q}_k - \alpha\mathbf{G}.$$

One may transform back from shape space to pre-shape space using:

$$\mathbf{X}_k = (\mathbf{T}_k + \alpha\mathbf{G})\mathbf{Q}'_k.$$

These are linear operations. When $\|\mathbf{G}\| = 1$, as it must if $\mathbf{g}$ is on the hypersphere, the squared length of $\mathbf{t}$ is $\mathbf{t}'\mathbf{t} = (1 - \alpha^2)$ whereas the Procrustes residual sum-of-squares is $\|\mathbf{X}_k\mathbf{Q}_k - \mathbf{G}\| = \|\mathbf{x} - \mathbf{g}\| = 2(1-\alpha)$. Provided the $\mathbf{X}_k\mathbf{Q}_k$ are close to $\mathbf{G}$, so that $\alpha = 1-\varepsilon$ for small ε, we have that $(1-\alpha^2) \sim 2(1-\alpha) = 2\varepsilon$, showing that then the metric properties of shape space will closely approximate those of pre-shape space. Furthermore, when fitting a scaling factor s_k we have $\|s_k\mathbf{X}_k\mathbf{Q}_k - \mathbf{G}\| = 1 - \alpha^2$, so that then the tangent space representation is exact. The shape space literature refers to *Partial Procrustes Analysis* (excluding scaling factors) and *Full Procrustes Analysis* (including scaling factors). The dimensionality of the tangent space is $(N-1)P - \frac{1}{2}P(P-1) - 1$, allowing for the rotation and a scaling parameter.

The group average of K optimally rotated configurations represented by points on the pre-shape hypersphere, will not be properly normalised and hence will lie in the interior of the hypersphere. This implies that if the K sets are random sample perturbations of a configuration $\mathbf{G}$, normalisation of the individual configurations results in the group average being a biased estimator of $\mathbf{G}$. It remains possible to project orthogonally onto the non unit-vector $\mathbf{g}$, giving for the kth configuration:

$$\mathbf{t}_k = \left(\mathbf{I} - \frac{\mathbf{g}\mathbf{g}'}{\mathbf{g}'\mathbf{g}}\right)\mathbf{x}_k = \mathbf{x}_k - (K/G)\alpha_k\mathbf{g}.$$

Now, $\sum_{k=1}^{K}\mathbf{t}'_k\mathbf{t}_k = K(1 - (1/G)\sum_1^K \alpha_k^2)$ contrasting with the residual sum-of-squares (Table 10.2) $\sum_{k=1}^{K}\|\mathbf{X}_k\mathbf{Q}_k - \mathbf{G}\| = K - G$ where $G = K\mathbf{g}'\mathbf{g} = \sum_1^K \alpha_k$. Again, for $\alpha_k = 1 - \varepsilon_k (k = 1, \ldots, K)$, both metrics are approximately equal to $\sum_1^K \varepsilon_k$.

Thus in the vicinity of the pole $\mathbf{G}$, the metric properties of the tangent space closely approximate the metric properties of the pre-shape hypersphere. In shape studies, the shapes are often quite similar, so the degree of approximation is acceptable.* The degree of approximation is less good in other fields of application.

To calculate the configurations $\mathbf{X}_k\mathbf{Q}_k$, we have to do the Procrustes matching, so may wonder what has been gained from erecting the rather imposing edifice of shape space. The gain comes from the ability to derive distributional results in the shape space. According to Kent and Mardia (1998) '*For shape data the Procrustes tangent projection is of fundamental importance because it reduces the problem of shape analysis to standard multivariate analysis*.' We translate this as meaning that, unlike with the pre-shape sphere, the distributional properties of points in the linear shape space may be handled by conventional multivariate methods based on multinormal assumptions.

Thus, we may test for a difference between the mean of K configurations and a specified target, or for the difference between the means of two sets of configurations, by using a variance ratio and F-statistic. The results are related to the analyses of variance discussed in Chapter 10 and include and extend the distributional results of Section 12.2. Goodall (1991) was the first to set these results in the shape space framework. If the distribution of error is not isotropic but has a diagonal variance covariance matrix, we may use Hottelling's T^2 statistic but the least-squares Procrustes analysis should be weighted as discussed in Section 8.1.2; a non-diagonal weighting matrix is also acceptable. Dryden and Mardia (1998) give full details.

It is also possible to do principal components analysis or a multidimensional scaling for the K points in the tangent space. We note that multidimensional scaling in shape space is similar to generalised pairwise Procrustes analysis (Section 9.3). We also note the possibility of fitting a multivariate mixture model in the tangent space, thus allowing the possibility of testing for two or more groups among the configurations. In principle, we can imagine configurations of points in the tangent space, each vertex of which represents a configuration $\mathbf{X}_{ik} (k = 1, \ldots, K)$. Then, when the K points can be presented as K_1 distinct sets each with the same number of points with the same labels, the use of Procrustes analysis is open to us. This would be similar to a Procrustes analysis of K_1 group averages.

Two-dimensional configurations may be represented in complex space, where multiplication by $e^{i\theta}$ is given by a rotation through the angle θ. Because rotation has this simple representation in complex space, the algebra of two-dimensional Procrustes problems in shape studies is often represented in terms of complex variables. Some find the complex representation 'elegant' but it has its counterpart in terms of $N \times 2$ and 2×2 matrices. Complex variable representations do not generalise to three or more dimensions and, therefore, we have not included them in the body of this book.

Kendall's shape spaces are not the only way of representing shapes as points in a high-dimensional Euclidean space. One alternative is to evaluate the matrix $\mathbf{D}$

* With similar shapes, one may wonder why there is so much concern with eliminating the remote possibility of requiring reflections to achieve a good match.

of squared-distances between all pairs of N points of the configuration $\mathbf{X}$. $\mathbf{D}$ has $\frac{1}{2}N(N-1)$ distinct values which may be represented as the coordinates of a point in a Euclidean space of that dimensionality. The set of such points form a convex cone. Translations and rotations (and also reflections) are eliminated in this representation so, apart from size, the cone may be regarded as a shape space. The standardisation $\mathbf{1}'\mathbf{D}\mathbf{1} = 1$ eliminates size differences, equivalent to the normalisation $\|\mathbf{X}\| = 1$, and defines a convex sub-region of the cone. This sub-region contains the average of any set of standardised distance matrices but, like the pre-shape hypersphere, does not contain the Procrustes group average. The distance-matrix shape space has many nice properties but a disadvantage is its inability to distinguish between rotations and reflections. An outline of the geometry is given by Gower (1995*b*) but whether it could be developed in anything like the degree that Kendall's shape spaces have, remains to be seen. Lele and Richtsmeier (2001) extensively use distance representations for analysing shape and have developed what they term *Euclidean Distance Matrix Analysis* (EDMA).

14.1.5 BIOMETRIC IDENTIFICATION AND WARPING

Closely related to shape analysis, the currently popular quest for methods of identifying a person accurately will probably continue to stimulate new research into image matching algorithms. Be it handwriting, authentication of signatures, a finger print, a face, an iris, or retina scan, all these images imply some Procrustean transformation of a *template*. Any image, when converted into a matrix of pixels, can serve as a two-dimensional configuration to be matched against another such configuration, possibly by some form of Procrustes analysis.

Quite general two-dimensional transformations of the kind $(x_1, x_2) \rightarrow u_1(x_1, x_2), u_2(x_1, x_2)$ are considered. These include as special cases the Procrustean transformations of translation, orthogonal rotation, and isotropic or anisotropic scaling but also much more general non-linear transformations, termed *image warping*. Before it can be decided if a sample fits a template, images have to be aligned (in this context, often termed *registration*). Identification has to be based on some measure of similarity of the template to the original. Matching may be done simultaneously with the above mentioned aligning, or a separate criterion may be evaluated subsequent to an affine aligning.

With the more general transformations, spatial contiguity is recognised and respected. For example, we may construct the DeLauney triangulation of the two-dimensional configurations and match piecewise while ensuring continuity across the triangle boundaries. Thin plate splines (Bookstein 1989, 1991 and Bookstein and Green 1993) are an example of this type of approach. Different parts of biological organisms are known to grow at different rates; the shapes of different species are known to vary from one part of an organism to another. Piecewise methods accommodate these possibilities in a way not available to the rigid body transformations of Procrustes analysis, even when supplemented by anisotropic scaling. With estimated functions $u_1(x_1, x_2), u_2(x_1, x_2)$, interpolation to non-observed points becomes possible, which seem to reflect the original D'Arcy Thompson (1917) transformation grid philosophy of shape change better than other proposals. Non-rigid body transformations are likely to get increasing interest. See Glasbey and Mardia (1998) for a review with a comprehensive bibliography of applications and the book by Glasbey and Horgan (1995); other examples are included in Dryden and Mardia (1998).

14.1.6 MOLECULAR BIOLOGY

Shapes of molecules can be matched to each other, or the shape of certain molecules could perhaps be matched to sites on a receptor. This may already take place in drug design or other physiological applications but we know of no explicit reference. When one of the molecules is DNA, a known genomic sequence can be matched to it using orthogonal Procrustes matching. The matching of genomic sequences to a DNA string can also call on Procrustean methodology. An introduction to this area is given by Crippen and Harvel (1981). Although the above types of problem are three-dimensional, the interesting sequencing information is essentially one-dimensional. It follows that interest will often be in matching sequences or in attempting to find regions on the molecules where sequences match. These types of problem are rather far from those discussed in this book. Thompson (1996) gives a recent review of statistical genetics.

14.1.7 IMAGE ANALYSIS

In recent years neuro imaging methods have increasingly been used following the development of powerful, non-invasive, scanning techniques, such as Magnetic Resonance Imaging (MRI). Both functional and structural MRI produce images that need to be compared with other images. Especially in functional MRI, with special on–off presentation schemes, the on and the off treatments are aggregated and each compared with the aggregate. Procrustean methods can be deployed as tools to compare images. Newer developments, such as Magneto Encephalography (MEG), also produce images that can be compared with other images by means of, for example, Procrustean methods.

Tomography is concerned with attempts to reconstruct three-dimensional objects from sets of two-dimensional views. Instrumentation produces images that may be regarded as some kind of projection of the original object. A simple Procrustean problem would be to estimate $\mathbf{X}$, the three-dimensional configuration, from known orthogonal projections $\mathbf{P}_k$ and two-dimensional images $\mathbf{X}_k (k = 1, 2, \ldots, K)$. Thus, we might estimate $\mathbf{X}$ by minimising:

$$\sum_{k=1}^{K} \|\mathbf{X}\mathbf{P}_k - \mathbf{X}_k\|.$$

This differs from the projection problems discussed in Chapter 5 but should not be difficult to solve. Of course, there is much more to tomography than this. A review, that gives further references to tomography and other image analysis applications, is given by Glasbey and Berman (1996).

14.1.8 PROCRUSTES METHODS AS A STATISTICAL TOOL

When comparing results from different statistical methods, Procrustes analysis naturally presents itself as a method for assessing the similarity of the results. This is especially true when results are presented spatially, as with multidimensional scaling and other methods of multidimensional data analysis. Procrustean rotation is also used in jackknifing situations where a jackknifed solution is compared with the complete solution (cf. Davis 1978, Sibson 1979, De Leeuw and Meulman 1986, Meyners *et al.* 2000, Martens and Martens 2001). Of course, it may also be used with other resampling methods (e.g. bootstrapping). An example of this use of Procrustes analysis is presented in Section 9.3.1 where different multivariate analyses of the same data are compared. Procrustes analysis may be used to assess samples of group averages derived by any of the methods discussed in Chapter 13, including Procrustes analysis itself!

14.2 Note on comparing methods

The comparison of different methods of multivariate analysis with other multivariate methods deserves some comment. The comparison often results in a competition for 'which method is better'. In this section we discuss how fraught with difficulty are comparisons of this kind. The situation is quite complex as we have to distinguish between (i) different criteria for assessing the optimum (e.g. least-squares, inner-product, robust criteria), (ii) different algorithms for arriving at the optimum (e.g. Algorithms 5.2 and 5.3), (iii) different implementations of the same algorithm and, lastly, (iv) comparing the performance of totally different models (e.g. Procrustes methods with INDSCAL). We shall discuss these in turn.

14.2.1 COMPARING CRITERIA

Given a particular model, there are many criteria that may be proposed for fitting the model. In general, different fitting-criteria are not comparable although their good and bad properties may be listed. Several criteria defining *best fit* have been mentioned in this book and we have given reasons why some are not always acceptable. Heuristic algorithms are often defined. For such methods the optimisation criterion is often obscure, or it is known but cannot be readily optimised. Then, it can be useful to compare the behaviour of different heuristics with respect to the known criterion or any other criterion of interest.

Ideally, we would like a proper inferential framework for Procrustes methods. What little that has been done is discussed in Chapter 12 and above in Section 14.1.4; more will be developed over the coming years. It seems that the best we can hope for are asymptotic distributional results and even this possibility depends on the choice of fitting-criterion. Least-squares criteria are the most amenable to methodological development and robust and non-linear criteria are the least amenable. This is a pity, as robust and non-linear methods are areas of increasing importance in describing the substantive differences and agreements among configurations (see Section 14.1.4 earlier). It is better to have a methodology that deals well with substantive effects than one that permits inferences based on unrealistic assumptions. Perhaps this should not be a cause for concern as we can always rely on the bootstrap and jackknifing. Our view is that it is nice to have distributional results in mathematical form but this is a secondary objective when comparing configurations.

14.2.2 COMPARING ALGORITHMS

The comparison of different algorithms for optimising the same criterion is a legitimate activity and we have reported some results along these lines. Interest is in speed, accuracy, and ability to find global rather than local optima. Most algorithms given in previous chapters have been of the simple alternating least-squares type and rather little is known of their performances. More work is required here. Recently, Trendafilov and colleagues (Chu and Trendavilov 1990, 2001, Trendavilov 1999) have investigated the use, for Procrustes analysis, of steepest descent algorithms developed by numerical analysts. In particular he has exploited what is known as the *Projected Gradient* method for constrained optimisation. In this method, a constraint such as that **T** be an orthognal matrix **Q** defines a manifold, in this case defined by $\frac{1}{2}P(P-1)$ parameters, embedded in a Euclidean space of $P \times P$ dimensions. The Procrustes criterion $\|\mathbf{X}_1\mathbf{Q} - \mathbf{X}_2\|$ is defined at all points of the manifold and the objective is to find the point *on the manifold* where the minimum is achieved. Starting at some valid point P_i, ordinary steepest descent methods will take one out of the manifold but, by projecting the gradient vector onto the tangent plane at P_i, it may be arranged to return to the manifold for the next point P_{i+1} in the iteration. The process requires knowledge of the differential geometry of the manifold and, at each step, the solution of a differential equation by numerical integration. Chu and Tredafilov (1990, 2001) give the details for $\mathbf{T} = \mathbf{Q}$ and $\mathbf{T} = \mathbf{P}$, respectively; Tredafilov (1999) handles the case $\mathbf{T} = \mathbf{C}$. Orthonormal matrices form what is known as the Steifel manifold of which the orthogonal matrix manifold is a special case. The basic numerical method is constant for all these applications (and for other non-Procrustean minimisations), all that changes is the geometry of the manifolds and the algebraic form of the criterion. Sophisticated as is this methodology is, the reported results are disappointing. For example, Chu and Trendafilov (1990) compare their algorithm (PG) with that of Koschat and Swayne (1991), which we refer to as (KS), showing that out of 200 random matrices, 66 (KS) and 69 (PG) converge to local optima. Of the 66 local optima found by KS the PG algorithm finds global minima in eight cases, while of the 69 local optima found by PG the KS algorithm finds global minima in 11 cases. Thus increased sophistication does not seem to have paid off. Furthermore, the PG algorithm is reported as being 'significantly slower' than KS, although it should be appreciated that relative speed is unimportant when we are concerned with two quick methods. Despite the disappointing outcome, we expect numerical analysis to contribute to a better understanding of the local minimum problem and to future improved algorithms.

14.2.3 COMPARING IMPLEMENTATIONS OF ALGORITHMS

Apart from the remarks made in (14.2.2), we have little to say on implementations, as so little is known. We draw attention to the danger of confounding comparisons between method/models with the implementations of their associated algorithms. A good implementation (including its documentation) may help promote a poor method, while a poor implementation may inhibit interest in a good method.

14.2.4 COMPARING MODELS

Comparing models is a minefield that many authors seem prepared to cross. Does it make sense to compare GOPA with, say, INDSCAL? We have already mentioned the possibilities of obfuscation given by using different fitting criteria and/or different qualities of algorithm. Although GOPA and INDSCAL both fit their models through optimising least-squares criteria, one does so with squared-residuals between configurations and the other with squared squared-residuals. This difference means that we are not comparing like with like and does not help with making valid comparisons. If both models were fitted with the same level of residuals, then we could compare their residual sum-of-squares and say, perhaps, that one model fitted better than the other, after making due allowance, if we could, for the differing numbers of free parameters in the two models. However, what would it mean to say that one method gives a bigger group average than the other? In our opinion, such statements have little validity, for if our objective is to maximise the group average then that is what to do. That is, we should fit a different model—perhaps GOPA itself or, if we are willing to specify the dimensionality of the group average, Peay's projected group average (Chapter 9, section on Maximal Group Average Projection Procrustes Analyses).

Thus, if we are going to compare models, we should make sure we are comparing like with like. Then, the ideal situation is when one model is nested within the other. That is one model contains a subset of the parameters of the other model. When we are comparing some common aspect of two models, such as a group average, then we should examine whether or not a third model, that focuses on modelling the chosen aspect should be preferred.

14.3 A note on software

Despite the wide range of application, few algorithms for Procrustes methods are available in mainstream statistical packages. Orthogonal Procrustes analysis is available as a Genstat macro (Arnold 1986) and in SAS (cf. Schlich 1993), but not in SPSS. In any case, orthogonal Procrustes analysis is easily implemented within any software package that includes SVD. Most of the algorithms that we have outlined are easily programmed by those with quite modest programming skills. Special purpose programs exist in the field of Sensory analysis (SensTools, OP&P Product Research, Utrecht, the Netherlands). In the fields of image analysis and molecular genetics, Procrustes methods are used and some special purpose programs probably exist there, which probably limit themselves to handling two- or three-dimensional cases. Dedicated programs for shape matching purposes, must exist, but they are not known to us.

Procrustes problems often arise as a part of other problems, for example, in matching the result from a jackknifed solution to the original solution, or as an initial step in many types of image-analysis. Here they may be part of dedicated programs, and not directly recognisable independent programs for performing a Procrustean matching.

Some of the other methods mentioned in Chapter 13 are implemented in software packages. There are special purpose programs available for STATIS, INDSCAL, etc. The HOMALS and OVERALS methods are implemented in SPSS (SPSS Inc.).

14.4 Conclusion

What is the future of Procrustes analysis? If we had asked this question of Mosier in 1939, it is likely that he would have foreseen few of the developments described in this book. Even if you had asked us this question over 10 years ago, when we were first planning this book, we would have got things sadly wrong. If recent trends continue, many new application areas will be explored and these will often raise their own special problems calling on the development of new Procrustean methodology. Orthogonal, projection, and oblique axis transformations are basic geometrical concepts that are bound to continue to find applications in the future, not least, in areas where computer-based visualisations are found useful. These basic transformations will be too inflexible for some applications, so more general types of matching transformations will continue to be developed. These transformations can be seen as truly violent methods for making the data fit the Procrustean bed and, in the spirit of the myth, may rightly be regarded as forms of Procrustes analysis.

Appendix A
Configurations

Geometrically, each row of an $N \times P$ data matrix can be represented as a point in a space spanned by the columns of the data matrix. The data matrix may be seen as a configuration of N points in a P-dimensional space. Figure A.1 shows the geometric configuration of the (4×2) data matrix

$$\begin{pmatrix} 2 & 3 \\ 4 & 1 \\ 5 & 5 \\ 6 & 4 \end{pmatrix}.$$

The four points (2, 3), (4, 1), (5, 5), and (6, 4) are drawn on two axes, labelled x_1 and x_2.

The configuration in Figure A.1 is two-dimensional; adding a third column to the data matrix would result in a configuration in three dimensions. The configuration

$$\begin{pmatrix} 2 & 3 & -2 \\ 4 & 1 & 0 \\ 5 & 5 & 1 \\ 6 & 4 & 1 \end{pmatrix}$$

is like the configuration in Figure A.1 but where the point (2, 3) lies two units below the plane of the page—actually it's the point (2, 3, −2) now, the point (4, 1, 0) lies in the plane, the points (5, 5, 1) and (6, 4, 1) lie one unit above the plane

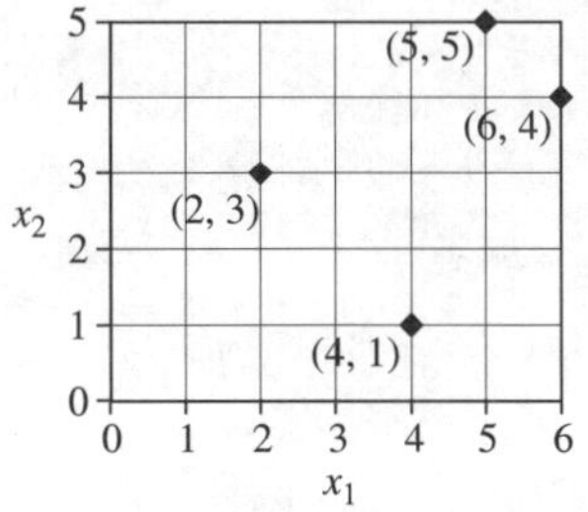

Fig. A.1. Example of a geometric representation of a data matrix.

of the page. Note that the number of columns in a data matrix does not necessarily dictate the dimensionality of the configuration. The points in the matrix

$$\begin{pmatrix} 2 & 3 & 3 \\ 4 & 1 & 3 \\ 3 & 2 & 3 \\ -1 & 6 & 3 \end{pmatrix}$$

can be drawn in one dimension on a straight line. To be more precise, when $N \geq P$ the N points lie in a space of at most P dimensions (e.g. $N = 4$ and $P = 3$ but the configuration is one-dimensional). When $P > N$, the configuration lies in at most $(N-1)$-dimensions (e.g. $N = 2$ points are collinear, however many columns there may be).

Configurations play an important part in the main text. We use the symbol $\mathbf{X}$ to denote data matrices of configurations. The ith row of $\mathbf{X}$ is denoted by $\mathbf{x}_i$, which is a row-vector giving the coordinates of the ith object. Thus, in the above data matrix $\mathbf{x}_3 = (3, 2, 3)$ are the coordinates of a point in three dimensions. In this book, coordinates are always written as row-vectors; column-vectors denote directions.

In the usual terminology the rows of the data set contain *objects*, and the columns contain *dimensions*. What the objects or the dimensions represent depends on context. There is a fundamental distinction, however, between two different cases:

(i) the data sets contain raw data and (ii) the data sets contain coordinates. The consequences of this distinction are discussed in detail in Chapter 2 of this book.

Appendix B
Rotations and reflections

Rotations and projections underpin many, but not all, Procrustes methods. In this section we examine the geometric notions concerned and establish the corresponding algebraic notation. Then some fundamental results are established. Projections are discussed in Appendix C.

B.1 Rotations

A rotation of a set of points is a transformation that fixes the origin and leaves all pairs of interpoint distances unchanged. Figure B.1(a) and (b) show two sets of 'rotations' in two dimensions.

Figure B.1(a) shows what is normally understood by a clockwise rotation through an angle θ about the origin O. The point A_1 rotates to B_1 and A_2 rotates to B_2. The distance between A_1 and A_2 is the same as the distance between B_1 and B_2. Clearly, we may introduce further points $A_3, A_4, \ldots$ rotating clockwise to $B_3, B_4, \ldots$ and at the same time preserving the equality of the distance between every A-pair and the corresponding B-pair. Figure B.1(b) represents a reflection in the line λ through the origin, also exhibiting the requirements of preserving distance and leaving the origin unchanged.

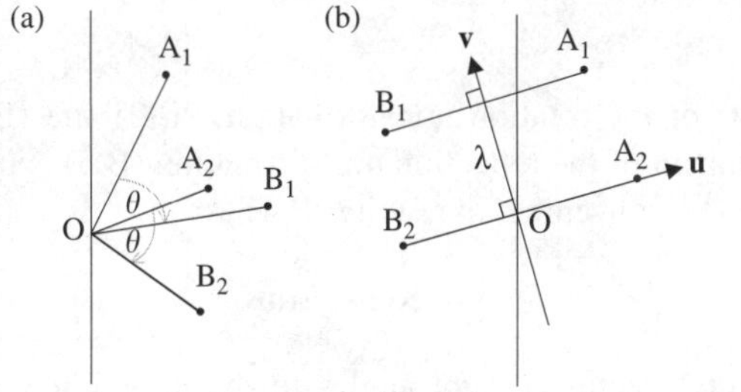

Fig. B.1. What is normally understood by a rotation through an angle θ is shown in (a); a reflection in the line λ is shown in (b). Interpoint distances are invariant to the transformations in both cases.

The transformations shown in Figure B.1 may be expressed algebraically in the form

$$\mathbf{y} = \mathbf{xQ} \tag{B.1}$$

where $\mathbf{x}$ represents the coordinates before transformation (i.e. of the A-set) and $\mathbf{y}$ represents the coordinates after transformation (i.e. of the B-set). We express coordinates of points as row-vectors. The precise form of $\mathbf{Q}$ depends on how the coordinates are represented. In the most simple form where we have orthogonal Cartesian axes in two dimensions, rotation takes the well-known form:

$$\mathbf{y} = \mathbf{x}\begin{pmatrix} \cos\theta & -\sin\theta \\ \sin\theta & \cos\theta \end{pmatrix} \tag{B.2}$$

giving the coordinates $\mathbf{y}$ of the rotated points in terms of their original coordinates $\mathbf{x}$. In this approach, the axes remain fixed. Equation (B.2) is known as a Jacobi plane rotation. We may also consider the points fixed and ask what are the coordinates of the points when the axes are rotated through an angle θ. This is equivalent to rotating the points through an angle $-\theta$ so now:

$$\mathbf{y} = \mathbf{x}\begin{pmatrix} \cos\theta & \sin\theta \\ -\sin\theta & \cos\theta \end{pmatrix}. \tag{B.3}$$

In (B.3) the transformation matrix is both the inverse and the transpose of (B.2).

If $\mathbf{v}$ in Figure B.1(b) denotes a unit column-vector in the direction of λ, then $\mathbf{x}+\mathbf{y}$ is a vector in the direction of λ of length $2\mathbf{xv}$. Thus, reflection in λ is given by:

$$\mathbf{y} = 2\mathbf{xvv}' - \mathbf{x}$$

or

$$\mathbf{y} = \mathbf{x}(2\mathbf{vv}' - \mathbf{I}). \tag{B.4}$$

If λ is inclined to the horizontal axis at an angle θ we have that $\mathbf{v}' = (\cos\theta, \sin\theta)$ and (B.4) may be written:

$$\mathbf{y} = \mathbf{x}\begin{pmatrix} \cos 2\theta & \sin 2\theta \\ \sin 2\theta & -\cos 2\theta \end{pmatrix}. \tag{B.5}$$

The determinants of the rotation transformations (B.2) and (B.3) are plus one, while the determinant of the reflection transformation (B.5) is minus one.

It turns out to be convenient to rewrite (B.4) as;

$$\mathbf{y} = \mathbf{x}(\mathbf{I} - 2\mathbf{uu}') \tag{B.6}$$

where $\mathbf{u}$ is the unit vector at right angles to the direction $\mathbf{v}$ (see Figure B.1). Then (B.6) represents a reflection in the space orthogonal to $\mathbf{u}$. In two dimensions the two forms (B.4) and (B.6) are equivalent, giving the same value of $\mathbf{y}$ for a given $\mathbf{x}$ but in higher dimensions, the two values of $\mathbf{y}$ would differ. We note that

$\det(\mathbf{I} - 2\mathbf{uu}') = -1$, whereas $\det(2\mathbf{vv}' - \mathbf{I}) = \pm 1$, depending on whether $\mathbf{x}$ is in a space of an odd or even number of dimensions. (B.6) is known as the Householder transform and gives a reflection in the *plane* orthogonal to $\mathbf{u}$.

Although Figure B.1 shows only two dimensions, it should be clear that the concepts of rotation and reflection may be generalised to any number, P, of dimensions. Thus, in (B.1) $\mathbf{x}$ and $\mathbf{y}$ are now P-dimensional row-vectors and $\mathbf{Q}$ is a $P \times P$ matrix. To preserve distance between two points $\mathbf{x}_1$ and $\mathbf{x}_2$ we must have $(\mathbf{x}_1 - \mathbf{x}_2)(\mathbf{x}_1 - \mathbf{x}_2)' = (\mathbf{x}_1 - \mathbf{x}_2)\mathbf{QQ}'(\mathbf{x}_1 - \mathbf{x}_2)'$ for all pairs of points. This implies that:

$$\mathbf{QQ}' = \mathbf{I} \tag{B.7}$$

so that $\mathbf{Q}^{-1} = \mathbf{Q}'$ and hence $\mathbf{Q}'\mathbf{Q} = \mathbf{I}$. Taking determinants of (B.7), we have that:

$$(\det \mathbf{Q})^2 = 1$$

and therefore $\det \mathbf{Q} = \pm 1$. Thus, these results, which we have already found for two dimensions, are universally valid. We also note that, after rotation, the inner-product $\mathbf{x}_1\mathbf{x}_2'$ becomes $\mathbf{x}_1\mathbf{QQ}'\mathbf{x}_2'$ so this too is invariant under the transformation, showing that angles subtended at the origin are unchanged. Square matrices which satisfy (B.7) are said to be orthogonal.

B.2 Some special cases

Constantine and Gower (1978) discuss special cases of elementary orthogonal matrices. These are orthogonal matrices representing rotations in the plane of two unit vectors $\mathbf{u}$, $\mathbf{v}$ (say), inclined at an angle γ given by $\mathbf{u}'\mathbf{v} = \cos\gamma$ or reflections in a $(P-1)$-dimensional plane with normal in the $(\mathbf{u}, \mathbf{v})$-plane. Consider the matrix:

$$\mathbf{Q} = \mathbf{I} - p\mathbf{uu}' - q\mathbf{vv}' + r\mathbf{uv}' + s\mathbf{vu}'. \tag{B.8}$$

Any vector in the $(\mathbf{u}, \mathbf{v})$-plane has the form $\lambda\mathbf{u}' + \mu\mathbf{v}'$ and $(\lambda\mathbf{u}' + \mu\mathbf{v}')\mathbf{Q}$ retains the same form. Further, if any vector $\mathbf{w}$ includes a component orthogonal to $\mathbf{u}$ and $\mathbf{v}$, then $\mathbf{w}'\mathbf{Q}$ retains the same component of $\mathbf{w}$ so is not coplanar with $\mathbf{u}$ and $\mathbf{v}$. Thus, (B.8) represents some transformation in the $(\mathbf{u}, \mathbf{v})$-plane. Constantine and Gower (1978) show that $\mathbf{Q}$ is orthogonal iff:

(i) $r = s$ and p, q are the roots of $t^2 - 2(1 + r\cos\gamma)t + r^2 = 0$

(ii) $p = q$ and r, s are the roots of $t^2 - 2(p\cos\gamma)t + (p^2 - 2p) = 0$.

Case (i) may be written as the Householder transform:

$$\mathbf{Q} = \mathbf{I} - (p^{1/2}\mathbf{u} - q^{1/2}\mathbf{v})(p^{1/2}\mathbf{u} - q^{1/2}\mathbf{v})',$$

representing reflections in the plane with normal $(p^{1/2}\mathbf{u} - q^{1/2}\mathbf{v})$. Case (ii) represents a rotation in the $(\mathbf{u}, \mathbf{v})$-plane through an angle θ given by:

$$\cos\theta = 1 - p\sin^2\gamma. \tag{B.9}$$

Using (B.9) allows the coefficients p, q, r, s of (ii) to be expressed entirely in terms of the two angles γ and θ. Different choices of γ allow rotations in the (**u**, **v**)-plane through an angle θ to be expressed in different ways, of which the following are of special interest:

Table B.1. Three basic forms for the elementary orthogonal rotation through an angle θ in the plane (**u**, **v**).

Case	γ	$p = q$	r	s
(a)	$\pm\pi/2$	$1 - \cos\theta$	$\pm\sin\theta$	$\pm\sin\theta$
(b)	$\theta/2$	2	$4\cos(\theta/2)$	0
(c)	θ	$(1 + \cos\theta)^{-1}$	$2 - (1 + \cos\theta)^{-1}$	$-(1 + \cos\theta)^{-1}$

In (a) **Q** is the classical Jacobi rotation matrix, here expressed in terms of *any* two orthogonal unit vectors **u** and **v** rather than unit vectors parallel to two coordinate axes. Case (b) may be written $\mathbf{Q} = (\mathbf{I} - 2\mathbf{u}\mathbf{u}')(\mathbf{I} - 2\mathbf{v}\mathbf{v}')$ showing the rotation as the product of two reflections; that is a reflection in the plane normal to **u** followed by a reflection in the plane normal to **v**. In case (c), $\theta = \gamma$ so $\mathbf{u}'\mathbf{Q} = \mathbf{v}'$.

B.3 General orthogonal matrices

The orthogonal matrices form a group under multiplication. This follows from noting that if $\mathbf{Q}_1$ and $\mathbf{Q}_2$ are orthogonal, then $\mathbf{Q}_2'\mathbf{Q}_1'\mathbf{Q}_1\mathbf{Q}_2 = \mathbf{Q}_2'\mathbf{I}\mathbf{Q}_2 = \mathbf{Q}_2'\mathbf{Q}_2 = \mathbf{I}$ so that the product $\mathbf{Q}_1\mathbf{Q}_2$ is also an orthogonal matrix. Clearly, **I** is the unit of the group and $\mathbf{Q}_1'$ is the inverse of $\mathbf{Q}_1$. The group of orthogonal matrices has a subgroup of all orthogonal matrices with unit positive determinant. These are often referred to as general rotation matrices while the orthogonal matrices with negative determinants, which do not form a subgroup, are referred to as general reflection matrices. Apart from the fact that this terminology is consistent with the two-dimensional geometry, discussed above, the following result, which we do not prove, supports the terminology. It is a standard result from algebra (see, for example, Gentmacher 1959, vol. 2, p. 23) that any orthogonal matrix **Q** may be written in the form:

$$\mathbf{Q} = \mathbf{U}\mathbf{D}\mathbf{U}'$$

where **U** is itself an orthogonal matrix and **D** is block-diagonal, formed from elementary orthogonal matrices of the types

$$\begin{pmatrix} \cos\theta & \sin\theta \\ -\sin\theta & \cos\theta \end{pmatrix}, (-1), \text{ and } (1)$$

which refer, respectively, to rotations in the plane determined by the two corresponding columns of **U**, reflection in the plane whose normal is given by the corresponding column of $\mathbf{U}$ or to a null effect. Note that

$$\begin{pmatrix} -1 & 0 \\ 0 & -1 \end{pmatrix}$$

is a rotation matrix with $\theta = \pi$ so that $\mathbf{Q}$ is a reflection matrix only when there are an odd number of (-1) terms on the diagonal of $\mathbf{D}$.

We may get a better understanding of the decomposition $\mathbf{Q} = \mathbf{UDU}'$ by writing:

$$\mathbf{Q} = \prod_{i=1}^{B} \mathbf{UD}_i\mathbf{U}' = \prod_{i=1}^{B} \mathbf{U}(\mathbf{I} + \mathbf{E}_i)\mathbf{U}' = \prod_{i=1}^{B} (\mathbf{I} + \mathbf{UE}_i\mathbf{U}') \tag{B.10}$$

where $\mathbf{D}_i$ consists of the ith of the B blocks of $\mathbf{D}$ with the remaining blocks being replaced by units on the diagonal, and $\mathbf{E}_i = \mathbf{D}_i - \mathbf{I}$. Thus, $\mathbf{E}_i$ is zero everywhere except for diagonal blocks of size 2×2 of the form

$$\begin{pmatrix} \cos\theta - 1 & \sin\theta \\ -\sin\theta & \cos\theta - 1 \end{pmatrix},$$

or of size one with forms either (-2) or (0). In the former case, $(\mathbf{I} + \mathbf{UE}_i\mathbf{U}') = \mathbf{J}_i$, say, reduces to the Jacobi rotation given as (a) in Table B.1 and where $\mathbf{u}$ and $\mathbf{v}$ of that table are the columns of $\mathbf{U}$ corresponding to the non-zero columns of $\mathbf{E}_i$. When the only non-zero element of $\mathbf{E}_i$ is -2 then $(\mathbf{I} + \mathbf{UE}_i\mathbf{U}') = \mathbf{H}_i$, say, reduces to a Householder transform in the plane normal to the corresponding column of $\mathbf{U}$. When $\mathbf{E}_i$ is zero then $(\mathbf{I} + \mathbf{UE}_i\mathbf{U}') = \mathbf{I}$. Putting these together shows that the decomposition (B.10) may be written:

$$\mathbf{Q} = \prod_{i=1}^{J} \mathbf{J}_i \prod_{i=1}^{H} \mathbf{H}_i \tag{B.11}$$

where J is the number of Jacobi blocks and H is the number of Householder blocks, so there are P–J–H unit-matrix blocks. Using (B.11), shows that any orthogonal transformation $\mathbf{xQ}$ may be represented as the product of a series of plane rotations and reflections. Because the order of the products in (B.10) is immaterial, so is it in (B.11).

We may also present the decomposition as a sum, as follows:

$$\mathbf{Q} = \sum_{i=1}^{J+H} \mathbf{UE}_i\mathbf{U}' = \sum_{i=1}^{J+H} (\mathbf{I} + \mathbf{UE}_i\mathbf{U}') - (J + H)\mathbf{I}$$

which may be written:

$$\mathbf{Q} = \sum_{i=1}^{J} \mathbf{J}_i + \sum_{i=1}^{H} \mathbf{H}_i - (J + H)\mathbf{I} \tag{B.12}$$

showing that any orthogonal transformation can be represented as the vector sum of a series of Jacobi rotations and Householder transforms, combined with a multiple of the original vector $\mathbf{x}$.

Finally, we note that:

$$\mathbf{Q} - \mathbf{I} = \mathbf{U}(\mathbf{D} - \mathbf{I})\mathbf{U}' = \mathbf{U}\,\mathrm{diag}(\mathbf{E}_1, \mathbf{E}_2, \ldots, \mathbf{E}_B)\mathbf{U}'$$

where now $\mathbf{E}_i$ represents just the non-zero diagonal blocks of the $N \times N$ representations of the previous $\mathbf{E}_i$. We may reexpress the Jacobi blocks as follows:

$$\begin{pmatrix} \cos\theta_i - 1 & \sin\theta_i \\ -\sin\theta_i & \cos\theta_i - 1 \end{pmatrix} = 2\sin\frac{\theta_i}{2}\begin{pmatrix} -\sin(\theta_i/2) & \cos(\theta_i/2) \\ -\cos(\theta_i/2) & -\sin(\theta_i/2) \end{pmatrix} = 2\sin\frac{\theta_i}{2}F_i, \text{ say.}$$

The matrix in $\theta_i/2$ is itself an orthogonal rotation matrix. Similarly if we write the Householder 'block' as $2(-1)$ then (-1) is itself a, very elementary, orthogonal matrix. We may therefore write:

$$\mathbf{Q} - \mathbf{I} = \mathbf{U}\mathbf{F}\boldsymbol{\Sigma}\mathbf{U}' \tag{B.13}$$

where $\mathbf{F} = \mathrm{diag}(\mathbf{F}_1, \mathbf{F}_2, \ldots, \mathbf{F}_B)$ and $\boldsymbol{\Sigma} = \mathrm{diag}(2\sin(\theta_i/2), \ldots, 2, \ldots, 0, \ldots)$. Now $\mathbf{F}$ is an orthogonal matrix and so, therefore, is $\mathbf{UF}$. It follows that (B.13) gives the singular value decomposition of $\mathbf{Q} - \mathbf{I}$, with singular values in $\boldsymbol{\Sigma}$. The singular values of 2 associated with the reflections are always dominant. The biggest value of $\sin(\theta_i)/2$ corresponds to the best rotation.

Appendix C
Orthogonal projections

The orthogonal projection B of a point A onto a linear space L is such that the direction AB is orthogonal to all directions in the space; it follows that B is the nearest point in L to A. Figure C.1 illustrates the geometry:

If A_1 has coordinates $\mathbf{x}$ and B_1 has coordinates $\mathbf{y}$ and the columns of ${}_P\mathbf{L}_R$ are independent and span the R-dimensional subspace L, then because A_1B_1 is orthogonal to L, $(\mathbf{x} - \mathbf{y})\mathbf{L} = 0$ and because $\mathbf{y} \in L$, there exists an R-dimensional column-vector $\mathbf{c}$ such that $\mathbf{y}' = \mathbf{Lc}$. It follows that:

$$\mathbf{c} = (\mathbf{L}'\mathbf{L})^{-1}\mathbf{L}'\mathbf{x}'$$

and

$$\mathbf{y} = \mathbf{xL}(\mathbf{L}'\mathbf{L})^{-1}\mathbf{L}'. \tag{C.1}$$

Equation (C.1) is the general algebraic form for representing orthogonal projections. However, if we choose the columns of $\mathbf{L}$ to be orthonormal, so that $\mathbf{L}'\mathbf{L} = \mathbf{I}$, we have the simplified form:

$$\mathbf{y} = \mathbf{xLL}'. \tag{C.2}$$

Equations (C.1) and (C.2) give the coordinates of the point B relative to the original P-dimensional axes. Relative to the R orthonormal axes in L, (C.2) becomes;

$$\mathbf{y} = \mathbf{xL}. \tag{C.3}$$

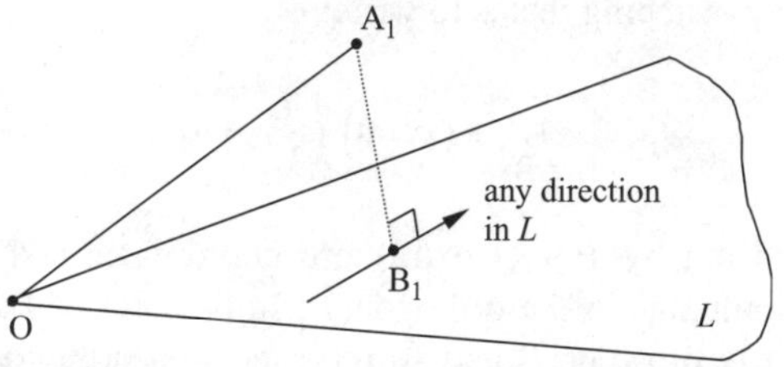

Fig. C.1. Illustration of the orthogonal projection of a point A_1 not lying in L onto a point B_1 in the subspace L.

Both the forms (C.2) and (C.3) will be required, and the reader is warned that there is often some confusion because it is not made sufficiently clear which is the reference coordinate system, the original one in P dimensions or a coordinate system embedded in the R-dimensional subspace.

When the columns of $\mathbf{L}$ are orthonormal we may regard them as the first R columns of some P-dimensional orthogonal matrix whose final P–R columns are irrelevant, so long as they are normal to each other and to the first R columns. Thus, if we partition an orthogonal matrix $\mathbf{Q} = (\mathbf{Q}_1, \mathbf{Q}_2)$, then we have:

$$\mathbf{xQ} = (\mathbf{xQ}_1, \mathbf{xQ}_2)$$

showing the rotation, or reflection, as the concatenation of two projections onto orthogonal subspaces.

We note that we must have $R \leq Pp$. We have seen that $R = P$ gives an orthogonal matrix. It is impossible to have $\mathbf{L}'\mathbf{L} = \mathbf{I}$ when $R > P$ because, geometrically, it would require the existence of R orthogonal directions in a space of fewer than R dimensions.

Returning to the general form (C.1) we shall write:

$$\mathbf{P} = \mathbf{L}(\mathbf{L}'\mathbf{L})^{-1}\mathbf{L}' \tag{C.4}$$

for the projection matrix. We note that $\mathbf{P}^2 = \mathbf{P}$, so that the matrix is idempotent. This means that once A is projected to B, repeated projection will leave the point B unchanged, which is obvious geometrically.

C.1 Orthonormal operators as rotations

Above, $\mathbf{xL}$ was interpreted as a projection from a P-dimensional space onto a subspace spanned by the Q orthogonal columns of $\mathbf{L}$. What might be the interpretation of $\mathbf{xL}'$? Because $Q < P$, this certainly means that a point in q dimensions is transformed into a point in a higher dimensional space of P dimensions. The columns of $\mathbf{L}'$ are not orthonormal, so $\mathbf{xL}'$ does not represent projection. We have that $(\mathbf{xL}')(\mathbf{xL}')' = \mathbf{xx}'$, so inner products are invariant to the transformation and, hence, so are distances. Unlike with projections, $\mathbf{xL}'$ preserves configurations and hence it must represent a generalised rotation of $\mathbf{x}$ into the higher dimensional space. Another way of seeing this is to write:

$$\mathbf{xL}' = (\mathbf{x} \quad \mathbf{0}) \begin{pmatrix} \mathbf{L}' \\ \mathbf{L}'_{\perp} \end{pmatrix}$$

where $\mathbf{x}$ is padded out by $P - Q$ extra zero coordinates and $\mathbf{L}_{\perp}$ represents a set of orthogonal columns in the orthogonal complement of $\mathbf{L}$. Now $(\mathbf{L}, \mathbf{L}_{\perp})$ is an orthogonal matrix of order P and so represents an orthogonal rotation in P dimensions.

Appendix D
Oblique axes

Let $\mathbf{x}$ be a row-vector giving the coordinates of a point relative to a set of P orthogonal Cartesian axes. Suppose now, we wish to refer this same point to a set of oblique axes whose directions relative to the Cartesian axes are given by the columns of a matrix $\mathbf{C}$. Thus the kth column $\mathbf{c}_k$ of $\mathbf{C}$ gives the direction cosines of the kth oblique axis. There are at least two ways of representing coordinates relative to oblique axes. These are shown in Figure D.1, below.

In the first version (a) one projects orthogonally onto the oblique axes, so that the oblique coordinates $\mathbf{y}$ are given by (C.3) as:

$$\mathbf{y} = \mathbf{x}\mathbf{C}. \tag{D.1}$$

In Figure D.1(a), $y_1 = 3$ and $y_2 = 3$.

In the second version (b), the *parallel axes system*, $\mathbf{x}$ is the vector-sum of the coordinates $(z_1, z_2, z_3, \ldots, z_P)$. Thus $\mathbf{x} = z_1\mathbf{c}_1 + z_2\mathbf{c}_2 + z_3\mathbf{c}_3 + \cdots + z_P\mathbf{c}_P$ so that $x_k = z_1 c_{1k} + z_2 c_{2k} + z_3 c_{3k} + \cdots + z_P c_{Pk}$ and $\mathbf{x} = \mathbf{z}\mathbf{C}'$ giving:

$$\mathbf{z} = \mathbf{x}(\mathbf{C}')^{-1}. \tag{D.2}$$

In Figure D.1(b), $z_1 = 2$ and $z_2 = 1$.

It follows from (D.1) and (D.2) that:

$$\mathbf{z} = \mathbf{y}(\mathbf{C}'\mathbf{C})^{-1} \quad \text{and} \quad \mathbf{y} = \mathbf{z}\mathbf{C}'\mathbf{C} \tag{D.3}$$

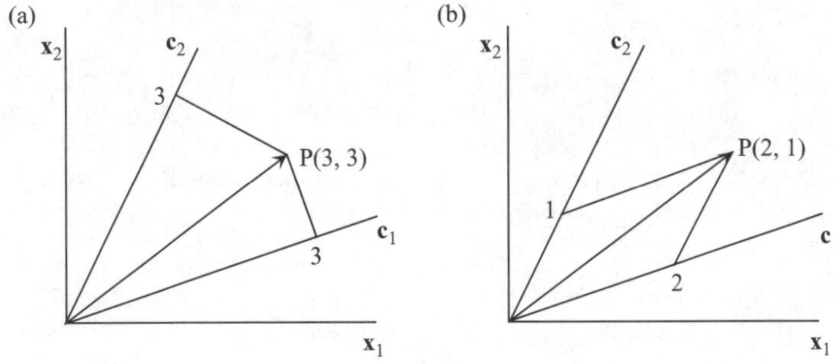

Fig. D.1. Oblique axes coordinate systems. In (a) the coordinates are given by orthogonal projection onto the oblique axes. In (b) the projections are parallel to the oblique axes.

give the transformations between the two oblique representations. The cosines of the angles between the oblique axes are the elements of $\mathbf{C}'\mathbf{C}$, which has a unit diagonal. When the oblique axes are themselves orthogonal then the cosines are zero so that $\mathbf{C}'\mathbf{C} = \mathbf{I}$, and $\mathbf{C}$ is orthogonal. In this case $\mathbf{y}$ and $\mathbf{z}$ coincide, as is obvious from the geometry of Figure D.1.

Using (D.1) we can evaluate the distance between two points given by their projection coordinates $\mathbf{y}_1$ and $\mathbf{y}_2$. From (D.1), their Cartesian coordinates are $\mathbf{y}_1\mathbf{C}^{-1}, \mathbf{y}_2\mathbf{C}^{-1}$ so that their squared distance is given by:

$$(\mathbf{y}_1 - \mathbf{y}_2)(\mathbf{C}'\mathbf{C})^{-1}(\mathbf{y}_1 - \mathbf{y}_2)'. \tag{D.4}$$

Similarly, from (D.2) the squared distance between two points given by their parallel axes coordinates $\mathbf{z}_1$ and $\mathbf{z}_2$ is:

$$(\mathbf{z}_1 - \mathbf{z}_2)(\mathbf{C}'\mathbf{C})(\mathbf{z}_1 - \mathbf{z}_2)'. \tag{D.5}$$

Appendix E
A minimisation problem

In Chapter 6 we need the minimum over $\mathbf{x}$ of $S = \|\mathbf{Xx} - \mathbf{y}\|$ subject to a constraint $\mathbf{x}'\mathbf{Bx} = k$, where $\mathbf{B}$ is symmetric but not necessarily definite and k is a constant. We may assume that k is either $+1$ or -1, because we may divide $\mathbf{B}$ by k. Note that $k = -1$ is not possible when $\mathbf{B}$ is p.s.d. and $k = +1$ is not possible when $\mathbf{B}$ is n.s.d. When $\mathbf{B}$ is not of full rank we can have $k = 0$ with $\mathbf{x}$ chosen to be in the null-space of $\mathbf{B}$. When $\mathbf{B}$ is indefinite, possibly of full rank, it is also possible to find a minimum with $k = 0$, where $\mathbf{x}$ is not in the null-space. We do not attempt a complete discussion for the case $k = 0$ but do cover the non-null-space case and, to indicate the possibilities, we give a brief discussion of null-space solutions at the end of this appendix.

Thus we require the minimum of:

$$S = \|\mathbf{Xx} - \mathbf{y}\| - \lambda(\mathbf{x}'\mathbf{Bx} - k),$$

where λ is a Lagrange multiplier. Writing $\mathbf{X}'\mathbf{X} = \mathbf{A}$ and expanding we have:

$$\mathrm{S} = \mathbf{x}'\mathbf{Ax} - 2\mathbf{x}'(\mathbf{X}'\mathbf{y}) + \mathbf{y}'\mathbf{y} - \lambda(\mathbf{x}'\mathbf{Bx} - k) \tag{E.1}$$

Differentiating (E.1) w.r.t $\mathbf{x}$ gives:

$$(\mathbf{A} - \lambda\mathbf{B})\mathbf{x} = \mathbf{X}'\mathbf{y} \tag{E.2}$$

so that

$$\mathbf{x} = (\mathbf{A} - \lambda\mathbf{B})^{-1}\mathbf{X}'\mathbf{y}. \tag{E.3}$$

Imposing the constraint shows that λ must satisfy:

$$\mathbf{y}'\mathbf{X}(\mathbf{A} - \lambda\mathbf{B})^{-1}\mathbf{B}(\mathbf{A} - \lambda\mathbf{B})^{-1}\mathbf{X}'\mathbf{y} = k. \tag{E.4}$$

(E.4) is essentially a polynomial of degree $2P$ in λ so we must decide which of its many roots minimises S. To make progress, we simplify (E.4) by using the two-sided eigenvalue decomposition:

$$\mathbf{AW\Gamma} = \mathbf{BW} \tag{E.5}$$

which requires $\mathbf{A}$ to be positive definite, which it is unless $\mathbf{X}$ has deficient rank, so we exclude this possibility. Because $\mathbf{A}$ is p.s.d. the eigenvectors may be normalised to satisfy $\mathbf{W}'\mathbf{AW} = \mathbf{I}$ so then $\mathbf{W}'\mathbf{BW} = \mathbf{\Gamma}$. It follows immediately from the identity $\mathbf{W}'\mathbf{B}\mathbf{W} = \mathbf{\Gamma}$ that the number of positive/negative eigenvalues of (E.5) is the positivity/negativity of $\mathbf{B}$. Also, if $\mathbf{w}$ is a null vector of $\mathbf{B}$, then $\gamma\mathbf{A}\mathbf{w} = \mathbf{B}\mathbf{w} = 0$ with $\gamma = 0$, so the number of zero eigenvalues γ is the same as the number of zero eigenvalues of $\mathbf{B}$.

We may write (E.2) as:

$$(\mathbf{W}'\mathbf{A}\mathbf{W} - \lambda\mathbf{W}'\mathbf{B}\mathbf{W})\mathbf{W}^{-1}\mathbf{x} = \mathbf{W}'\mathbf{X}'\mathbf{y}$$

or

$$(\mathbf{I} - \lambda\mathbf{\Gamma})\mathbf{u} = \mathbf{z} \qquad \text{(E.6)}$$

where $\mathbf{u} = \mathbf{W}^{-1}\mathbf{x}$ and $\mathbf{z} = \mathbf{W}'\mathbf{X}'\mathbf{y}$. Using the normalisation relationships, (E.4) simplifies to:

$$\mathbf{z}'(\mathbf{I} - \lambda\mathbf{\Gamma})^{-1}\mathbf{\Gamma}(\mathbf{I} - \lambda\mathbf{\Gamma})^{-1}\mathbf{z} = k \qquad \text{(E.7)}$$

or, in non-matrix form:

$$f(\lambda) = \sum_{i=1}^{P} \frac{\gamma_i z_i^2}{(1 - \lambda\gamma_i)^2} - k = 0. \qquad \text{(E.8)}$$

Thus we are concerned with the roots of (E.8). To determine which of these corresponds to a minimum of S requires the second differential of (E.1). Differentiating (E.2) gives the second differential as $\mathbf{A} - \lambda\mathbf{B}$ (the Hessian) which may be written:

$$\mathbf{A} - \lambda\mathbf{B} = (\mathbf{W}')^{-1}(\mathbf{I} - \lambda\mathbf{\Gamma})\mathbf{W}^{-1}.$$

A minimum occurs when $\mathbf{A} - \lambda\mathbf{B}$ is positive definite, which is the same as requiring that the Hessian

$$\mathbf{I} - \lambda\mathbf{\Gamma} \qquad \text{(E.9)}$$

be positive definite. This is a diagonal matrix so we require its diagonal values $1 - \lambda\gamma_i$ to be positive for all i. Suppose the eigenvalues are given in increasing order:

$$\gamma_1 < \gamma_2 < \cdots < \gamma_- < 0 < \gamma_+ < \cdots < \gamma_{P-1} < \gamma_P$$

where γ_- is the biggest negative eigenvalue and γ_+ is the smallest positive eigenvalue. With this ordering, when γ_i is positive, $1 - \lambda\gamma_i > 0$ implies that $\lambda < 1/\gamma_i$ so that for a minimum we have that $\lambda < 1/\gamma_P$. Similarly, when γ_i is negative, positivity of the diagonal value requires that $\lambda > 1/\gamma_i$. This shows that the minimum of S occurs for a value λ_0 of λ in the range:

$$\frac{1}{\gamma_1} < \lambda_0 < \frac{1}{\gamma_P}. \qquad \text{(E.10)}$$

We refer to (E.10) as defining the *permissible range* for a minimum.

The above assumes that there is at least one positive and at least one negative eigenvalue. The case when this is not so, is considered below in Section E.1.

Next, we show that there is just one minimum in the range (E.10). The individual terms $(\gamma_i z_i^2)/(1-\lambda\gamma_i)^2$ of (E.8) define a function which has a vertical asymptote at $\lambda = 1/\gamma_i$. If $z_i = 0$, then the term vanishes and may normally be disregarded—but see Section E.1. When γ_i is positive, the right-hand branch of the function (i.e. $\lambda > 1/\gamma_i$) is monotone decreasing and to the left it is monotone increasing. When γ_i is negative, the opposite applies and, the right-hand branch of the function is monotone increasing and to the left it is monotone decreasing. It follows that in the interval (E.10), all the component parts of (E.8) are monotone increasing, so that (E.8) is itself monotone increasing in the interval. Equation (E.8) is negatively infinite at $\lambda = 1/\gamma_1$ and is positively infinite at $\lambda = 1/\gamma_P$ and is monotone increasing in between, so (E.8) is zero at only one point in the interval. Thus we have shown that S has a single minimum λ_0 in the interval (E.10), *whatever the value of k*. That there is a unique minimum of S is surprising but useful, as it shows that we need have no concern with the possibility of local minima. However, in intervals bounded by other adjacent asymptotes, S may have maxima, saddle-points etc. formed from combinations of monotonic increasing and monotonic decreasing components of (E.8). Thus, it is best to devise algorithms for computing λ_0 that operate only on values of λ within the range (E.10), otherwise if one moves outside the interval there is the possibility of divergence. We note the possibility that minimisation algorithms based on differentiation of $f(\lambda)$ might take one onto the wrong branch. Indeed, just this kind of difficulty is reported by Cramer (1974) when using the Newton–Raphson method as recommended by Browne (1967); we refer to gradient methods further in Section E.2.

In any case, the use of gradient methods is unnecessary, as the value λ_0 at the minimum is easily and safely found by bisecting in the interval (E.10), thus remaining in the required range throughout; this is the method we use in Algorithm E.2.1. Bisection methods improve the accuracy of the root by about one bit per iteration, so three decimal figure accuracy requires about 10 iterations ($2^{10} \sim 1000$). When λ_0 has been found, it may be substituted into (E.3) to obtain $\mathbf{x}$, or by substituting into (E.6) to obtain $\mathbf{u}$, whence $\mathbf{x} = \mathbf{Wu}$. The value of the minimum follows from (E.1) and (E.2) as:

$$S_{\min} = \mathbf{y}'\mathbf{y} - \mathbf{x}'(\mathbf{X}'\mathbf{y}) + \lambda_0 k$$

Note that when $\lambda = 0$, then (E.2) gives $\mathbf{x} = (\mathbf{X}'\mathbf{X})^{-1}\mathbf{X}'\mathbf{y}$ confirming that $S_{\min}$ is then the classical (multiple regression) unconstrained least-squares minimum $\mathbf{y}'[\mathbf{I} - \mathbf{X}(\mathbf{X}'\mathbf{X})^{-1}\mathbf{X}']\mathbf{y}$.

E.1 Zero values of $\mathbf{z}_i$

Each term of (E.8) pairs a value of γ_i with a value of z_i. In deriving the above, we wrote that when $z_i = 0$, then the corresponding term of (E.8) vanishes and may normally be disregarded. There is one pathological case, a particular instance of which was first noted by Cramer (1974) and analysed in detail by ten Berge and Nevels (1977), where more care has to be taken. This is when $z_i = 0$ pairs with

γ_1 and/or γ_P and possibly with adjacent values of γ_i. Then, the asymptotes vanish (we refer to *phantom asymptotes*) at these values and the admissible range for λ_0 extends to:

$$\frac{1}{\gamma_a} < \lambda_0 < \frac{1}{\gamma_b} \tag{E.11}$$

where $z_i = 0$ pairs with $\gamma_1, \gamma_2, \ldots, \gamma_{a-1}$ and with $\gamma_{b+1}, \gamma_{b+2}, \ldots, \gamma_P$. Thus, γ_a is the smallest negative eigenvalue paired with a non-zero z_i and γ_b is the biggest positive eigenvalue paired with a non-zero z_i. The admissible range for a minimum remains as given by (E.10)—see Figure E.1. Then, writing $\mathbf{\Gamma}_1 = \text{diag}(\gamma_1, \gamma_2, \ldots, \gamma_{a-1})$, $\mathbf{\Gamma}_2 = \text{diag}(\gamma_a, \gamma_{a+1}, \ldots, \gamma_b)$ and $\mathbf{\Gamma}_3 = \text{diag}(\gamma_{b+1}, \gamma_{b+2}, \ldots, \gamma_P)$, (E.6) splits into three parts:

$$\begin{aligned}(\mathbf{I} - \lambda\mathbf{\Gamma}_1)\mathbf{u}_1 &= 0\\ (\mathbf{I} - \lambda\mathbf{\Gamma}_2)\mathbf{u}_2 &= \mathbf{z}_2\\ (\mathbf{I} - \lambda\mathbf{\Gamma}_3)\mathbf{u}_3 &= 0,\end{aligned} \tag{E.12}$$

where the first equation derives from the $a - 1$ zero elements of $\mathbf{z}$ paired with $\mathbf{\Gamma}_1$ and the third equation derives from the $P - b + 1$ zero elements of $\mathbf{z}$ paired with $\mathbf{\Gamma}_3$. The vector $\mathbf{z}_2$ represents the non-zero part of $\mathbf{z}$, although all that is required is that its first and last elements are non-zero. There are two types of solution to (E.12):

(i) Type 1 occurs when the first and third equations of (E.12) give $\mathbf{u}_1 = \mathbf{0}$ and $\mathbf{u}_3 = \mathbf{0}$,

(ii) Type 2 occurs when one of the first and third equations of (E.12) satisfies $\lambda\gamma_i = 1$, in which case u_i is arbitrary and the remaining elements of $\mathbf{u}_1$ and $\mathbf{u}_3$ are zero.

We now examine these solutions in more detail.

With the type 1, solution where $\mathbf{u}_1 = \mathbf{0}$ and $\mathbf{u}_3 = \mathbf{0}$, the middle equation of (E.12) may be solved for $\mathbf{u}_2$ as before, under the constraint $\mathbf{u}_2'\mathbf{\Gamma}_2\mathbf{u}_2 = \mathbf{u}'\mathbf{\Gamma}\mathbf{u} = k$. The previous results apply when λ_0 turns out to be in the range (E.10) and a condition for this is that $f(1/\gamma_P) > 0$ and $f(1/\gamma_1) < 0$ (see the curve through point A in Figure E.1). When λ_0 is outside the range (E.10) and inside the range (E.11) then the solution for λ_0 is either at a point like B or a point like C of Figure E.1. Taking point B we have $\lambda_0 - 1/\gamma_P > 0$ so that $1 - \lambda_0\gamma_P$ is negative and the Hessian (E.9) determines a saddle point and not a minimum. Similarly, for point C, $\lambda_0 - 1/\gamma_1 < 0$ and because γ_1 is negative, $1 - \lambda_0\gamma_1$ is again negative, so that C also corresponds to a saddle point solution. Saddle point solutions are not acceptable, so we have to look for solutions to (E.12) additional to our setting of $\mathbf{u}_1 = \mathbf{0}$ and $\mathbf{u}_3 = \mathbf{0}$.

With type 2 solutions, the only permissible values are given by $1 - \lambda\gamma_1 = 0$ and/or $1 - \lambda\gamma_P = 0$; other choices of γ would give values of λ_0 outside the permissible range (E.10). Taking the case $1 - \lambda\gamma_P = 0$, yields $\lambda_0 = 1/\gamma_P$ (point D of Figure E.1) which may be substituted into the middle equation of (E.12) to give $\mathbf{u}_2$. The values of $\mathbf{u}_1$ and $\mathbf{u}_3$ are all zero except for the term corresponding to

γ_P which is arbitrary, say u_P. We may use u_P to satisfy the constraint $\mathbf{u}'\mathbf{\Gamma u} = k$. Thus, we have:

$$\sum_{i=a}^{b} \frac{\gamma_i z_i^2}{(1 - \gamma_P^{-1}\gamma_i)^2} - k + \gamma_P u_P^2 = 0$$

or

$$f\left(\frac{1}{\gamma_P}\right) + \gamma_P u_P^2 = 0. \qquad \text{(E.13)}$$

which determines u_P, apart from its sign.

A similar argument for the solution based on $\lambda_0 = 1/\gamma_1$ (point E of Figure E.1) and arbitrary u_1 leads to:

$$f\left(\frac{1}{\gamma_1}\right) + \gamma_1 u_1^2 = 0. \qquad \text{(E.14)}$$

There is a question as to whether either or both of (E.13) and (E.14) can give real solutions for u_1 and u_P. For (E.13) to have a real solution we must have

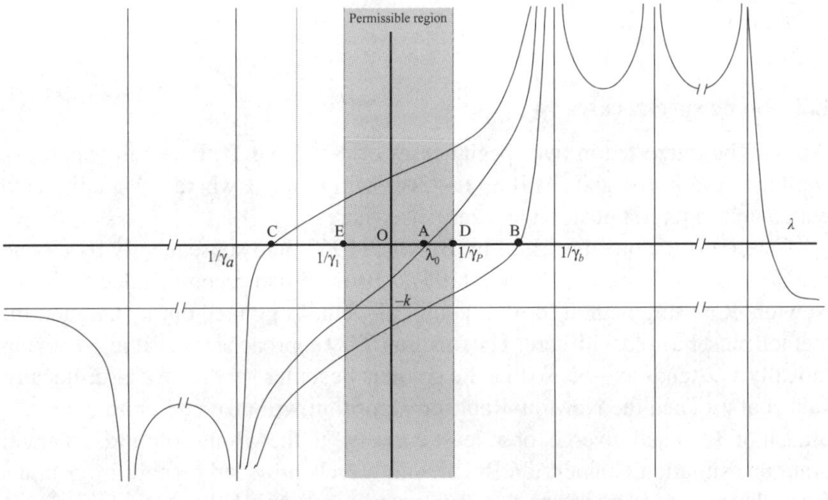

Fig. E.1. Various possibilities for the performance of $f(\lambda)$ when $z_i = 0$. The asymptotes at $\lambda = 1/\gamma_1$ and $1/\gamma_P$ correspond to zero values of z_1 and z_P giving vanishing terms in (E.8) and hence to phantom asymptotes. Nevertheless, they continue to define the boundaries of the region giving permissible minima of S. Other phantom asymptotes may exist in the region bounded by $\lambda = 1/\gamma_a$ and $1/\gamma_b$, one of which is shown between $\lambda = 1/\gamma_a$ and $\lambda = 1/\gamma_1$. The solution at A is within the permissible region but those at B and C give saddle point solutions and not minima. When there is no solution like A, then one of the solutions at D or E is available. O is the origin.

$f(1/\gamma_p) < 0$ and for (E.14) to have a real solution we must have $f(1/\gamma_1) > 0$. Not both can be true or the function $f(\lambda)$ would not be monotone increasing. When both are false is the condition for a type 1 solution. Hence, either we have a type 1 solution or a type 2 solution for which one of (E.13) and (E.14) holds. For a saddle point solution, as at B, between $1/\gamma_P$ and $1/\gamma_b$, $f(1/\gamma_P)$ is negative, so we have solution D given by (E.13). For a saddle point solution, as at C, between $1/\gamma_a$ and $1/\gamma_1$, $f(1/\gamma_1)$ is positive, so we have solution E given by (E.14). Note that the solutions are the points in the permissible region nearest to the corresponding saddle point solutions.

Thus, when there is no type 1 solution we have a possible type 2 solution at D or E in Figure E.1 and these are at the extremities of the permissible region. With either of these settings, the Hessian (E.9) has positive diagonal elements apart from a zero in either the first or last position. It is not immediate that these solutions pertain to a minimum of S. However, S being a sum-of-squares, must have a minimum unless the constraint is inadmissible, such as a requirement that $\mathbf{x}'\mathbf{B}\mathbf{x} < 0$ for a p.s.d. matrix. There being no other putative minimum available, it follows that the extreme solutions at points D and E of Figure E.1 may be accepted as true minima when the type 1 solution at A is unavailable. Thus, there is a unique* real minimum corresponding to one of three mutually exclusive possibilities: (i) $f(1/\gamma_P) > 0$ and $f(1/\gamma_1) < 0$, (ii) $f(1/\gamma_P) < 0$ and $f(1/\gamma_1) < 0$, and (iii) $f(1/\gamma_P) > 0$ and $f(1/\gamma_1) > 0$, corresponding respectively to points A, D, and E in Figure E.1.

E.2 Some special cases

We shall be interested in two special cases of the above: (i) $\mathbf{B} = \mathbf{I}$ in which case we must have $k = 1$ and (ii) $\mathbf{B} = \mathbf{I} - 2\mathbf{e}\mathbf{e}'$ with $k = -1$ where $\mathbf{e}$ is a unit vector with a unit in its last position and zero elsewhere.

Case (i) is the one discussed by Browne (1967) and subsequently by Cramer (1974) and ten Berge and Nevels (1977). Browne had recommended using the Newton–Raphson method for determining λ_0 but, as pointed out by Cramer, this method must be used with care. The function $f(\lambda)$ approaches the value -1 asymptotically as λ tends to $-\infty$, so that the gradient becomes very flat for large negative values of λ. Then the Newton–Raphson algorithm will take one on to the wrong branch of $f(\lambda)$ and diverge, or at least converge to the wrong solution. To ameliorate this situation Cramer, ten Berge, and Nevels provided a good lower bound for λ_0 but it is not clear to us that they proved that their lower bound guarantees

* This statement is at variance with the claim by ten Berge and Nevels (1977) that, for $k = 1$, $\mathbf{B} = \mathbf{I}$, there are an infinite number of type 2 solutions. Our analysis allows two closely related solutions, depending on the sign given to u_p (or u_1). However, if the leading, or trailing, eigenvalues are equal *and* pair with zero values of z_i, then $(\ldots, u_{p-1}, u_p)$ or $(u_1, u_2, \ldots)$ are arbitrary and there are indeed an infinite number of solutions.

convergence under all circumstances. We prefer the simple bisection method which avoids these difficulties but it remains efficient to confine the bisection process to a small range, so good lower bounds on λ_0 remain of interest.

In case (i) there are no negative values of γ_i so the permissible region of (E.10) becomes $-\infty < \lambda_0 < 1/\gamma_P$ and the bisection algorithm to find the root λ_0 of $f(\lambda) = 0$ starts with an infinite lower bound. This bound may be improved as follows. Let $z^2 = \max(z_i^2/\gamma_i)$, then:

$$f(\lambda) + 1 = \sum \frac{\gamma_i z_i^2}{(1 - \lambda\gamma_i)^2} = \sum \frac{z_i^2/\gamma_i}{(1/\gamma_i - \lambda)^2}$$

$$\leq \sum \frac{z^2}{(1/\gamma_i - \lambda)^2} \leq \sum \frac{z^2}{(1/\gamma_P - \lambda)^2} = \frac{Pz^2}{(1/\gamma_P - \lambda)^2}.$$

Now, the function (E.8) is positive at $\lambda = 1/\gamma_P$ so to get a lower bound on λ_0 it is sufficient to choose $\lambda < 1/\gamma_P$ so that $Pz^2/(1/\gamma_P - \lambda)^2 - 1$ is negative. This is achieved when $1/\gamma_P - z\sqrt{P} < \lambda$. Hence, with this value of λ (E.8) is negative. Thus, we have shown that when **B** is p.d:

$$1/\gamma_P - z\sqrt{P} < \lambda_0 < 1/\gamma_P$$

gives bounds for initiating the bisection that may replace (E.10). The permissible region continues to extend to $-\infty$ and apart from type 1 solutions, only the type 2 solution given by (E.13) is available should γ_P be paired with zero z_P.

When $\mathbf{B} = -\mathbf{I}$ and there are no positive eigenvalues, a similar argument leads to the interval:

$$1/\gamma_1 < \lambda_0 < 1/\gamma_1 + z\sqrt{P}$$

where $-z^2 = \min(z_i^2/\gamma_i)$. Now the only available type 2 solution is given by (E.14).

In case (ii) **B** has one negative eigen value, so there is one negative eigenvalue, γ_1, and there are no special problems unless $z_1 = 0$, when $f(\lambda)$ is always positive and there is no solution.

When there are zero values of z_i, as discussed in E.1, the above limits may be improved further. In the first case p may be replaced by b and in the second by $P - a + 1$.

All the above considerations are incorporated in Algorithm E.2.1.

Algorithm E.2.1

Step 1 Solve the two-sided eigenvalue problem $\mathbf{AW\Gamma} = \mathbf{BW}$, normalised so that $\mathbf{W}'\mathbf{AW} = \mathbf{I}$.

Step 2 Determine a, where γ_a is the smallest negative eigenvalue coupled with a non-zero z_i and b, where γ_b is the biggest

Step 3 Set $\lambda_- = 1/\gamma_a$, $\lambda_+ = 1/\gamma_b$, $\mu_- = 1/\gamma_1$, $\mu_+ = 1/\gamma_P$.

Step 4 Set bounds in special cases:

If $\lambda_- > 0$ then $\mathbf{B}$ is p.s.d. Set $\lambda_- = \mu_- = 1/\gamma_b - z\sqrt{b}$ where $z^2 = \max(z_i^2/\gamma_i)$

If $\lambda_+ < 0$ then $\mathbf{B}$ is n.s.d. Set $\lambda_+ = \mu_+ = 1/\gamma_a + z\sqrt{(P-a+1)}$ where $-z^2 = \min(z_i^2/\gamma_i)$

Step 5 *Bisection step.* Repeat:

(1) $\lambda_0 = \frac{1}{2}(\lambda_- + \lambda_+)$

(2) if $f(\lambda_0) > 0$, set $\lambda_+ = \lambda_0$ else $\lambda_- = \lambda_0$

(3) if $|f(\lambda_0)| > \varepsilon$, go to (1)

Use (E.6) to give $\mathbf{u}$.

Step 6 If $\mu_- < \lambda_0 < \mu_+$ go to *Step 7*, else:

If $\lambda_0 \geq \mu_+$

then (i) set $\lambda_0 = 1/\gamma_p$, (ii) set $u_P = \sqrt{-f(\lambda_0)/\gamma_P}$

else (i) set $\lambda_0 = 1/\gamma_1$, (ii) set $u_1 = \sqrt{-f(\lambda_0)/\gamma_1}$

Step 7 Then $\mathbf{x} = \mathbf{Wu}$, $S_{\min} = \mathbf{y}'\mathbf{y} - \mathbf{x}'\mathbf{X}'\mathbf{y} + \lambda_0 k$.

E.3 Some remarks on the case $k = 0$

The above algorithm is designed for $k = -1$ or $k = +1$; however, it should also work for $k = 0$ when $\mathbf{B}$ is indefinite, that is it has positive and negative roots (and, perhaps, also zero roots). We have not needed such solutions but ten Berge (1983b) considers a regression problem which leads to the same minimisation problem with the constraint $\mathbf{x}'\mathbf{Bx} = 0$ for $P = 3$ and where $b_{13} = b_{31} = 2$ and $b_{22} = -1$. Returning to the general problem, when $\mathbf{B}$ is not of full rank we can have $\mathbf{x}'\mathbf{Bx} = 0$ for any vector $\mathbf{x}$ in the null space of $\mathbf{B}$; these null space solutions have been considered neither in our analysis nor in the algorithm. However, solutions in the null space are easily found. Suppose that $\mathbf{BV} = \mathbf{0}$, so that the columns of $\mathbf{V}$, assumed to be independent, determine the null space of $\mathbf{B}$. Then any vector $\mathbf{x}$ in the null space is a linear combination of the form $\mathbf{x} = \mathbf{Va}$. Then, in our least-squares criterion, the Lagrangian term vanishes and (E.1) becomes:

$$S = \|\mathbf{Xx} - \mathbf{y}\| = \mathbf{a}'(\mathbf{V}'\mathbf{AV})\mathbf{a} - 2\mathbf{a}'(\mathbf{V}'\mathbf{X}'\mathbf{y}) + \mathbf{y}'\mathbf{y}$$

which is minimal when:

$$\mathbf{a} = (\mathbf{V}'\mathbf{AV})^{-1}\mathbf{V}'\mathbf{X}'\mathbf{y}$$

from which $\mathbf{x} = \mathbf{V}(\mathbf{V}'\mathbf{AV})^{-1}\mathbf{V}'\mathbf{X}'\mathbf{y}$
and $S_{\min} = \mathbf{y}'\mathbf{y} - \mathbf{y}'\mathbf{X}'\mathbf{V}(\mathbf{V}'\mathbf{AV})^{-1}\mathbf{V}'\mathbf{X}'\mathbf{y}$.
When $\mathbf{B}$ is semi-definite this solution gives the minimum of the criterion for $k = 0$. When $\mathbf{B}$ is indefinite and not of full rank, it is unclear whether or not this null-space minimum can be less than that given by (E.8) and the Algorithm E.2.1 with $k = 0$. To summarise the possibilities:

Type of **B**	Permissible values of k
Positive definite	$+1$
Negative definite	-1
Positive semi-definite	$+1, 0$ ($\mathbf{x}$ in null space)
Negative semi-definite	$-1, 0$ ($\mathbf{x}$ in null space)
Indefinite, full rank	$+1, -1, 0$ ($\mathbf{x}$ not in null space)
Indefinite, not full rank	$+1, -1, 0$ ($\mathbf{x}$ not in null space), or 0 ($\mathbf{x}$ in null space)

An exhaustive algorithm should take account of all these possibilities and check for violations of the nominated value of k against permissible values allowed for the actual form of $\mathbf{B}$.

E.4 Note on an alternative parameterisation

We note that if $\mathbf{X}'\mathbf{X} = \mathbf{LL}'$ is a Choleski decomposition in terms of a lower-triangular matrix $\mathbf{L}$, then:

$$\begin{aligned}\|\mathbf{Xs} - \mathbf{y}\| &= \mathbf{s}'\mathbf{X}'\mathbf{Xs} - 2\mathbf{y}'\mathbf{Xs} + \mathbf{y}'\mathbf{y} = \mathbf{s}'\mathbf{LL}' - \mathbf{s} \\ &\quad - 2\mathbf{y}'\mathbf{X}(\mathbf{L}')^{-1}(\mathbf{L}'\mathbf{s}) + \mathbf{y}'\mathbf{y} = \|\mathbf{t} - \mathbf{L}^{-1}\mathbf{X}'\mathbf{y}\| + \text{constant},\end{aligned}$$

where $\mathbf{t} = \mathbf{L}'\mathbf{s}$.

Thus, minimising $\|\mathbf{Xs} - \mathbf{y}\|$ subject to $\mathbf{s}'\mathbf{Bs} = k$ is the same as minimising $\|\mathbf{t} - \mathbf{z}\|$ subject to $\mathbf{t}'\mathbf{L}^{-1}\mathbf{B}(\mathbf{L}')^{-1}\mathbf{t} = k$ where $\mathbf{z} = \mathbf{L}^{-1}\mathbf{X}'\mathbf{y}$. This parameterisation is simpler than the one we have adopted and allows the ordinary spectral decomposition of $\mathbf{L}^{-1}\mathbf{B}(\mathbf{L}')^{-1}$ to be used rather than the two-sided eigen-decomposition used above. However, it leads to precisely the same polynomial (E.8) and its subsequent analysis. We prefer the form $\|\mathbf{Xs} - \mathbf{y}\|$ as this includes the case where $\mathbf{s} = \mathbf{c}$, which is then shown explicitly.

Appendix F
Symmetric matrix products

In this section we derive a result that is needed in Section 7.5 for the development of double Procrustes problems. Suppose $\mathbf{A}$ and $\mathbf{B}$ are two $p \times q$ matrices such that the products $\mathbf{B}'\mathbf{A}$ and $\mathbf{A}\mathbf{B}'$ are both symmetric. We shall show that necessary and sufficient conditions for this to hold are that $\mathbf{A}$ and $\mathbf{B}$ share the same singular vectors.

Sufficiency is easily established for if the SVDs are given by $\mathbf{A} = \mathbf{X\Phi Y}'$ and $\mathbf{B} = \mathbf{X\Psi Y}'$ then $\mathbf{AB}' = \mathbf{X\Phi\Psi X}'$ and $\mathbf{B}'\mathbf{A} = \mathbf{Y\Psi\Phi Y}'$ which are both patently symmetric.

Necessity is harder to establish. We begin from the spectral decompositions:

$$\mathbf{B}'\mathbf{A} = \mathbf{V\Gamma V}' \quad \mathbf{AB}' = \mathbf{U\Delta U}' \tag{F.1}$$

where $\mathbf{U}$ has r columns and $\mathbf{V}$ has s columns, with $\mathbf{U}'\mathbf{U} = \mathbf{I}$ and $\mathbf{V}'\mathbf{V} = \mathbf{I}$. Thus, $\mathbf{\Gamma}$ and $\mathbf{\Delta}$ are diagonal matrices of non-zero eigenvalues and the columns of $\mathbf{V}$ and $\mathbf{U}$ are the corresponding eigenvectors. We first show that $r = s$ and that $\mathbf{\Delta} = \mathbf{\Gamma}$, possibly after some permutation of rows and columns.

Consider the eigenvalue/vector equation:

$$\mathbf{B}'\mathbf{Av} = \gamma\mathbf{v}.$$

Pre-multiplication by $\mathbf{A}$ gives:

$$(\mathbf{AB}')\mathbf{Av} = \gamma\mathbf{Av},$$

showing that $(\gamma, \mathbf{Av})$ is an eigenvalue/vector pair of $\mathbf{AB}'$ and so γ must correspond to some eigenvalue δ, and $\mathbf{Av}$ to an eigenvector $\mathbf{u}$ which is a column of $\mathbf{U}$. Similarly, considering $\mathbf{AB}'\mathbf{u} = \delta\mathbf{u}$, which on pre-multiplying by $\mathbf{B}'$ gives:

$$(\mathbf{B}'\mathbf{A})\mathbf{B}'\mathbf{u} = \delta\mathbf{B}'\mathbf{u},$$

shows that $(\delta, \mathbf{B}'\mathbf{u})$ is an eigenvalue/vector pair of $\mathbf{B}'\mathbf{A}$ and so δ must correspond to some eigenvalue γ, and $\mathbf{B}'\mathbf{u}$ to an eigenvector $\mathbf{v}$ which is a column of $\mathbf{V}$.

It follows that to every δ, there corresponds a γ, and conversely. Hence the non-zero eigenvalues contained in $\mathbf{\Delta}$ and $\mathbf{\Gamma}$ are equal and therefore $r = s$. We have not shown that $\gamma_i = \delta_i (i = 1, 2, \ldots, r)$, only that there is a one-to-one correspondence (bijection) between the two sets. However, without loss of generality, we may assume that the values of δ are permuted into the same order as those of γ, provided the columns of $\mathbf{U}$ are permuted similarly. Then, $\mathbf{\Gamma} = \mathbf{\Delta}$. We shall assume that this ordering has been done. Also, $\mathbf{AV}$ must be an unnormalised version of $\mathbf{U}$ and $\mathbf{B}'\mathbf{U}$ must be an unnormalised version of $\mathbf{V}$. Thus:

$$\mathbf{AV} = \mathbf{U\Phi} \quad \text{and} \quad \mathbf{B}'\mathbf{U} = \mathbf{V\Psi} \tag{F.2}$$

where $\mathbf{\Phi}$ and $\mathbf{\Psi}$ are diagonal matrices of normalising constants. This gives:

$$\mathbf{U}'\mathbf{AV} = \mathbf{\Phi} \quad \mathbf{U}'\mathbf{BV} = \mathbf{\Psi}.$$

From (F.2), $\mathbf{V}'\mathbf{B}'\mathbf{AV} = \mathbf{V}'\mathbf{B}'\mathbf{U\Phi} = \mathbf{\Psi\Phi}$, while from (F.1) $\mathbf{V}'\mathbf{B}'\mathbf{AV} = \mathbf{\Gamma}$. Hence, $\mathbf{\Gamma} = \mathbf{\Psi\Phi}$.

The above has shown that $\mathbf{B}'\mathbf{A}$ and $\mathbf{AB}'$ both have rank r and the same non-zero eigenvalues $\mathbf{\Gamma}$. The matrices $\mathbf{A}$ and $\mathbf{B}$ themselves may have rank greater than r. Additional singular vectors must lie in the orthogonal complements of $\mathbf{U}$ and $\mathbf{V}$, which we shall write as $\mathbf{U}_\perp, \mathbf{V}_\perp$. Thus:

$$\mathbf{A}(\mathbf{V}, \mathbf{V}_\perp) = (\mathbf{U}, \mathbf{U}_\perp)\begin{pmatrix} \Phi & \\ & \Phi_\perp \end{pmatrix}$$

corresponding to the SVD:

$$\mathbf{A} = \mathbf{U\Phi V}' + \mathbf{U}_\perp \mathbf{\Phi}_\perp \mathbf{V}'_\perp,$$

where $\mathbf{\Phi}_\perp$ may contain zero singular values. Writing $\mathbf{\Phi}_\mathrm{A}$, $\mathbf{\Psi}_\mathrm{B}$ for the non-zero singular values of $\mathbf{\Phi}_\perp$, $\mathbf{\Psi}_\perp$ and the corresponding vectors as $\mathbf{U}_\mathrm{A} \in \mathbf{U}_\perp$ and $\mathbf{V}_\mathrm{A} \in \mathbf{V}_\perp$, the above becomes:

$$\left.\begin{aligned} \mathbf{A} &= \mathbf{U\Phi V}' + \mathbf{U}_\mathrm{A}\mathbf{\Phi}_\mathrm{A}\mathbf{V}'_\mathrm{A} \\ \text{Similarly,} \quad \mathbf{B} &= \mathbf{U\Psi V}' + \mathbf{U}_\mathrm{B}\mathbf{\Psi}_\mathrm{B}\mathbf{V}'_\mathrm{B} \end{aligned}\right\}. \tag{F.3}$$

Thus, we have shown that $\mathbf{A}$ and $\mathbf{B}$ share the singular vectors $\mathbf{U}$ and $\mathbf{V}$, given as the eigenvectors of (F.1), plus possible additional components $\mathbf{U}_\mathrm{A}$, $\mathbf{U}_\mathrm{B}$, $\mathbf{V}_\mathrm{A}$, and $\mathbf{V}_\mathrm{B}$ in the null-spaces of $\mathbf{B}'\mathbf{A}$ and $\mathbf{AB}'$. We now show that these additional components must be independent.
From (F.3):

$$\mathbf{B}'\mathbf{A} = \mathbf{V\Phi\Psi V}' + (\mathbf{V}_\mathrm{B}\mathbf{\Psi}_\mathrm{B}\mathbf{U}'_\mathrm{B})(\mathbf{U}_\mathrm{A}\mathbf{\Phi}_\mathrm{A}\mathbf{V}'_\mathrm{A}).$$

But $\mathbf{\Psi\Phi} = \mathbf{\Gamma}$, and so from (F.1) we have that:

$$(\mathbf{V}_\mathrm{B}\mathbf{\Psi}_\mathrm{B}\mathbf{U}'_\mathrm{B})(\mathbf{U}_\mathrm{A}\mathbf{\Phi}_\mathrm{A}\mathbf{V}'_\mathrm{A}) = \mathbf{0}.$$

Pre-multiplying by $\mathbf{\Psi}_\mathrm{B}^{-1}\mathbf{V}'_\mathrm{B}$ and post-multiplying by $\mathbf{V}_\mathrm{A}\mathbf{\Phi}_\mathrm{A}^{-1}$ gives:

$$\mathbf{U}'_\mathrm{B}\mathbf{U}_\mathrm{A} = \mathbf{0}.$$

Similarly, from the expansion of $\mathbf{AB}'$ we may show that $\mathbf{V}_B'\mathbf{V}_A = 0$. This shows that the additional vectors occupy independent subspaces of the null-spaces of **A** and **B**.

There remain possible vectors $\mathbf{V}_C$ and $\mathbf{U}_C$ that are null-vectors of both **A** and **B**. Thus, the full SVDs are:

$$\mathbf{A} = (\mathbf{U} \quad \mathbf{U}_A \quad \mathbf{U}_B \quad \mathbf{U}_C)\begin{pmatrix} \mathbf{\Phi} & & & \\ & \mathbf{\Phi}_A & & \\ & & \mathbf{0} & \\ & & & \mathbf{0} \end{pmatrix}\begin{pmatrix} \mathbf{V}' \\ \mathbf{V}_A' \\ \mathbf{V}_B' \\ \mathbf{V}_C' \end{pmatrix}$$

$$\mathbf{B} = (\mathbf{U} \quad \mathbf{U}_A \quad \mathbf{U}_B \quad \mathbf{U}_C)\begin{pmatrix} \mathbf{\Psi} & & & \\ & \mathbf{0} & & \\ & & \mathbf{\Psi}_B & \\ & & & \mathbf{0} \end{pmatrix}\begin{pmatrix} \mathbf{V}' \\ \mathbf{V}_A' \\ \mathbf{V}_B' \\ \mathbf{V}_C' \end{pmatrix}.$$

When $\mathbf{\Gamma}$ has repeated eigenvalues, then corresponding columns of **U** and **V** are not unique. Nevertheless, to every admissible choice of **V** corresponds a unique **U**, given by $\mathbf{U} = \mathbf{AV}$. Of course, this needs the normalisation $\mathbf{V}'\mathbf{A}'\mathbf{AV} = \mathbf{\Phi}^2$, from which it follows that $\mathbf{U} = \mathbf{AV\Phi}^{-1}$. Similarly, when $\mathbf{U}_B$, $\mathbf{V}_B$ correspond to multiple zero singular values of **A** there is indeterminacy, but admissible unique settings may be obtained from the SVD of **B**, unless $\mathbf{\Psi}_B$ includes repeated singular values. In that case, any orthogonal vectors corresponding to repeated values in $\mathbf{\Psi}_B$ may also be chosen for the SVD of **A**. Similar remarks apply to repeated values in $\mathbf{\Phi}_A$ and any corresponding indeterminancies in $\mathbf{U}_A$, $\mathbf{V}_A$.

Thus, the SVDs of **A** and **B** necessarily share the same singular vectors, with the provisos just discussed concerning repeated singular vectors, in which case they share subspaces of vectors. The fundamental result may be stated as follows: *For $\mathbf{B}'\mathbf{A}$ and $\mathbf{AB}'$ both to be symmetric, it is necessary and sufficient that they have the SVDs (F.3), where $\mathbf{U}_A'\mathbf{U}_B = 0$ and $\mathbf{V}_A'\mathbf{V}_B = 0$.*

A simple way of thinking of things is that the non-null spaces of **A** and **B** share some singular vectors, while each may possibly have additional singular vectors in orthogonal subspaces of the null-spaces of $\mathbf{AB}'$ (equivalently $\mathbf{B}'\mathbf{A}$).

F.1 Special case

When **A** and **B** are square and symmetric, and **AB** is symmetric, then $\mathbf{B}'\mathbf{A} = \mathbf{BA} = \mathbf{AB} = \mathbf{AB}'$, so the conditions of our result are satisfied. Singular vectors become eigenvectors; it follows that **A** and **B** share the same eigenvectors and that this is the condition for two symmetric matrices to commute, a result that is occasionally found in algebra texts (see, for example, Wedderburn 1934).

Comments (i) It is clear that if $\mathbf{\Pi}$ is any permutation matrix, then $\mathbf{U\Psi V}' = \mathbf{U\Pi}'(\mathbf{\Pi\Psi\Pi}')\mathbf{\Pi V}'$. Thus, although we have shown that the singular vectors **U**, **V**

common to $\mathbf{A}$ and $\mathbf{B}$ are the same, we have not shown that they are necessarily presented in the same order.

(ii) The result established here is used in theoretical contexts. For practical purposes, it is easier to evaluate $\mathbf{AB}'$ and $\mathbf{B}'\mathbf{A}$ and inspect for symmetry, or more likely, for approximate symmetry. If one insisted on examining the SVD's of $\mathbf{A}$ and $\mathbf{B}$, one would proceed as follows.

(a) Evaluate the SVDs $\mathbf{A} = \mathbf{U}_1\boldsymbol{\Gamma}_1\mathbf{V}_1'$ and $\mathbf{B} = \mathbf{U}_2\boldsymbol{\Gamma}_2\mathbf{V}_2'$. Should $\mathbf{A}$ and/or $\mathbf{B}$ have repeated singular values, choose arbitrary admissible corresponding orthogonal singular vectors $\mathbf{U}_1$, $\mathbf{V}_1$, and/or $\mathbf{U}_2$, $\mathbf{V}_2$.

(b) Examine $\mathbf{V}_1$ and $\mathbf{V}_2$ for pairs of vectors $(\mathbf{v}_{1i}, \mathbf{v}_{2j})$ that agree in all respects. Let $\mathbf{V}$ be the set of all such pairs. Then set $\mathbf{V}_{\mathrm{A}} = \mathbf{V}_{1\perp}\mathbf{V}$, $\mathbf{V}_{\mathrm{B}} = \mathbf{V}_{2\perp}\mathbf{V}$. Check, $\mathbf{V}_{\mathrm{A}}'\mathbf{V}_{\mathrm{B}} = \mathbf{0}$.

(c) Repeat (b) similarly for $\mathbf{U}_1$ and $\mathbf{U}_2$ forming $\mathbf{U}_{\mathrm{A}} = \mathbf{U}_{1\perp}\mathbf{U}$, $\mathbf{U}_{\mathrm{B}} = \mathbf{U}_{2\perp}\mathbf{U}$. Check, $\mathbf{U}_{\mathrm{A}}'\mathbf{U}_{\mathrm{B}} = \mathbf{0}$.

(d) If the checks in (b) and (c) both succeed, then $\mathbf{AB}'$ and $\mathbf{B}'\mathbf{A}$ are symmetric.

(iii) We note that if $\mathbf{B}'\mathbf{A}$ and $\mathbf{AB}'$ do not share the same singular vectors, it is possible for one of the products to be symmetric and the other not. To see this, consider the situation where $\mathbf{C}$ and $\mathbf{D}$ do satisfy the condition. Then, for any $q \times q$ non-singular matrix $\mathbf{T}$, $(\mathbf{D}'\mathbf{T})(\mathbf{T}^{-1}\mathbf{C})$ is unaltered, so remains symmetric. However $(\mathbf{T}^{-1}\mathbf{C})(\mathbf{D}'\mathbf{T}) = \mathbf{T}^{-1}(\mathbf{CD}')\mathbf{T}$ is not symmetric even though $\mathbf{CD}'$ is symmetric. Thus, $\mathbf{A} = \mathbf{T}^{-1}\mathbf{C}$ and $\mathbf{B} = \mathbf{T}'\mathbf{D}$ are two matrices for which $\mathbf{B}'\mathbf{A}$ is symmetric and $\mathbf{AB}'$ is not. We have not pursued general conditions for this property to hold.

References

Andersson, L.E. and Elving, T. (1997). A constrained Procrustes problem. *SIAM J. Matrix Anal. Appl.* **18**(1), 124–139.

Andrews, D.F., Bickel, P.J., Hample, F.R., Huber, P.J., Rogers, W.F. and Tukey, J.W. (1972). *Robust Estimates of Location: Survey and Advances*. Princeton: Princeton University Press.

Arabie, P., Carroll, J.D. and De Sarbo, W.S. (1987). *Three-way Scaling and Clustering*. Newbury Park, CA: Sage.

Arnold, G.M. (1986). A generalised Procrustes macro for sensory analysis. *Genstat Newslett.* **18**, 61–80.

Arnold, G.M. (1990). Problems with Procrustes analysis. *J. Appl. Statist.*, Letters to the Editor **17**, 449–451.

Arnold, G.M. (1992). Scaling factors in generalised Procrustes analysis. In: Yadolah Dodge and Joe Whittaker (eds.) *Computational Statistics (Compstat 92)*. Heidelberg: Physica Verlag.

Arnold, G.M. and Williams, A.A. (1986). The use of generalised Procrustes analysis in sensory analysis. In: Piggott, J.R. (ed.), *Statistical Procedures in Food Research*. London: Elsevier Applied Science.

Banfield, C.F. and Harries, J.M. (1975). A technique for comparing judges' performance in sensory tests. *Journal of Food Technology* **10**, 1–10.

Bookstein, F.L. (1989). Principal warps: thin plate splines and the decomposition of deformations. *IEEE Transactions on Pattern Anal. Machine Intelligence* **11**, 567–585.

Bookstein, F.L. (1991). *Morphometric Tools for Landmark Data: Geometry and Biology*. Cambridge: Cambridge University Press.

Bookstein, F.L. and Green, W.D.K. (1993). A thin-plate spline for deformations with specified derivatives. In: *Mathematical Models in Medical Imaging*.

Borg, I. and Groenen, P.J.F. (1997). *Modern Multidimensional Scaling: Theory and Applications*. Springer Series in Statistics, Heidelberg: Springer Verlag.

Brokken, F.B. (1983). Orthogonal Procrustes rotation maximising congruence. *Psychometrika* **48**, 343–352.

Browne, M.W. (1967). On oblique Procrustes rotation. *Psychometrika* **32**, 125–132.

Browne, M.W. and Kristof, W. (1969). On the oblique rotation of a factor matrix to a specified pattern. *Psychometrika* **34**, 237–248.

Byrne, D.V., O'Sullivan, M.G., Bredie, W.L.P. and Martens, M. (2003). Descriptive sensory profiling and physical/chemical analyses of warmed-overflavour in meat patties from carriers and non-carriers of the RN-allele. *Meat Sci.* **63**, 211–224.

Byrne, D.V., O'Sullivan, M.G., Dijksterhuis, G.B., Bredie, W.L.P. and Martens, M. (2001). Sensory panel consistency during development of a vocabulary for warmed-over flavour. *Food Quality Pref.* **12**(3), 171–187.

Carroll, J.D. (1968). Generalization of canonical correlation analysis to three or more sets of variables. In: *Proceedings of the 76th Annual Convention of the American Psychological Association*, 227–228.

Carroll, J.D. and Chang, J.J. (1970). Analysis of individual differences in multidimensional scaling via n-way generalization of 'Eckhart-Young' decomposition. *Psychometrika* **35**, 283–319.

Carroll, J.D. and Arabie, P. (1980). Multidimensional scaling. *Annu. Rev. Psychol.* **31**, 607–649.

Chu, M.T. and Trendafilov, N.T. (1990). On a differential equation approach to the weighted orthogonal Procrustes problem. *Statist. Comput.* **8**, 125–133.

Chu, M.T. and Trendafilov, N.T. (2001). The orthogonally constrained regression revisited. *J. Comput. Graphical Statist.* **10**(4), 746–771.

Cliff, N. (1966). Orthogonal rotation to congruence. *Psychometrika* **31**, 33–42.

Collins, A.J. (1991). The use of generalised Procrustes techniques in sensory analysis. In: *Agro-Industrie et Methodes Statistiques*, 2èmes journeé europèenes. Nantes, 43–54.

Commandeur, J.J.F. (1991). *Matching Configurations*. Leiden: DSWO Press.

Constantine, A.C. and Gower, J.C. (1978). Some properties and applications of simple orthogonal matrices. *J. Inst. Maths. Applics.* **21**, 445–454.

Constantine, A.C. and Gower, J.C. (1982). Models for the analysis of inter-regional migration. *Environ. Planning 'A'* **14**, 477–497.

Coppi, R. and Bolasco, S. (eds) (1989). *Multiway Data Analysis*. Amsterdam: Elsevier/North Holland.

Cox, T.F. and Cox, M.A.A. (2001). *Multidimensional Scaling*, 2nd edn. Chapman and Hall: Monographs on Statistics and Applied Probability 88.

Cramer, E.M. (1974). On Browne's solution for oblique Procrustes rotation. *Psychometrika* **39**, 139–163.

Crippen, G. and Harvel, T. (1988). *Distance Geometry and Molecular Conformation*. Baldock, U.K: Research Studies Press.

Davis, A.W. (1978). On the asymptotic distribution of Gower's m^2 goodness-of-fit criterion in a particular case. *Ann. Inst. Statist. Math.* **30**, 71–79.

De Leeuw, J. and Meulman, J. (1986). A special jackknife for multidimensional scaling. *J. Classification* **3**, 97–112.

Dempster, A.P., Laird, N.M. and Rubin, D.B. (1977). Maximum likelihood from incomplete data via the *EM* algorithm (with discussion). *J. R. Statist. Soc. B* **39**, 1–38.

Dijksterhuis, G.B. (1994). Procrustes analysis in studying sensory-instrumental relations. *Food Quality Pref.* **5**, 115–120.

Dijksterhuis, G.B. (1996). Procrustes analysis in sensory research. In: Næs, T. and Risvik, E. (eds.) *Multivariate Analysis of Data in Sensory Science*. Amsterdam: Elsevier Science Publishers.

Dijksterhuis, G.B. (1997). *Multivariate Data Analysis in Sensory and Consumer Science*. Trumbull, CT: Food and Nutrition Press.

Dijksterhuis, G.B. and Gower, J.C. (1991/1992). The interpretation of generalised Procrustes analysis and allied methods. *Food Quality Pref.* **3**, 67–87.

Dijksterhuis, G.B. and Heiser, W.J. (1995). The role of permutation tests in multivariate data analysis. *Food Quality Pref.* **6**, 263–270.

Dijksterhuis, G.B. and Punter, P.H. (1990). Interpreting generalized Procrustes analysis 'Analysis of Variance' tables. *Food Quality Pref.* **2**, 255–265.

Dryden, I.L. and Mardia, K.V. (1998). *Statistical Shape Analysis*. Chichester: Wiley.

Eckart, C. and Young, G. (1936). The approximation of one matrix by another of lower rank. *Psychometrika* **1**, 211–218.

Edgington, E.S. (1987). *Randomization Tests*. New York: Marcel Dekker, Inc.

Ekblom, H. (1987). The L_1-estimate as limiting case of an L_p- or Huber-estimate. In: Dodge, Y. (ed.) *Statistical Data Analysis Based on the L_1-Norm*. Amsterdam: Elsevier (North Holland).

Escofier, B. and Pages, J. (1988). *Analyse factorielles Multiples*. Paris: Dunod.

Escoufier, Y. (1985). Objetives et procédures de l'analyse conjoint de plusiers tableaux. *Statist. et analyse de données* **10**, 1–10.

Everitt, B.J. and Gower, J.C. (1981). Plotting the optimal positions of an array of cortical electric phosphenes. In: Barnett, V. (ed.) *Interpreting Multivariate Data*. Chichester: Wiley.

Fletcher, R. and Powell, M.J.D. (1963). A rapidly convergent descent method for minimization. *Comput. J.* **2**, 163–168.

Flury, B. (1988). *Common Principal Components Analysis and Related Multivariate Models*. New York: Wiley.

Flury, B. (1995). Developments in principal components analysis. In: Krzanowski, W.J. (ed.) *Recent Advances in Descriptive Multivariate Analysis*. Oxford: Clarendon Press, pp. 14–33.

Frewer, L.J., Salter, B. and Lambert, N. (2001). Understanding patients' preferences for treatment: the need for innovative methodologies. *Quality in Health Care* **10**, (Suppl. I), i50–i54.

Gibson, W.A. (1962). On the least-squares orthogonalization of an oblique transformation. *Psychometrika* **27**, 193–196.

Gifi, A. (1990). *Nonlinear Multivariate Analysis*. Chichester: John Wiley and Sons.

Glasbey, C.A. and Berman, M. (1995). *A review of image analysis in biometry*. In: Armitage, P. and David, H.A. (eds.) *Advances in Biometry*. Chichester: Wiley.

Glasbey, C.A. and Horgan, G.W. (1995). *Image Analysis for the Biological Sciences*. Chichester: Wiley.

Glasbey, C.A. and Mardia, K.V. (1998). A review of image warping methods. *J. Appl. Statist.* **25**, 155–171.

Goodall, C. (1991). Procrustes methods in the statistical analysis of shape. *J. R. Statist. Soc. B* **53**, 285–339.

Gordon, A.D. (1999). *Classification*. 2nd edn. Monographs on Statistics and Applied Probability 82. London: Chapman and Hall.

Gower, J.C. (1966). Some distance properties of latent root and vector methods used in multivariate analysis. *Biometrika* **53**(3/4), 325–338.

Gower, J.C. (1971). Statistical methods for comparing different multivariate analyses of the same data. In: Hodson, J.R., Kendall, D.G. and Tautu, P. (eds.) *Mathematics in the Archaeological and Historical Sciences*. Edinburgh: Edinburgh University Press, pp. 138–149.

Gower, J.C. (1975). Generalized Procrustes analysis. *Psychometrika* **40**, 33–51.

Gower, J.C. (1984). Multivariate analysis: ordination, multidimensional scaling and allied topics. In: Lloyd, E.H. (ed.) *Handbook of Applicable Mathematics: Vol. VI, Statistics*, Chichester: John Wiley and Sons, pp. 727–781.

Gower, J.C. (1989). Generalised canonical analysis. In: Coppi, R. and Bolasco, S. (eds.) *Multiway Data Analysis*. Amsterdam: North Holland, pp. 221–232.

Gower, J.C. (1992). Generalized biplots. *Biometrika* **79**, 475–493.

Gower, J.C. (1995*a*). Orthogonal and projection Procrustes analysis. In: Krzanowski, W. (ed.) *Recent Advances in Descriptive Multivariate Analysis*. Oxford: Clarendon Press.

Gower, J.C. (1995*b*). Distance-geometry and shape. In: Mardia, K.V. and Gill, C.A. (eds.) *Proceedings in Current Issues in Statistical Shape Analysis*. Leeds: Leeds University Press, pp. 11–17.

Gower, J.C. (1998). The role of constraints in determining optimal scores. *Statist. Med.* **17**, 2709–2721.

Gower, J.C. (2002). Categories and quantities. In: Nishisato, S., Baba, Y., Bozdogan, H. and Kanefuji, K. (eds.) *Measurement and Mutivariate Analysis*. Tokyo: Springer-Verlag. pp. 1–12.

Gower, J.C. and Banfield, C.F. (1975). Goodness-of-fit criteria for hierarchical classification and their empirical distributions. In: Corsten, L.C.A. and Postelnicu, T. (eds.) *Proceedings of the Eighth International Biometrics Conference*. Bucharest: Editura Acadamiei Republicii Socialiste Romania, pp. 347–361.

Gower, J.C. and Dijksterhuis, G.B. (1994). Coffee images: a study in the simultaneous display of multivariate quantitative and qualitative variables for several assessors. *Quality Quantity* **28**, 165–184.

Gower, J.C. and Hand, D.J. (1996). Biplots. Monographs on Statistics and Applied Probability 54. London: Chapman and Hall.

Gower, J.C. and Harding, S.A. (1988). Nonlinear biplots. *Biometrika* **75**, 445–455.

Gower, J.C. and Harding, S.A. (1998). Prediction regions for categorical variables. In: Blasius, J. and Greenacre, M.J. (eds.) *Vizualisation of Categorical Variables*. Academic Press, London, pp. 405–419.

Gower, J.C., Meulman, J.J. and Arnold, G.M. (1999). Nonmetric linear biplots. *J. Classification* **16**, 181–196.

Green, B.F. (1969). Best linear composites with a specified structure. *Psychometrika* **34**, 301–318.

Green, B.F. and Gower, J.C. (1979). A problem with congruence. Paper presented at the Annual Meeting of the Psychometric Society, Monterey, CA.

Gruvaeus, T.T. (1970). A general approach to Procrustes pattern rotation. *Psychometrika* **35**, 493–505.

Hand, D.J. and Taylor, C.C. (1987). *Multivariate Analysis of Variance and Repeated Measures*. London: Chapman and Hall.

Hardy, G.H., Littlewood, J.E. and Polya, G. (1952). *Inequalities*, 2nd edn. Cambridge: Cambridge University Press.

Harries, J.M. and MacFie, H.J.H. (1976). The use of a rotational fitting technique in the interpretation of sensory scores for different characteristics. *J. Food Technol.* **11**, 449–456.

Harman, H.H. (1976). *Modern Factor Analysis*, 3rd edition. Chicago: University of Chicago Press.

Healy, M.J.R. and Goldstein, H. (1966). An approach to the scaling of categorised attributers. *Biometrika* **63**, 219–229.

Heiser, W. and de Leeuw, J. (1979). *How to use SMACOFF-1, A Program for Metric Multidimensional Scaling*. Department of Datatheory, Faculty of Social Sciences, University of Leiden, Wassenaarseweg 80, Leiden, The Netherlands.

Hotelling, H. (1936). Relations between two sets of variates. *Biometrika* **28**, 321–327.

Huber, P.J. (1981). *Robust Statistics*. New York: Wiley.

Huitson, A. (1989). Problems with Procrustes analysis. *J. Appl. Statist.* **16**, 39–45.

Huitson, A. (1990). Comments on: 'problems with Procrustes analysis' by Gillian M. Arnold. *J. Appl. Statist.*, Letters to the Editor **17**, 451–452.

Hurley, J.R. and Cattell, R.B. (1962). The Procrustes program: producing direct rotation to test a hypothesized factor structure. *Beh. Sci.* **7**, 258–262.

Jennrich, R.I. (2001). A simple general procedure for orthogonal rotation. *Psychometrika* **66**, 289–306.

Jolliffe, I.T. (1986). *Principal Component Analysis*. Springer-Verlag.

Jöreskog, K.G. (1967). Some contributions to maximum likelihood factor analysis. *Psychometrika* **23**, 443–482.

Kaiser, H.F. (1958). The varimax criterion for analytical rotation in factor analysis. *Psychometrika* **32**, 187–200.

Kendall, D.G. (1984). Shape manifolds, procrustean metrics, and complex projective spaces. *Bull. London Math. Soc.* **16**, 81–121.

Kendall, D.G. (1988). A survey of the statisitical theory of shape. *Statist. Sci.* **4**, 87–120.

Kendall, D.G., Barden, D., Carne, T.K. and Le, H. (1999). *Shape and Shape Theory*. Chichester: John Wiley and sons Ltd.

Kent, J.T. and Mardia, K.V. (1997). Consistency of Procrustes estimators. *J. R. Statist. Soc. B* **59**, 281–290.

Kent, J.T. and Mardia, K.V. (2001). Shape, Procrustes tangent projections and bilateral symmetry. *Biometrika* **88**, 469–485.

Kettenring, J.R. (1971). Canonical analysis of several sets of variables. *Biometrika* **56**, 433–451.

Kiers, H.A.L. and ten Berge, J.M.F. (1992). Minimization of a class of matrix trace functions by means of refined majorization. *Psychometrika* **57**, 371–382.

King, B.M. and Arents, P. (1991). A statistical test of consensus obtained from generalized Procrustes analysis of sensory data. *J. Sens. Stud.* **6**, 37–48.

Koschat, M.A. and Swayne, D.F. (1991). A weighted Procrustes criterion. *Psychometrika* **56**, 229–239.

Kristof, W. and Wingersky, B. (1971). Generalization of the orthogonal Procrustes rotation procedure to more than two matrices. *Proceedings of the 79th Annual Convention of the American Psychological Association* **6**, 89–90.

Kruskal, J.B. and Wish, M. (1978). *Multidimensional Scaling*. Newbury Park, CA: Sage.

Krzanowski, W.J. (1979). Between-groups comparison of principal components. *J. Am. Statist. Assoc.* **74**, 703–707. Correction note: (1981), 76, 1022.

Langeheine, R. (1982). Statistical evaluation of measures of fit in the Lingoes-Borg Procrustean individual differences scaling. *Psychometrika* **47**, 427–442.

Langron, S.P. (1981). The statistical treatment of sensory analysis data. Ph.D. thesis, University of Bath.

Langron, S.P. and Collins, A.J. (1985). Perturbation theory for generalized Procrustes analysis. *J.R. Statist. Soc. B* **47**, 277–284.

Lavit, C. (1988). Analyse Conjointe de Tableaux Quantitatifs. Paris: Masson.

Law, H.G., Snyder, C.W., Hattie, J.A. and McDonald, R.P. (1984). *Research Methods for Multimode Data Analysis*. New York: Praeger.

Lebart, L., Morineau, A. and Piron, M. (1995). *Statstique exploratoire multidimensionelle*. Dunod: Paris.

Lele, S.R. and Richtsmeier, J.T. (2001). *An Invariant Approach to the Statistical Analysis of Shapes*. Chapman and Hall CRC Boca Raton.

Lingoes, J.C. and Borg, I. (1978). A direct approach to individual differences scaling using increasingly complex transformations. *Psychometrika* **43**, 491–519.

Lissitz, R.W., Shönemann, P.H. and Lingoes, J.C. (1976). A solution of the weighted Procrustes problem in which the transformation is in agreement with the loss function. *Psychometrika* **41**, 547–550.

Mardia, K.V., Bookstein, F.L. and Moreton, I.J. (2000). Statistical assessment of bilateral symmetry of shape. *Biometrika* **87**, 285–300.

Martens, H. and Martens, M. (2001). *Multivariate Analysis of Quality. An Introduction*. Chichester: John Wiley and Sons Ltd.

McCullagh, P. and Nelder, J.A. (1989). Generalized linear models. Monographs on Statistics and Applied Probability 37. London: Chapman and Hall.

Meredith, W. (1964). Rotation to achieve factorial invariance. *Psychometrika* **29**, 187–206.

Meulman, J.J. (1992). The integration of multidimensional scaling and multivariate analyses with optimal transformations. *Psychometrika* **57**, 239–565.

Meyners, M., Kunert, J. and Qannari, E.M. (2000). Comparing generalized procrustes analysis and statis. *Food Quality Pref.* **11**, 77–83.

Mosier, C.I. (1939). Determinig a simple structure when loadings for certain tests are known. *Psychometrika* **4**, 149–162.

Mosteller, F. and Tukey, J.W. (1977). *Data Analysis and Regression*. Massachusetts: Addison-Wesley.

Neuhaus, J.O. and Wrigley, C. (1954). The quartimax method. An analytical approach to orthogonal simple structure. *Br. J. Math. Statist. Psychol.* **7**, 81–91.

Peay, E.R. (1988). Multidimensional rotation and scaling of configurations to optimal agreement. *Psychometrika* **53**, 199–208.

Schönemann, P.H. (1966). A generalized solution of the orthogonal Procrustes problem. *Psychometrika* **31**(1), 1–10.

Schönemann, P.H. (1968). On two sided orthogonal Procrustes problems. *Psychometrika* **33**, 19–33.

Schönemann, P.H. and Carroll, R.M. (1970). Fitting one matrix to another under choice of a central dilation and a rigid motion. *Psychometrika* **35**, 245–255.

Schönemann, P.H., James, W.L. and Carter, F.S. (1979). Statistical inference in multidimensional scaling: a method for fitting and testing Horan's model. In: Lingoes, J.C., Roskam, E.E. and Borg, I. (eds.) *Geometric Representations of Relational Data. Readings in Multidimensional Scaling*. Ann Arbor: Mathesis Press.

Schott, P.H. (1991). Some tests for common principal component subspaces in several groups. *Biometrika* **78**, 177–184.

Sibson, R. (1978). Studies in the robustness of multidimensional scaling: Procrustes statistics. *J. R. Statist. Soc. B* **40**, 234–238.

Sibson, R. (1979). Studies in the robustness of multidimensional scaling: perturbational analysis of classical scaling. *J. R. Statist. Soc. B* **41**, 217–229.

Siegel, A. and Benson, R. (1982). A robust comparison of biological shapes. *Biometrics* **38**, 341–350.

Slater, P. (1976). *The Measurement of Intrapersonal Space by Grid Technique. Volume 1: Explorations of Intrapersonal Space*. London: John Wiley and Sons.

Slater, P. (1977). *The Measurement of Intrapersonal Space by Grid Technique. Volume 2: Dimensions of Intrapersonal Space*. London: John Wiley and Sons.

Steenkamp, J.-B.E.M., Trijp, J.C.M. van, and Berge, J.M.F. ten (1994). Perceptual mapping based on idiosyncratic sets of attributes. *J. Marketing Res.* **XXXI**, 15–27.

Stone, H. and Sidel, J.L. (1993). *Sensory Evaluation Practices*, 2nd edn. New York: Academic Press Inc.

Stoyan, D. and Stoyan, H. (1990). A further application of D.G. Kendall's Procrustes analysis. *Biometrical J.* **32**, 293–301, Berlin: Akademie-Verlag.

ten Berge, J.M.F. (1977*a*). Orthogonal Procrustes rotation for two or more matrices. *Psychometrika* **42**, 267–276.

ten Berge, J.M.F. (1977*b*). Optimising factorial invariance. Unpublished Doctoral Dissertation. University of Groningen.

ten Berge, J.M.F. and Nevels, K. (1977). A general solution for Mosier's oblique Procrustes problem. *Psychometrika* **42**, 593–600.

ten Berge, J.M.F. (1979). On the equivalence of two oblique congruence rotation methods, and orthogonal approximations. *Psychometrika* **44**, 359–364.

ten Berge, J.M.F. (1983*a*). On Green's best linear composites with a specified structure, and oblique estimates of factor scores. *Psychometrika* **48**, 371–375.

ten Berge, J.M.F. (1983*b*). A generalization of Verhelst's solution for a constrained regression problem in ALSCAL and related MDS-algorithms. *Psychometrika* **48**, 631–638.

ten Berge, J.M.F. and Knol, D. (1984). Orthogonal rotations to maximal agreement for two or more matrices of different column orders. *Psychometrika* **49**, 49–55.

ten Berge, J.M.F., Kiers, H.A.L. and Commandeur, J.J.F. (1993). Orthogonal Procrustes rotation for matrices with missing values. *Br. J. Math. Statist. Psychol.* **46**, 119–134.

Thompson, D.W. (1917). *On Growth and Form.* Cambridge: Cambridge University press.

Thompson, E.A. (1995). *Statistical Genetics*. In: Armitage, P. and David, H.A. (eds.) *Advances in Biometry*. Chichester: Wiley.

Trendafilov, N.T. (1999). A continuous time approach to the oblique Procrustes problem. *Behaviormetrika* **26**, 167–181.

Tucker, L.R. (1951). A method for synthesis of factor analysis studies (Personnel research section report no. 984. Washington D.C.: Department of the Army).

van Buuren, S. and Dijksterhuis, G.B. (1988). Procrustes analysis of discrete data. In: Jansen, M.G.H. and van Schuur, W.H. (eds.) *The Many Faces of Multivariate Analysis, Proceedings of the SMABS-88 Conference*, Vol. 1. Groningen: RION, Institute for Educational Research, pp. 53–66.

van der Burg and De Leeuw (1983). Non-linear canonical correlation. *Br. J. Math. Statist. Psychol.* **36**, 54–80.

van der Burg, E. (1988). *Nonlinear Canonical Correlation and Some Related Techniques.* Leiden: DSWO press.

Verboon, P. and Heiser, W. (1992). Resistant orthogonal Procrustes analysis. *J. Classification* **9**, 237–256.

Verboon, P. and Gabriel, K.R. (1995). Generalized Procrustes analysis with iterative weighting to achieve resistance. *Br. J. Math. Statist. Psychol.* **48**, 57–74.

Verdegaal, R. (1986). OVERALS. Leiden: University of Leiden, Department of Data Theory, UG-96-01.

Wakeling, I.N., Raats, M.M. and MacFie, H.J.H. (1992). A new significance test for consensus in generalized Procrustes analysis. *J. Sens. Stud.* **7**, 91–96.

Wedderburn, J.H.M. (1934). *Lectures on matrices.* Colloquium publications of the American Mathematical Society, 17. New York: American Mathematical Society.

Williams, A.A. and Langron, S.P. (1984). The use of free-choice profiling for the evaluation of commercial ports. *J. Sci. Food Agric.* **35**, 558–568.

Young, F., De Leeuw, J. and Takane, Y. (1976). Regression with qualitative and quantitative variables: an alternating least squares method with optimal scaling features. *Psychometrika* **41**, 505–529.

Young, F., De Leeuw, J. and Takane, Y. (1978). The principal components of mixed measurement level multivariate data: an alternating least squares method with optimal scaling features. *Psychometrika* **43**, 279–281.

Index